S0-AEL-330

An Introduction to
Categorical Data Analysis

THE WILEY BICENTENNIAL–KNOWLEDGE FOR GENERATIONS

Each generation has its unique needs and aspirations. When Charles Wiley first opened his small printing shop in lower Manhattan in 1807, it was a generation of boundless potential searching for an identity. And we were there, helping to define a new American literary tradition. Over half a century later, in the midst of the Second Industrial Revolution, it was a generation focused on building the future. Once again, we were there, supplying the critical scientific, technical, and engineering knowledge that helped frame the world. Throughout the 20th Century, and into the new millennium, nations began to reach out beyond their own borders and a new international community was born. Wiley was there, expanding its operations around the world to enable a global exchange of ideas, opinions, and know-how.

For 200 years, Wiley has been an integral part of each generation's journey, enabling the flow of information and understanding necessary to meet their needs and fulfill their aspirations. Today, bold new technologies are changing the way we live and learn. Wiley will be there, providing you the must-have knowledge you need to imagine new worlds, new possibilities, and new opportunities.

Generations come and go, but you can always count on Wiley to provide you the knowledge you need, when and where you need it!

WILLIAM J. PESCE
PRESIDENT AND CHIEF EXECUTIVE OFFICER

PETER BOOTH WILEY
CHAIRMAN OF THE BOARD

An Introduction to Categorical Data Analysis

Second Edition

ALAN AGRESTI

Department of Statistics
University of Florida
Gainesville, Florida

WILEY-INTERSCIENCE
A JOHN WILEY & SONS, INC., PUBLICATION

Published by John Wiley & Sons, Inc., Hoboken, New Jersey
Published simultaneously in Canada

For general information on our other products and services or for technical support, please contact our Customer Care Department within the United States at (800) 762-2974, outside the United States at (317) 572-3993 or fax (317) 572-4002.

Wiley also publishes its books in a variety of electronic formats. Some content that appears in print may not be available in electronic formats. For more information about Wiley products, visit our web site at www.wiley.com.

Library of Congress Cataloging-in-Publication Data

Agresti, Alan
An introduction to categorical data analysis / Alan Agresti.
p. cm.
Includes bibliographical references and index.
ISBN 978-0-471-22618-5
1. Multivariate analysis. I. Title.

QA278.A355 1996
519.5'35 - - dc22 2006042138

Printed in the United States of America.

10 9 8

Contents

Preface to the Second Edition

In recent years, the use of specialized statistical methods for categorical data has increased dramatically, particularly for applications in the biomedical and social sciences. Partly this reflects the development during the past few decades of sophisticated methods for analyzing categorical data. It also reflects the increasing methodological sophistication of scientists and applied statisticians, most of whom now realize that it is unnecessary and often inappropriate to use methods for continuous data with categorical responses.

This book presents the most important methods for analyzing categorical data. It summarizes methods that have long played a prominent role, such as chi-squared tests. It gives special emphasis, however, to modeling techniques, in particular to logistic regression.

The presentation in this book has a low technical level and does not require familiarity with advanced mathematics such as calculus or matrix algebra. Readers should possess a background that includes material from a two-semester statistical methods sequence for undergraduate or graduate nonstatistics majors. This background should include estimation and significance testing and exposure to regression modeling.

This book is designed for students taking an introductory course in categorical data analysis, but I also have written it for applied statisticians and practicing scientists involved in data analyses. I hope that the book will be helpful to analysts dealing with categorical response data in the social, behavioral, and biomedical sciences, as well as in public health, marketing, education, biological and agricultural sciences, and industrial quality control.

The basics of categorical data analysis are covered in Chapters 1–8. Chapter 2 surveys standard descriptive and inferential methods for contingency tables, such as odds ratios, tests of independence, and conditional vs marginal associations. I feel that an understanding of methods is enhanced, however, by viewing them in the context of statistical models. Thus, the rest of the text focuses on the modeling of categorical responses. Chapter 3 introduces generalized linear models for binary data and count data. Chapters 4 and 5 discuss the most important such model for binomial (binary) data, logistic regression. Chapter 6 introduces logistic regression models for multinomial responses, both nominal and ordinal. Chapter 7 discusses loglinear models for Poisson (count) data. Chapter 8 presents methods for matched-pairs data.

I believe that logistic regression is more important than loglinear models, since most applications with categorical responses have a single binomial or multinomial response variable. Thus, I have given main attention to this model in these chapters and in later chapters that discuss extensions of this model. Compared with the first edition, this edition places greater emphasis on logistic regression and less emphasis on loglinear models.

I prefer to teach categorical data methods by unifying their models with ordinary regression and ANOVA models. Chapter 3 does this under the umbrella of generalized linear models. Some instructors might prefer to cover this chapter rather lightly, using it primarily to introduce logistic regression models for binomial data (Sections 3.1 and 3.2).

The main change from the first edition is the addition of two chapters dealing with the analysis of clustered correlated categorical data, such as occur in longitudinal studies with repeated measurement of subjects. Chapters 9 and 10 extend the matched-pairs methods of Chapter 8 to apply to clustered data. Chapter 9 does this with marginal models, emphasizing the generalized estimating equations (GEE) approach, whereas Chapter 10 uses random effects to model more fully the dependence. The text concludes with a chapter providing a historical perspective of the development of the methods (Chapter 11) and an appendix showing the use of SAS for conducting nearly all methods presented in this book.

The material in Chapters 1–8 forms the heart of an introductory course in categorical data analysis. Sections that can be skipped if desired, to provide more time for other topics, include Sections 2.5, 2.6, 3.3 and 3.5, 5.3–5.5, 6.3, 6.4, 7.4, 7.5, and 8.3–8.6. Instructors can choose sections from Chapters 9–11 to supplement the basic topics in Chapters 1–8. Within sections, subsections labelled with an asterisk are less important and can be skipped for those wanting a quick exposure to the main points.

This book is of a lower technical level than my book *Categorical Data Analysis* (2nd edition, Wiley, 2002). I hope that it will appeal to readers who prefer a more applied focus than that book provides. For instance, this book does not attempt to derive likelihood equations, prove asymptotic distributions, discuss current research work, or present a complete bibliography.

Most methods presented in this text require extensive computations. For the most part, I have avoided details about complex calculations, feeling that computing software should relieve this drudgery. Software for categorical data analyses is widely available in most large commercial packages. I recommend that readers of this text use software wherever possible in answering homework problems and checking text examples. The Appendix discusses the use of SAS (particularly PROC GENMOD) for nearly all methods discussed in the text. The tables in the Appendix and many of the data sets analyzed in the book are available at the web site http://www.stat.ufl.edu/~aa/intro-cda/appendix.html. The web site http://www.stat.ufl.edu/~aa/cda/software.html contains information about the use of other software, such as S-Plus and R, Stata, and SPSS, including a link to an excellent free manual prepared by Laura Thompson showing how to use R and S-Plus to conduct nearly all the examples in this book and its higher-level companion. Also listed at the text website are known typos and errors in early printings of the text.

I owe very special thanks to Brian Marx for his many suggestions about the text over the past 10 years. He has been incredibly generous with his time in providing feedback based on using the book many times in courses. He and Bernhard Klingenberg also very kindly reviewed the draft for this edition and made many helpful suggestions. I also thank those individuals who commented on parts of the manuscript or who made suggestions about examples or material to cover. These include Anna Gottard for suggestions about Section 7.4, Judy Breiner, Brian Caffo, Allen Hammer, and Carla Rampichini. I also owe thanks to those who helped with the first edition, especially Patricia Altham, James Booth, Jane Brockmann, Brent Coull, Al DeMaris, Joan Hilton, Peter Imrey, Harry Khamis, Svend Kreiner, Stephen Stigler, and Larry Winner. Thanks finally to those who helped with material for my more advanced text (*Categorical Data Analysis*) that I extracted here, especially Bernhard Klingenberg, Yongyi Min, and Brian Caffo. Many thanks to Stephen Quigley at Wiley for his continuing interest, and to the Wiley staff for their usual high-quality support.

As always, most special thanks to my wife, Jacki Levine, for her advice and encouragement. Finally, a truly nice byproduct of writing books is the opportunity to teach short courses based on them and spend research visits at a variety of institutions. In doing so, I have had the opportunity to visit about 30 countries and meet many wonderful people. Some of them have become valued friends. It is to them that I dedicate this book.

ALAN AGRESTI

London, United Kingdom
January 2007

CHAPTER 1

Introduction

From helping to assess the value of new medical treatments to evaluating the factors that affect our opinions on various controversial issues, scientists today are finding myriad uses for methods of analyzing categorical data. It's primarily for these scientists and their collaborating statisticians – as well as those training to perform these roles – that this book was written. The book provides an introduction to methods for analyzing categorical data. It emphasizes the ideas behind the methods and their interpretations, rather than the theory behind them.

This first chapter reviews the probability distributions most often used for categorical data, such as the *binomial distribution*. It also introduces *maximum likelihood*, the most popular method for estimating parameters. We use this estimate and a related *likelihood function* to conduct statistical inference about proportions. We begin by discussing the major types of categorical data and summarizing the book's outline.

1.1 CATEGORICAL RESPONSE DATA

Let us first define categorical data. A *categorical* variable has a measurement scale consisting of a set of categories. For example, political philosophy may be measured as "liberal," "moderate," or "conservative"; choice of accommodation might use categories "house," "condominium," "apartment"; a diagnostic test to detect e-mail spam might classify an incoming e-mail message as "spam" or "legitimate e-mail."

Categorical scales are pervasive in the social sciences for measuring attitudes and opinions. Categorical scales also occur frequently in the health sciences, for measuring responses such as whether a patient survives an operation (yes, no), severity of an injury (none, mild, moderate, severe), and stage of a disease (initial, advanced).

Although categorical variables are common in the social and health sciences, they are by no means restricted to those areas. They frequently occur in the behavioral

---

An Introduction to Categorical Data Analysis, Second Edition. By Alan Agresti

sciences (e.g., categories "schizophrenia," "depression," "neurosis" for diagnosis of type of mental illness), public health (e.g., categories "yes" and "no" for whether awareness of AIDS has led to increased use of condoms), zoology (e.g., categories "fish," "invertebrate," "reptile" for alligators' primary food choice), education (e.g., categories "correct" and "incorrect" for students' responses to an exam question), and marketing (e.g., categories "Brand A," "Brand B," and "Brand C" for consumers' preference among three leading brands of a product). They even occur in highly quantitative fields such as engineering sciences and industrial quality control, when items are classified according to whether or not they conform to certain standards.

1.1.1 Response/Explanatory Variable Distinction

Most statistical analyses distinguish between *response* variables and *explanatory* variables. For instance, regression models describe how the distribution of a continuous response variable, such as annual income, changes according to levels of explanatory variables, such as number of years of education and number of years of job experience. The response variable is sometimes called the *dependent variable* or *Y variable*, and the explanatory variable is sometimes called the *independent variable* or *X variable*.

The subject of this text is the analysis of categorical response variables. The categorical variables listed in the previous subsection are response variables. In some studies, they might also serve as explanatory variables. Statistical models for categorical response variables analyze how such responses are influenced by explanatory variables. For example, a model for political philosophy could use predictors such as annual income, attained education, religious affiliation, age, gender, and race. The explanatory variables can be categorical or continuous.

1.1.2 Nominal/Ordinal Scale Distinction

Categorical variables have two main types of measurement scales. Many categorical scales have a natural ordering. Examples are attitude toward legalization of abortion (disapprove in all cases, approve only in certain cases, approve in all cases), appraisal of a company's inventory level (too low, about right, too high), response to a medical treatment (excellent, good, fair, poor), and frequency of feeling symptoms of anxiety (never, occasionally, often, always). Categorical variables having ordered scales are called *ordinal* variables.

Categorical variables having unordered scales are called *nominal* variables. Examples are religious affiliation (categories Catholic, Jewish, Protestant, Muslim, other), primary mode of transportation to work (automobile, bicycle, bus, subway, walk), favorite type of music (classical, country, folk, jazz, rock), and favorite place to shop (local mall, local downtown, Internet, other).

For nominal variables, the order of listing the categories is irrelevant. The statistical analysis should not depend on that ordering. Methods designed for nominal variables give the same results no matter how the categories are listed. Methods designed for ordinal variables utilize the category ordering. Whether we list the categories from low to high or from high to low is irrelevant in terms of substantive conclusions, but results of ordinal analyses would change if the categories were reordered in any other way.

Methods designed for ordinal variables *cannot* be used with nominal variables, since nominal variables do not have ordered categories. Methods designed for nominal variables *can* be used with nominal or ordinal variables, since they only require a categorical scale. When used with ordinal variables, however, they do not use the information about that ordering. This can result in serious loss of power. It is usually best to apply methods appropriate for the actual scale.

Categorical variables are often referred to as *qualitative*, to distinguish them from numerical-valued or *quantitative* variables such as weight, age, income, and number of children in a family. However, we will see it is often advantageous to treat ordinal data in a quantitative manner, for instance by assigning ordered scores to the categories.

1.1.3 Organization of this Book

Chapters 1 and 2 describe some standard methods of categorical data analysis developed prior to about 1960. These include basic analyses of association between two categorical variables.

Chapters 3–7 introduce models for categorical responses. These models resemble regression models for continuous response variables. In fact, Chapter 3 shows they are special cases of a generalized class of linear models that also contains the usual normal-distribution-based regression models. The main emphasis in this book is on *logistic regression* models. Applying to response variables that have two outcome categories, they are the focus of Chapters 4 and 5. Chapter 6 presents extensions to multicategory responses, both nominal and ordinal. Chapter 7 introduces *loglinear* models, which analyze associations among multiple categorical response variables.

The methods in Chapters 1–7 assume that observations are independent. Chapters 8–10 discuss logistic regression models that apply when some observations are correlated, such as with repeated measurement of subjects in longitudinal studies. An important special case is matched pairs that result from observing a categorical response for the same subjects at two separate times. The book concludes (Chapter 11) with a historical overview of categorical data methods.

Most methods for categorical data analysis require extensive computations. The Appendix discusses the use of SAS statistical software. A companion website for the book, http://www.stat.ufl.edu/~aa/intro-cda/software.html, discusses other software.

1.2 PROBABILITY DISTRIBUTIONS FOR CATEGORICAL DATA

Inferential statistical analyses require assumptions about the probability distribution of the response variable. For regression and analysis of variance (ANOVA) models for continuous data, the normal distribution plays a central role. This section presents the key distributions for categorical data: the *binomial* and *multinomial* distributions.

1.2.1 Binomial Distribution

Often, categorical data result from n independent and identical trials with two possible outcomes for each, referred to as "success" and "failure." These are generic labels, and the "success" outcome need not be a preferred result. *Identical trials* means that the probability of success is the same for each trial. *Independent trials* means the response outcomes are independent random variables. In particular, the outcome of one trial does not affect the outcome of another. These are often called *Bernoulli trials*. Let π denote the probability of success for a given trial. Let Y denote the number of successes out of the n trials.

Under the assumption of n independent, identical trials, Y has the *binomial distribution* with index n and parameter π. You are probably familiar with this distribution, but we review it briefly here. The probability of outcome y for Y equals

$$P(y) = \frac{n!}{y!(n-y)!}\pi^y(1-\pi)^{n-y}, \quad y = 0, 1, 2, \ldots, n \tag{1.1}$$

To illustrate, suppose a quiz has 10 multiple-choice questions, with five possible answers for each. A student who is completely unprepared randomly guesses the answer for each question. Let Y denote the number of correct responses. The probability of a correct response is 0.20 for a given question, so $n = 10$ and $\pi = 0.20$. The probability of $y = 0$ correct responses, and hence $n - y = 10$ incorrect ones, equals

$$P(0) = [10!/(0!10!)](0.20)^0(0.80)^{10} = (0.80)^{10} = 0.107.$$

The probability of 1 correct response equals

$$P(1) = [10!/(1!9!)](0.20)^1(0.80)^9 = 10(0.20)(0.80)^9 = 0.268.$$

Table 1.1 shows the entire distribution. For contrast, it also shows the distributions when $\pi = 0.50$ and when $\pi = 0.80$.

The binomial distribution for n trials with parameter π has mean and standard deviation

$$E(Y) = \mu = n\pi, \quad \sigma = \sqrt{n\pi(1-\pi)}$$

The binomial distribution in Table 1.1 has $\mu = 10(0.20) = 2.0$ and $\sigma = \sqrt{[10(0.20)(0.80)]} = 1.26$.

The binomial distribution is always symmetric when $\pi = 0.50$. For fixed n, it becomes more skewed as π moves toward 0 or 1. For fixed π, it becomes more

Table 1.1. Binomial Distribution with $n = 10$ and $\pi = 0.20, 0.50$, and 0.80. The Distribution is Symmetric when $\pi = 0.50$

y	$P(y)$ when $\pi = 0.20$	$P(y)$ when $\pi = 0.50$	$P(y)$ when $\pi = 0.80$
0	0.107	0.001	0.000
1	0.268	0.010	0.000
2	0.302	0.044	0.000
3	0.201	0.117	0.001
4	0.088	0.205	0.005
5	0.027	0.246	0.027
6	0.005	0.205	0.088
7	0.001	0.117	0.201
8	0.000	0.044	0.302
9	0.000	0.010	0.268
10	0.000	0.001	0.107

bell-shaped as n increases. When n is large, it can be approximated by a normal distribution with $\mu = n\pi$ and $\sigma = \sqrt{[n\pi(1-\pi)]}$. A guideline is that the expected number of outcomes of the two types, $n\pi$ and $n(1-\pi)$, should both be at least about 5. For $\pi = 0.50$ this requires only $n \geq 10$, whereas $\pi = 0.10$ (or $\pi = 0.90$) requires $n \geq 50$. When π gets nearer to 0 or 1, larger samples are needed before a symmetric, bell shape occurs.

1.2.2 Multinomial Distribution

Some trials have more than two possible outcomes. For example, the outcome for a driver in an auto accident might be recorded using the categories "uninjured," "injury not requiring hospitalization," "injury requiring hospitalization," "fatality." When the trials are independent with the same category probabilities for each trial, the distribution of counts in the various categories is the *multinomial.*

Let c denote the number of outcome categories. We denote their probabilities by $\{\pi_1, \pi_2, \ldots, \pi_c\}$, where $\sum_j \pi_j = 1$. For n independent observations, the multinomial probability that n_1 fall in category 1, n_2 fall in category 2, ..., n_c fall in category c, where $\sum_j n_j = n$, equals

$$P(n_1, n_2, \ldots, n_c) = \left(\frac{n!}{n_1!n_2!\cdots n_c!}\right)\pi_1^{n_1}\pi_2^{n_2}\cdots\pi_c^{n_c}$$

The binomial distribution is the special case with $c = 2$ categories. We will not need to use this formula, as we will focus instead on sampling distributions of useful statistics computed from data assumed to have the multinomial distribution. We present it here merely to show how the binomial formula generalizes to several outcome categories.

The multinomial is a multivariate distribution. The marginal distribution of the count in any particular category is binomial. For category j, the count n_j has mean $n\pi_j$ and standard deviation $\sqrt{[n\pi_j(1-\pi_j)]}$. Most methods for categorical data assume the binomial distribution for a count in a single category and the multinomial distribution for a set of counts in several categories.

1.3 STATISTICAL INFERENCE FOR A PROPORTION

In practice, the parameter values for the binomial and multinomial distributions are unknown. Using sample data, we estimate the parameters. This section introduces the estimation method used in this text, called *maximum likelihood*. We illustrate this method for the binomial parameter.

1.3.1 Likelihood Function and Maximum Likelihood Estimation

The parametric approach to statistical modeling assumes a family of probability distributions, such as the binomial, for the response variable. For a particular family, we can substitute the observed data into the formula for the probability function and then view how that probability depends on the unknown parameter value. For example, in $n = 10$ trials, suppose a binomial count equals $y = 0$. From the binomial formula (1.1) with parameter π, the probability of this outcome equals

$$P(0) = [10!/(0!)(10!)]\pi^0(1-\pi)^{10} = (1-\pi)^{10}$$

This probability is defined for all the potential values of π between 0 and 1.

The probability of the observed data, expressed as a function of the parameter, is called the *likelihood function*. With $y = 0$ successes in $n = 10$ trials, the binomial likelihood function is $\ell(\pi) = (1-\pi)^{10}$. It is defined for π between 0 and 1. From the likelihood function, if $\pi = 0.40$ for instance, the probability that $Y = 0$ is $\ell(0.40) = (1-0.40)^{10} = 0.006$. Likewise, if $\pi = 0.20$ then $\ell(0.20) = (1-0.20)^{10} = 0.107$, and if $\pi = 0.0$ then $\ell(0.0) = (1-0.0)^{10} = 1.0$. Figure 1.1 plots this likelihood function.

The *maximum likelihood estimate* of a parameter is the parameter value for which the probability of the observed data takes its greatest value. It is the parameter value at which the likelihood function takes its maximum. Figure 1.1 shows that the likelihood function $\ell(\pi) = (1-\pi)^{10}$ has its maximum at $\pi = 0.0$. Thus, when $n = 10$ trials have $y = 0$ successes, the maximum likelihood estimate of π equals 0.0. This means that the result $y = 0$ in $n = 10$ trials is more likely to occur when $\pi = 0.00$ than when π equals any other value.

In general, for the binomial outcome of y successes in n trials, the maximum likelihood estimate of π equals $p = y/n$. This is the sample proportion of successes for the n trials. If we observe $y = 6$ successes in $n = 10$ trials, then the maximum likelihood estimate of π equals $p = 6/10 = 0.60$. Figure 1.1 also plots the

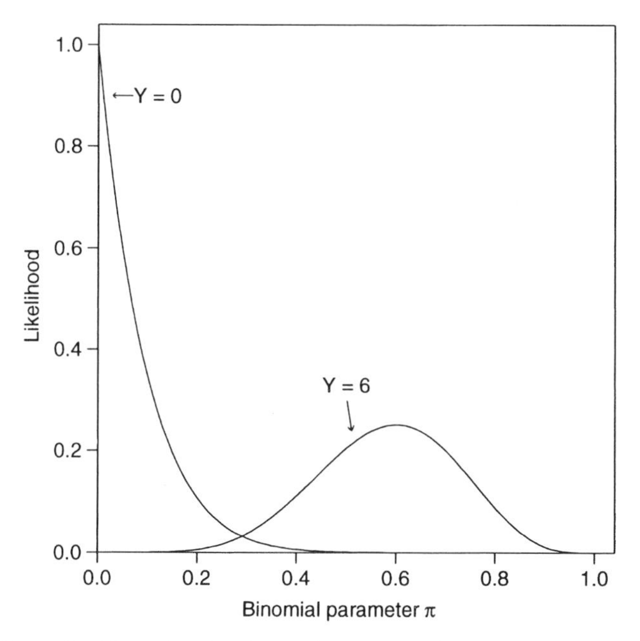

Figure 1.1. Binomial likelihood functions for $y = 0$ successes and for $y = 6$ successes in $n = 10$ trials.

likelihood function when $n = 10$ with $y = 6$, which from formula (1.1) equals $\ell(\pi) = [10!/(6!)(4!)]\pi^6(1-\pi)^4$. The maximum value occurs when $\pi = 0.60$. The result $y = 6$ in $n = 10$ trials is more likely to occur when $\pi = 0.60$ than when π equals any other value.

Denote each success by a 1 and each failure by a 0. Then the sample proportion equals the sample mean of the results of the individual trials. For instance, for four failures followed by six successes in 10 trials, the data are 0,0,0,0,1,1,1,1,1,1, and the sample mean is

$$p = (0+0+0+0+1+1+1+1+1+1)/10 = 0.60.$$

Thus, results that apply to sample means with random sampling, such as the Central Limit Theorem (large-sample normality of its sampling distribution) and the Law of Large Numbers (convergence to the population mean as n increases) apply also to sample proportions.

The abbreviation *ML* symbolizes the term *maximum likelihood*. The ML estimate is often denoted by the parameter symbol with a ˆ (a "hat") over it. The ML estimate of the binomial parameter π, for instance, is often denoted by $\hat{\pi}$, called *pi-hat*.

Before we observe the data, the value of the ML estimate is unknown. The estimate is then a variate having some sampling distribution. We refer to this variate as an *estimator* and its value for observed data as an *estimate*. Estimators based on the method of maximum likelihood are popular because they have good large-sample behavior. Most importantly, it is not possible to find good estimators that are more precise, in terms of having smaller large-sample standard errors. Also, large-sample distributions of ML estimators are usually approximately normal. The estimators reported in this text use this method.

1.3.2 Significance Test About a Binomial Proportion

For the binomial distribution, we now use the ML estimator in statistical inference for the parameter π. The ML estimator is the sample proportion, p. The sampling distribution of the sample proportion p has mean and standard error

$$E(p) = \pi, \quad \sigma(p) = \sqrt{\frac{\pi(1-\pi)}{n}}$$

As the number of trials n increases, the standard error of p decreases toward zero; that is, the sample proportion tends to be closer to the parameter value π. The sampling distribution of p is approximately normal for large n. This suggests large-sample inferential methods for π.

Consider the null hypothesis H_0: $\pi = \pi_0$ that the parameter equals some fixed value, π_0. The test statistic

$$z = \frac{p - \pi_0}{\sqrt{\frac{\pi_0(1-\pi_0)}{n}}} \tag{1.2}$$

divides the difference between the sample proportion p and the null hypothesis value π_0 by the *null standard error* of p. The null standard error is the one that holds under the assumption that the null hypothesis is true. For large samples, the null sampling distribution of the z test statistic is the standard normal – the normal distribution having a mean of 0 and standard deviation of 1. The z test statistic measures the number of standard errors that the sample proportion falls from the null hypothesized proportion.

1.3.3 Example: Survey Results on Legalizing Abortion

Do a majority, or minority, of adults in the United States believe that a pregnant woman should be able to obtain an abortion? Let π denote the proportion of the American adult population that responds "yes" to the question, "Please tell me whether or not you think it should be possible for a pregnant woman to obtain a legal abortion if she is married and does not want any more children." We test H_0: $\pi = 0.50$ against the two-sided alternative hypothesis, H_a: $\pi \neq 0.50$.

This item was one of many included in the 2002 General Social Survey. This survey, conducted every other year by the National Opinion Research Center (NORC) at the University of Chicago, asks a sample of adult American subjects their opinions about a wide variety of issues. (It is a multistage sample, but has characteristics similar to a simple random sample.) You can view responses to surveys since 1972 at http://sda.berkeley.edu/GSS. Of 893 respondents to this question in 2002, 400 replied "yes" and 493 replied "no".

The sample proportion of "yes" responses was $p = 400/893 = 0.448$. For a sample of size $n = 893$, the null standard error of p equals $\sqrt{[(0.50)(0.50)/893]} = 0.0167$. The test statistic is

$$z = (0.448 - 0.50)/0.0167 = -3.1$$

The two-sided P-value is the probability that the absolute value of a standard normal variate exceeds 3.1, which is $P = 0.002$. There is strong evidence that, in 2002, $\pi < 0.50$, that is, that fewer than half of Americans favored legal abortion in this situation. In some other situations, such as when the mother's health was endangered, an overwhelming majority favored legalized abortion. Responses depended strongly on the question wording.

1.3.4 Confidence Intervals for a Binomial Proportion

A significance test merely indicates whether a particular value for a parameter (such as 0.50) is plausible. We learn more by constructing a confidence interval to determine the range of plausible values. Let SE denote the estimated standard error of p. A large-sample $100(1 - \alpha)\%$ confidence interval for π has the formula

$$p \pm z_{\alpha/2}(SE), \quad \text{with } SE = \sqrt{p(1-p)/n} \tag{1.3}$$

where $z_{\alpha/2}$ denotes the standard normal percentile having right-tail probability equal to $\alpha/2$; for example, for 95% confidence, $\alpha = 0.05$, $z_{\alpha/2} = z_{0.025} = 1.96$. This formula substitutes the sample proportion p for the unknown parameter π in $\sigma(p) = \sqrt{[\pi(1-\pi)/n]}$.

For the attitudes about abortion example just discussed, $p = 0.448$ for $n = 893$ observations. The 95% confidence interval equals

$$0.448 \pm 1.96\sqrt{(0.448)(0.552)/893}, \quad \text{which is } 0.448 \pm 0.033, \quad \text{or } (0.415, 0.481)$$

We can be 95% confident that the population proportion of Americans in 2002 who favored legalized abortion for married pregnant women who do not want more children is between 0.415 and 0.481.

Formula (1.3) is simple. Unless π is close to 0.50, however, it does not work well unless n is very large. Consider its actual coverage probability, that is, the probability that the method produces an interval that captures the true parameter value. This may be quite a bit less than the nominal value (such as 95%). It is especially poor when π is near 0 or 1.

A better way to construct confidence intervals uses a duality with significance tests. This confidence interval consists of all values π_0 for the null hypothesis parameter that are judged plausible in the z test of the previous subsection. A 95% confidence interval contains all values π_0 for which the two-sided P-value exceeds 0.05. That is, it contains all values that are "not rejected" at the 0.05 significance level. These are the null values that have test statistic z less than 1.96 in absolute value. This alternative method does not require estimation of π in the standard error, since the standard error in the test statistic uses the null value π_0.

To illustrate, suppose a clinical trial to evaluate a new treatment has nine successes in the first 10 trials. For a sample proportion of $p = 0.90$ based on $n = 10$, the value $\pi_0 = 0.596$ for the null hypothesis parameter leads to the test statistic value

$$z = (0.90 - 0.596)/\sqrt{(0.596)(0.404)/10} = 1.96$$

and a two-sided P-value of $P = 0.05$. The value $\pi_0 = 0.982$ leads to

$$z = (0.90 - 0.982)/\sqrt{(0.982)(0.018)/100} = -1.96$$

and also a two-sided P-value of $P = 0.05$. (We explain in the following paragraph how to find 0.596 and 0.982.) All π_0 values between 0.596 and 0.982 have $|z| < 1.96$ and $P > 0.05$. So, the 95% confidence interval for π equals (0.596, 0.982). By contrast, the method (1.3) using the *estimated* standard error gives confidence interval $0.90 \pm 1.96\sqrt{[(0.90)(0.10)/10]}$, which is (0.714, 1.086). However, it works poorly to use the sample proportion as the midpoint of the confidence interval when the parameter may fall near the boundary values of 0 or 1.

For given p and n, the π_0 values that have test statistic value $z = \pm 1.96$ are the solutions to the equation

$$\frac{| p - \pi_0 |}{\sqrt{\pi_0(1 - \pi_0)/n}} = 1.96$$

for π_0. To solve this for π_0, squaring both sides gives an equation that is quadratic in π_0 (see Exercise 1.18). The results are available with some software, such as an R function available at http://www.stat.ufl.edu/~aa/cda/software.html.

Here is a simple alternative interval that approximates this one, having a similar midpoint in the 95% case but being a bit wider: Add 2 to the number of successes and 2 to the number of failures (and thus 4 to n) and then use the ordinary formula (1.3) with the estimated standard error. For example, with nine successes in 10 trials, you find $p = (9 + 2)/(10 + 4) = 0.786$, $SE = \sqrt{[0.786(0.214)/14]} = 0.110$, and obtain confidence interval (0.57, 1.00). This simple method, sometimes called the *Agresti–Coull confidence interval*, works well even for small samples.[1]

[1] A. Agresti and B. Coull, *Am. Statist.*, **52**: 119–126, 1998.

1.4 MORE ON STATISTICAL INFERENCE FOR DISCRETE DATA

We have just seen how to construct a confidence interval for a proportion using an estimated standard error or by inverting results of a significance test using the null standard error. In fact, there are three ways of using the likelihood function to conduct inference (confidence intervals and significance tests) about parameters. We finish the chapter by summarizing these methods. They apply to any parameter in a statistical model, but we will illustrate using the binomial parameter.

1.4.1 Wald, Likelihood-Ratio, and Score Inference

Let β denote an arbitrary parameter. Consider a significance test of H_0: $\beta = \beta_0$ (such as H_0: $\beta = 0$, for which $\beta_0 = 0$).

The simplest test statistic uses the large-sample normality of the ML estimator $\hat{\beta}$. Let SE denote the standard error of $\hat{\beta}$, evaluated by substituting the ML estimate for the unknown parameter in the expression for the true standard error. (For example, for the binomial parameter π, $SE = \sqrt{[p(1-p)/n]}$.) When H_0 is true, the test statistic

$$z = (\hat{\beta} - \beta_0)/SE$$

has approximately a standard normal distribution. Equivalently, z^2 has approximately a chi-squared distribution with $df = 1$. This type of statistic, which uses the standard error evaluated at the ML estimate, is called a *Wald statistic*. The z or chi-squared test using this test statistic is called a *Wald test*.

You can refer z to the standard normal table to get one-sided or two-sided P-values. Equivalently, for the two-sided alternative H_0: $\beta \neq \beta_0$, z^2 has a chi-squared distribution with $df = 1$. The P-value is then the right-tail chi-squared probability above the observed value. The two-tail probability beyond $\pm z$ for the standard normal distribution equals the right-tail probability above z^2 for the chi-squared distribution with $df = 1$. For example, the two-tail standard normal probability of 0.05 that falls below -1.96 and above 1.96 equals the right-tail chi-squared probability above $(1.96)^2 = 3.84$ when $df = 1$.

An alternative test uses the likelihood function through the ratio of two maximizations of it: (1) the maximum over the possible parameter values that assume the null hypothesis, (2) the maximum over the larger set of possible parameter values, permitting the null or the alternative hypothesis to be true. Let ℓ_0 denote the maximized value of the likelihood function under the null hypothesis, and let ℓ_1 denote the maximized value more generally. For instance, when there is a single parameter β, ℓ_0 is the likelihood function calculated at β_0, and ℓ_1 is the likelihood function calculated at the ML estimate $\hat{\beta}$. Then ℓ_1 is always at least as large as ℓ_0, because ℓ_1 refers to maximizing over a larger set of possible parameter values.

The *likelihood-ratio* test statistic equals

$$-2 \log(\ell_0/\ell_1)$$

In this text, we use the *natural log*, often abbreviated on calculators by LN. If the maximized likelihood is much larger when the parameters are not forced to satisfy H_0, then the ratio ℓ_0/ℓ_1 is far below 1. The test statistic $-2\log(\ell_0/\ell_1)$ must be nonnegative, and relatively small values of ℓ_0/ℓ_1 yield large values of $-2\log(\ell_0/\ell_1)$ and strong evidence against H_0. The reason for taking the log transform and doubling is that it yields an approximate chi-squared sampling distribution. Under H_0: $\beta = \beta_0$, the likelihood-ratio test statistic has a large-sample chi-squared distribution with $df = 1$. Software can find the maximized likelihood values and the likelihood-ratio test statistic.

A third possible test is called the *score test*. We will not discuss the details except to say that it finds standard errors under the assumption that the null hypothesis holds. For example, the z test (1.2) for a binomial parameter that uses the standard error $\sqrt{[\pi_0(1-\pi_0)/n]}$ is a score test.

The Wald, likelihood-ratio, and score tests are the three major ways of constructing significance tests for parameters in statistical models. For ordinary regression models assuming a normal distribution for Y, the three tests provide identical results. In other cases, for large samples they have similar behavior when H_0 is true.

When you use any of these tests, the P-value that you find or software reports is an approximation for the true P-value. This is because the normal (or chi-squared) sampling distribution used is a large-sample approximation for the true sampling distribution. Thus, when you report a P-value, it is overly optimistic to use many decimal places. If you are lucky, the P-value approximation is good to the second decimal place. So, for a P-value that software reports as 0.028374, it makes more sense to report it as 0.03 (or, at best, 0.028) rather than 0.028374. An exception is when the P-value is zero to many decimal places, in which case it is sensible to report it as $P < 0.001$ or $P < 0.0001$. In any case, a P-value merely summarizes the strength of evidence against the null hypothesis, and accuracy to two or three decimal places is sufficient for this purpose.

Each method has a corresponding confidence interval. This is based on inverting results of the significance test: The 95% confidence interval for a parameter β is the set of β_0 values for the significance test of H_0: $\beta = \beta_0$ such that the P-value is larger than 0.05. For example, the 95% *Wald confidence interval* is the set of β_0 values for which $z = (\hat{\beta} - \beta_0)/SE$ has $|z| < 1.96$. It equals $\hat{\beta} \pm 1.96(SE)$. For a binomial proportion, the score confidence interval is the one discussed in Section 1.3.4 that has endpoints that are π_0 values having P-value 0.05 in the z-test using the null standard error.

1.4.2 Wald, Score, and Likelihood-Ratio Inference for Binomial Parameter

We illustrate the Wald, likelihood-ratio, and score tests by testing H_0: $\pi = 0.50$ against H_a: $\pi \neq 0.50$ for the example mentioned near the end of Section 1.3.4 of a clinical trial that has nine successes in the first 10 trials. The sample proportion is $p = 0.90$ based on $n = 10$.

For the Wald test of H_0: $\pi = 0.50$, the estimated standard error is $SE = \sqrt{[0.90(0.10)/10]} = 0.095$. The z test statistic is

$$z = (0.90 - 0.50)/0.095 = 4.22$$

The corresponding chi-squared statistic is $(4.22)^2 = 17.8$ $(df = 1)$. The P-value <0.001.

For the score test of H_0: $\pi = 0.50$, the null standard error is $\sqrt{[0.50(0.50)/10]} = 0.158$. The z test statistic is

$$z = (0.90 - 0.50)/0.158 = 2.53$$

The corresponding chi-squared statistic is $(2.53)^2 = 6.4$ $(df = 1)$. The P-value $= 0.011$.

Finally, consider the likelihood-ratio test. When H_0: $\pi = 0.50$ is true, the binomial probability of the observed result of nine successes is $\ell_0 = [10!/9!1!]$ $(0.50)^9(0.50)^1 = 0.00977$. The likelihood-ratio test compares this to the value of the likelihood function at the ML estimate of $p = 0.90$, which is $\ell_1 = [10!/9!1!](0.90)^9(0.10)^1 = 0.3874$. The likelihood-ratio test statistic equals

$$-2\log(\ell_0/\ell_1) = -2\log(0.00977/0.3874) = -2\log(0.0252) = 7.36$$

From the chi-squared distribution with $df = 1$, this statistic has P-value $= 0.007$.

When the sample size is small to moderate, the Wald test is the least reliable of the three tests. We should not trust it for such a small n as in this example $(n = 10)$. Likelihood-ratio inference and score-test based inference are better in terms of actual error probabilities coming close to matching nominal levels. A marked divergence in the values of the three statistics indicates that the distribution of the ML estimator may be far from normality. In that case, small-sample methods are more appropriate than large-sample methods.

1.4.3 Small-Sample Binomial Inference

For inference about a proportion, the large-sample two-sided z score test and the confidence interval based on that test (using the null hypothesis standard error) perform reasonably well when $n\pi \geq 5$ and $n(1 - \pi) \geq 5$. When π_0 is not near 0.50 the normal P-value approximation is better for the test with a two-sided alternative than for a one-sided alternative; a probability that is "too small" in one tail tends to be approximately counter-balanced by a probability that is "too large" in the other tail.

For small sample sizes, it is safer to use the binomial distribution directly (rather than a normal approximation) to calculate P-values. To illustrate, consider testing H_0: $\pi = 0.50$ against H_a: $\pi > 0.50$ for the example of a clinical trial to evaluate a new treatment, when the number of successes $y = 9$ in $n = 10$ trials. The exact P-value, based on the right tail of the null binomial distribution with $\pi = 0.50$, is

$$P(Y \geq 9) = [10!/9!1!](0.50)^9(0.50)^1 + [10!/10!0!](0.50)^{10}(0.50)^0 = 0.011$$

For the two sided alternative H_a: $\pi \neq 0.50$, the P-value is

$$P(Y \geq 9 \text{ or } Y \leq 1) = 2 \times P(Y \geq 9) = 0.021$$

1.4.4 Small-Sample Discrete Inference is Conservative*

Unfortunately, with discrete probability distributions, small-sample inference using the ordinary P-value is *conservative*. This means that when H_0 is true, the P-value is ≤ 0.05 (thus leading to rejection of H_0 at the 0.05 significance level) not *exactly* 5% of the time, but typically *less* than 5% of the time. Because of the discreteness, it is usually not possible for a P-value to achieve the desired significance level exactly. Then, the actual probability of type I error is less than 0.05.

For example, consider testing H_0: $\pi = 0.50$ against H_a: $\pi > 0.50$ for the clinical trial example with $y = 9$ successes in $n = 10$ trials. Table 1.1 showed the binomial distribution with $n = 10$ and $\pi = 0.50$. Table 1.2 shows it again with the corresponding P-values (right-tail probabilities) for this one-sided alternative. The P-value is ≤ 0.05 when $y = 9$ or 10. This happens with probability $0.010 + 0.001 = 0.011$. Thus, the probability of getting a P-value ≤ 0.05 is only 0.011. For a desired significance level of 0.05, the actual probability of type I error is 0.011. The actual probability of type I error is much smaller than the intended one.

This illustrates an awkward aspect of significance testing when the test statistic has a discrete distribution. For test statistics having a *continuous* distribution, the P-value has a *uniform* null distribution over the interval [0, 1]. That is, when H_0 is true, the P-value is equally likely to fall anywhere between 0 and 1. Then, the probability that the P-value falls below 0.05 equals exactly 0.05, and the expected value of the P-value is exactly 0.50. For a test statistic having a *discrete* distribution, the null distribution of the P-value is discrete and has an expected value greater than 0.50.

For example, for the one-sided test summarized above, the P-value equals 1.000 with probability $P(0) = 0.001$, it equals 0.999 with probability $P(1) = 0.010, \ldots,$ and it equals 0.001 with probability $P(10) = 0.001$. From the table, the null expected

Table 1.2. Null Binomial Distribution and One-Sided P-values for Testing H_0: $\pi = 0.50$ against H_a: $\pi > 0.50$ with $n = 10$

y	$P(y)$	P-value	Mid P-value
0	0.001	1.000	0.9995
1	0.010	0.999	0.994
2	0.044	0.989	0.967
3	0.117	0.945	0.887
4	0.205	0.828	0.726
5	0.246	0.623	0.500
6	0.205	0.377	0.274
7	0.117	0.172	0.113
8	0.044	0.055	0.033
9	0.010	0.011	0.006
10	0.001	0.001	0.0005

value of the P-value is

$$\sum P \times \text{Prob}(P) = 1.000(0.001) + 0.999(0.010) + \cdots + 0.001(0.001) = 0.59$$

In this average sense, P-values for discrete distributions tend to be too large.

1.4.5 Inference Based on the Mid P-value*

With small samples of discrete data, many statisticians prefer to use a different type of P-value. Called the *mid P-value*, it adds only *half* the probability of the observed result to the probability of the more extreme results. To illustrate, in the above example with $y = 9$ successes in $n = 10$ trials, the ordinary P-value for H_a: $\pi > 0.50$ is $P(9) + P(10) = 0.010 + 0.001 = 0.011$. The mid P-value is $P(9)/2 + P(10) = 0.010/2 + 0.001 = 0.006$. Table 1.2 also shows the mid P-values for the possible y values when $n = 10$.

Tests using the mid P-value are, on the average, less conservative than tests using the ordinary P-value. The mid P-value has a null expected value of 0.50, the same as the regular P-value for continuous variates. Also, the two separate one-sided mid P-values sum to 1.0. For example, for $y = 9$ when $n = 10$, for H_a: $\pi > 0.50$ the ordinary P-value is

$$\text{right-tail } P\text{-value} = P(9) + P(10) = 0.011$$

and for H_a: $\pi < 0.50$ it is

$$\text{left-tail } P\text{-value} = P(0) + P(1) + \cdots + P(9) = 0.999$$

That is, $P(9)$ gets counted in each tail for each P-value. By contrast, for H_a: $\pi > 0.50$, the mid P-value is

$$\text{right-tail mid } P\text{-value} = P(9)/2 + P(10) = 0.006$$

and for H_a: $\pi < 0.50$ it is

$$\text{left-tail mid } P\text{-value} = P(0) + P(1) + \cdots + P(9)/2 = 0.994$$

and these one-sided mid P-values sum to 1.0.

The two-sided P-value for the large-sample z score test approximates the two-sided mid P-value in the small-sample binomial test. For example, with $y = 9$ in $n = 10$ trials for H_0: $\pi = 0.50$, $z = (0.90 - 0.50)/\sqrt{[0.50(0.50)/10]} = 2.53$ has two-sided P-value $= 0.0114$. The two-sided mid P-value is $2[P(9)/2 + P(10)] = 0.0117$.

For small samples, one can construct confidence intervals by inverting results of significance tests that use the binomial distribution, rather than a normal approximation. Such inferences are very conservative when the test uses the ordinary P-value. We recommend inverting instead the binomial test using the mid P-value. The mid-P confidence interval is the set of π_0 values for a two-sided test in which the mid P-value using the binomial distribution exceeds 0.05. This is available in some software, such as an R function (written by A. Gottard) at http://www.stat.ufl.edu/~aa/cda/software.html.

1.4.6 Summary

This chapter has introduced the key distributions for categorical data analysis: the binomial and the multinomial. It has also introduced maximum likelihood estimation and illustrated its use for proportion data using Wald, likelihood-ratio, and score methods of inference. The rest of the text uses ML inference for binomial and multinomial parameters in a wide variety of contexts.

PROBLEMS

1.1 In the following examples, identify the response variable and the explanatory variables.

a. Attitude toward gun control (favor, oppose), Gender (female, male), Mother's education (high school, college).

b. Heart disease (yes, no), Blood pressure, Cholesterol level.

c. Race (white, nonwhite), Religion (Catholic, Jewish, Protestant), Vote for president (Democrat, Republican, Other), Annual income.

d. Marital status (married, single, divorced, widowed), Quality of life (excellent, good, fair, poor).

1.2 Which scale of measurement is most appropriate for the following variables – nominal, or ordinal?

a. Political party affiliation (Democrat, Republican, unaffiliated).

b. Highest degree obtained (none, high school, bachelor's, master's, doctorate).

c. Patient condition (good, fair, serious, critical).

d. Hospital location (London, Boston, Madison, Rochester, Toronto).

e. Favorite beverage (beer, juice, milk, soft drink, wine, other).

f. How often feel depressed (never, occasionally, often, always).

1.3 Each of 100 multiple-choice questions on an exam has four possible answers but one correct response. For each question, a student randomly selects one response as the answer.

a. Specify the distribution of the student's number of correct answers on the exam.

b. Based on the mean and standard deviation of that distribution, would it be surprising if the student made at least 50 correct responses? Explain your reasoning.

1.4 A coin is flipped twice. Let Y = number of heads obtained, when the probability of a head for a flip equals π.

a. Assuming $\pi = 0.50$, specify the probabilities for the possible values for Y, and find the distribution's mean and standard deviation.

b. Find the binomial probabilities for Y when π equals (i) 0.60, (ii) 0.40.

c. Suppose you observe $y = 1$ and do not know π. Calculate and sketch the likelihood function.

d. Using the plotted likelihood function from (c), show that the ML estimate of π equals 0.50.

1.5 Refer to the previous exercise. Suppose $y = 0$ in 2 flips. Find the ML estimate of π. Does this estimate seem "reasonable"? Why? [The *Bayesian* estimator is an alternative one that combines the sample data with your prior beliefs about the parameter value. It provides a nonzero estimate of π, equaling $(y + 1)/(n + 2)$ when your prior belief is that π is equally likely to be anywhere between 0 and 1.]

1.6 Genotypes AA, Aa, and aa occur with probabilities (π_1, π_2, π_3). For $n = 3$ independent observations, the observed frequencies are (n_1, n_2, n_3).

a. Explain how you can determine n_3 from knowing n_1 and n_2. Thus, the multinomial distribution of (n_1, n_2, n_3) is actually two-dimensional.

b. Show the set of all possible observations, (n_1, n_2, n_3) with $n = 3$.

c. Suppose $(\pi_1, \pi_2, \pi_3) = (0.25, 0.50, 0.25)$. Find the multinomial probability that $(n_1, n_2, n_3) = (1, 2, 0)$.

d. Refer to (c). What probability distribution does n_1 alone have? Specify the values of the sample size index and parameter for that distribution.

1.7 In his autobiography *A Sort of Life*, British author Graham Greene described a period of severe mental depression during which he played Russian Roulette. This "game" consists of putting a bullet in one of the six chambers of a pistol, spinning the chambers to select one at random, and then firing the pistol once at one's head.

a. Greene played this game six times, and was lucky that none of them resulted in a bullet firing. Find the probability of this outcome.

b. Suppose one kept playing this game until the bullet fires. Let Y denote the number of the game on which the bullet fires. Argue that the probability of the outcome y equals $(5/6)^{y-1}(1/6)$, for $y = 1, 2, 3, \ldots$. (This is called the *geometric distribution*.)

1.8 When the 2000 General Social Survey asked subjects whether they would be willing to accept cuts in their standard of living to protect the environment, 344 of 1170 subjects said "yes."

a. Estimate the population proportion who would say "yes."

b. Conduct a significance test to determine whether a majority or minority of the population would say "yes." Report and interpret the P-value.

c. Construct and interpret a 99% confidence interval for the population proportion who would say "yes."

1.9 A sample of women suffering from excessive menstrual bleeding have been taking an analgesic designed to diminish the effects. A new analgesic is claimed to provide greater relief. After trying the new analgesic, 40 women reported greater relief with the standard analgesic, and 60 reported greater relief with the new one.

a. Test the hypothesis that the probability of greater relief with the standard analgesic is the same as the probability of greater relief with the new analgesic. Report and interpret the P-value for the two-sided alternative. (Hint: Express the hypotheses in terms of a single parameter. A test to compare matched-pairs responses in terms of which is better is called a *sign test*.)

b. Construct and interpret a 95% confidence interval for the probability of greater relief with the new analgesic.

1.10 Refer to the previous exercise. The researchers wanted a sufficiently large sample to be able to estimate the probability of preferring the new analgesic to within 0.08, with confidence 0.95. If the true probability is 0.75, how large a sample is needed to achieve this accuracy? (Hint: For how large an n does a 95% confidence interval have margin of error equal to about 0.08?)

1.11 When a recent General Social Survey asked 1158 American adults, "Do you believe in Heaven?", the proportion who answered yes was 0.86. Treating this as a random sample, conduct statistical inference about the true proportion of American adults believing in heaven. Summarize your analysis and interpret the results in a short report of about 200 words.

1.12 To collect data in an introductory statistics course, recently I gave the students a questionnaire. One question asked whether the student was a vegetarian. Of 25 students, 0 answered "yes." They were not a random sample, but let us use these data to illustrate inference for a proportion. (You may wish to refer to Section 1.4.1 on methods of inference.) Let π denote the population proportion who would say "yes." Consider H_0: $\pi = 0.50$ and H_a: $\pi \neq 0.50$.

a. What happens when you try to conduct the "Wald test," for which $z = (p - \pi_0)/\sqrt{[p(1-p)/n]}$ uses the *estimated* standard error?

b. Find the 95% "Wald confidence interval" (1.3) for π. Is it believable? (When the observation falls at the boundary of the sample space, often Wald methods do not provide sensible answers.)

c. Conduct the "score test," for which $z = (p - \pi_0)/\sqrt{[\pi_0(1-\pi_0)/n]}$ uses the *null* standard error. Report the P-value.

d. Verify that the 95% score confidence interval (i.e., the set of π_0 for which $|z| < 1.96$ in the score test) equals (0.0, 0.133). (Hint: What do the z test statistic and P-value equal when you test H_0: $\pi = 0.133$ against H_a: $\pi \neq 0.133$.)

1.13 Refer to the previous exercise, with $y = 0$ in $n = 25$ trials.

a. Show that ℓ_0, the maximized likelihood under H_0, equals $(1 - \pi_0)^{25}$, which is $(0.50)^{25}$ for H_0: $\pi = 0.50$.

b. Show that ℓ_1, the maximum of the likelihood function over all possible π values, equals 1.0. (Hint: This is the value at the ML estimate value of 0.0.)

c. For H_0: $\pi = 0.50$, show that the likelihood-ratio test statistic, $-2\log(\ell_0/\ell_1)$, equals 34.7. Report the P-value.

d. The 95% likelihood-ratio confidence interval for π is (0.000, 0.074). Verify that 0.074 is the correct upper bound by showing that the likelihood-ratio test of H_0: $\pi = 0.074$ against H_a: $\pi \neq 0.074$ has chi-squared test statistic equal to 3.84 and P-value = 0.05.

1.14 Sections 1.4.4 and 1.4.5 found binomial P-values for a clinical trial with $y = 9$ successes in 10 trials. Suppose instead $y = 8$. Using the binomial distribution shown in Table 1.2:

a. Find the P-value for (i) H_a: $\pi > 0.50$, (ii) H_a: $\pi < 0.50$.

b. Find the mid P-value for (i) H_a: $\pi > 0.50$, (ii) H_a: $\pi < 0.50$.

c. Why is the sum of the one-sided P-values greater than 1.0 for the ordinary P-value but equal to 1.0 for the mid P-value?

1.15 If Y is a variate and c is a positive constant, then the standard deviation of the distribution of cY equals $c\sigma(Y)$. Suppose Y is a binomial variate, and let $p = Y/n$.

a. Based on the binomial standard deviation for Y, show that $\sigma(p) = \sqrt{[\pi(1-\pi)/n]}$.

b. Explain why it is easier to estimate π precisely when it is near 0 or 1 than when it is near 0.50.

1.16 Using calculus, it is easier to derive the maximum of the log of the likelihood function, $L = \log \ell$, than the likelihood function ℓ itself. Both functions have maximum at the same value, so it is sufficient to do either.

a. Calculate the log likelihood function $L(\pi)$ for the binomial distribution (1.1).

b. One can usually determine the point at which the maximum of a log likelihood L occurs by solving the *likelihood equation*. This is the equation resulting from differentiating L with respect to the parameter, and setting the derivative equal to zero. Find the likelihood equation for the binomial distribution, and solve it to show that the ML estimate equals $p = y/n$.

1.17 Suppose a researcher routinely conducts significance tests by rejecting H_0 if the P-value satisfies $P \leq 0.05$. Suppose a test using a test statistic T and right-tail probability for the P-value has null distribution $P(T = 0) = 0.30$, $P(T = 3) = 0.62$, and $P(T = 9) = 0.08$.

a. Show that with the usual P-value, the actual probability of type I error is 0 rather than 0.05.

b. Show that with the mid P-value, the actual probability of type I error equals 0.08.

c. Repeat (**a**) and (**b**) using $P(T = 0) = 0.30$, $P(T = 3) = 0.66$, and $P(T = 9) = 0.04$. Note that the test with mid P-value can be "conservative" [having actual P(type I error) below the desired value] or "liberal" [having actual P(type I error) above the desired value]. The test with the ordinary P-value cannot be liberal.

1.18 For a given sample proportion p, show that a value π_0 for which the test statistic $z = (p - \pi_0)/\sqrt{[\pi_0(1 - \pi_0)/n]}$ takes some fixed value z_0 (such as 1.96) is a solution to the equation $(1 + z_0^2/n)\pi_0^2 + (-2p - z_0^2/n)\pi_0 + p^2 = 0$. Hence, using the formula $x = [-b \pm \sqrt{(b^2 - 4ac)}]/2a$ for solving the quadratic equation $ax^2 + bx + c = 0$, obtain the limits for the 95% confidence interval in Section 1.3.4 for the probability of success when a clinical trial has nine successes in 10 trials.

CHAPTER 2

Contingency Tables

Table 2.1 cross classifies a sample of Americans according to their gender and their opinion about an afterlife. For the females in the sample, for example, 509 said they believed in an afterlife and 116 said they did not or were undecided. Does an association exist between gender and belief in an afterlife? Is one gender more likely than the other to believe in an afterlife, or is belief in an afterlife independent of gender?

Table 2.1. Cross Classification of Belief in Afterlife by Gender

	Belief in Afterlife	
Gender	Yes	No or Undecided
Females	509	116
Males	398	104

Source: Data from 1998 General Social Survey.

Analyzing associations is at the heart of multivariate statistical analysis. This chapter deals with associations between categorical variables. We introduce parameters that describe the association and we present inferential methods for those parameters.

2.1 PROBABILITY STRUCTURE FOR CONTINGENCY TABLES

For a single categorical variable, we can summarize the data by counting the number of observations in each category. The sample proportions in the categories estimate the category probabilities.

Suppose there are two categorical variables, denoted by X and Y. Let I denote the number of categories of X and J the number of categories of Y. A rectangular table having I rows for the categories of X and J columns for the categories of Y has *cells* that display the IJ possible combinations of outcomes.

A table of this form that displays counts of outcomes in the cells is called a *contingency table*. A table that cross classifies two variables is called a *two-way contingency table*; one that cross classifies three variables is called a *three-way contingency table*, and so forth. A two-way table with I rows and J columns is called an $I \times J$ (read I–by–J) table. Table 2.1 is a 2×2 table.

2.1.1 Joint, Marginal, and Conditional Probabilities

Probabilities for contingency tables can be of three types – *joint*, *marginal*, or *conditional*. Suppose first that a randomly chosen subject from the population of interest is classified on X and Y. Let $\pi_{ij} = P(X = i, Y = j)$ denote the probability that (X, Y) falls in the cell in row i and column j. The probabilities $\{\pi_{ij}\}$ form the *joint distribution* of X and Y. They satisfy $\sum_{i,j} \pi_{ij} = 1$.

The *marginal distributions* are the row and column totals of the joint probabilities. We denote these by $\{\pi_{i+}\}$ for the row variable and $\{\pi_{+j}\}$ for the column variable, where the subscript "+" denotes the sum over the index it replaces. For 2×2 tables,

$$\pi_{1+} = \pi_{11} + \pi_{12} \quad \text{and} \quad \pi_{+1} = \pi_{11} + \pi_{21}$$

Each marginal distribution refers to a single variable.

We use similar notation for samples, with Roman p in place of Greek π. For example, $\{p_{ij}\}$ are cell proportions in a sample joint distribution. We denote the cell counts by $\{n_{ij}\}$. The marginal frequencies are the row totals $\{n_{i+}\}$ and the column totals $\{n_{+j}\}$, and $n = \sum_{i,j} n_{ij}$ denotes the total sample size. The sample cell proportions relate to the cell counts by

$$p_{ij} = n_{ij}/n$$

In many contingency tables, one variable (say, the column variable, Y) is a response variable and the other (the row variable, X) is an explanatory variable. Then, it is informative to construct a separate probability distribution for Y at each level of X. Such a distribution consists of *conditional probabilities* for Y, given the level of X. It is called a *conditional distribution*.

2.1.2 Example: Belief in Afterlife

Table 2.1 cross classified $n = 1127$ respondents to a General Social Survey by their gender and by their belief in an afterlife. Table 2.2 illustrates the cell count notation for these data. For example, $n_{11} = 509$, and the related sample joint proportion is $p_{11} = 509/1127 = 0.45$.

Table 2.2. Notation for Cell Counts in Table 2.1

	Belief in Afterlife		
Gender	Yes	No or Undecided	Total
Females	$n_{11} = 509$	$n_{12} = 116$	$n_{1+} = 625$
Males	$n_{21} = 398$	$n_{22} = 104$	$n_{2+} = 502$
Total	$n_{+1} = 907$	$n_{+2} = 220$	$n = 1127$

In Table 2.1, belief in the afterlife is a response variable and gender is an explanatory variable. We therefore study the conditional distributions of belief in the afterlife, given gender. For females, the proportion of "yes" responses was $509/625 = 0.81$ and the proportion of "no" responses was $116/625 = 0.19$. The proportions $(0.81, 0.19)$ form the sample conditional distribution of belief in the afterlife. For males, the sample conditional distribution is $(0.79, 0.21)$.

2.1.3 Sensitivity and Specificity in Diagnostic Tests

Diagnostic testing is used to detect many medical conditions. For example, the mammogram can detect breast cancer in women, and the prostate-specific antigen (PSA) test can detect prostate cancer in men. The result of a diagnostic test is said to be *positive* if it states that the disease is present and *negative* if it states that the disease is absent.

The accuracy of diagnostic tests is often assessed with two conditional probabilities: Given that a subject has the disease, the probability the diagnostic test is positive is called the *sensitivity*. Given that the subject does not have the disease, the probability the test is negative is called the *specificity*. Let X denote the true state of a person, with categories 1 = diseased, 2 = not diseased, and let Y = outcome of diagnostic test, with categories 1 = positive, 2 = negative. Then,

$$\text{sensitivity} = P(Y = 1|X = 1), \quad \text{specificity} = P(Y = 2|X = 2)$$

The higher the sensitivity and specificity, the better the diagnostic test.

In practice, if you get a positive result, what is more relevant is $P(X = 1|Y = 1)$. Given that the diagnostic test says you have the disease, what is the probability you truly have it? When relatively few people have the disease, this probability can be low even when the sensitivity and specificity are high. For example, breast cancer is the most common form of cancer in women. Of women who get mammograms at any given time, it has been estimated that 1% truly have breast cancer. Typical values reported for mammograms are sensitivity = 0.86 and specificity = 0.88. If these are true, then given that a mammogram has a positive result, the probability that the woman truly has breast cancer is only 0.07. This can be shown with Bayes theorem (see Exercise 2.2).

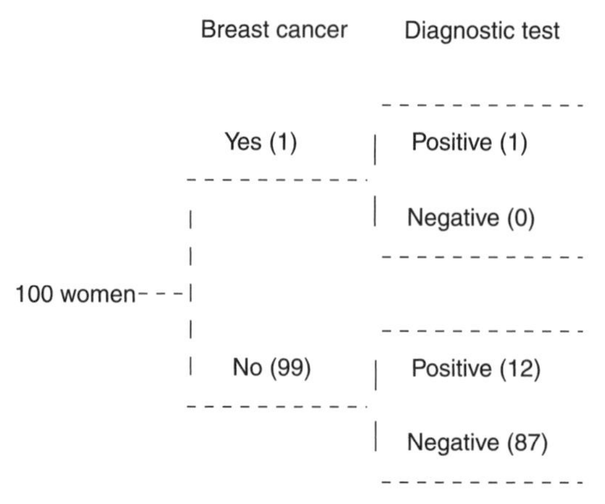

Figure 2.1. Tree diagram showing results of 100 mammograms, when sensitivity = 0.86 and specificity = 0.88.

How can $P(X = 1|Y = 1)$ be so low, given the relatively good sensitivity and specificity? Figure 2.1 is a tree diagram that shows results for a typical sample of 100 women. The first set of branches shows whether a woman has breast cancer. Here, one of the 100 women have it, 1% of the sample. The second set of branches shows the mammogram result, given the disease status. For a woman with breast cancer, there is a 0.86 probability of detecting it. So, we would expect the one woman with breast cancer to have a positive result, as the figure shows. For a woman without breast cancer, there is a 0.88 probability of a negative result. So, we would expect about $(0.88)99 = 87$ of the 99 women without breast cancer to have a negative result, and $(0.12)99 = 12$ to have a positive result. Figure 2.1 shows that of the 13 women with a positive test result, the proportion $1/13 = 0.08$ actually have breast cancer. The small proportion of errors for the large majority of women who do not have breast cancer swamps the large proportion of correct diagnoses for the few women who have it.

2.1.4 Independence

Two variables are said to be *statistically independent* if the population conditional distributions of Y are identical at each level of X. When two variables are independent, the probability of any particular column outcome j is the same in each row. Belief in an afterlife is independent of gender, for instance, if the actual probability of believing in an afterlife equals 0.80 both for females and for males.

When both variables are response variables, we can describe their relationship using their joint distribution, or the conditional distribution of Y given X, or the conditional distribution of X given Y. Statistical independence is, equivalently, the property that all joint probabilities equal the product of their marginal probabilities,

$$\pi_{ij} = \pi_{i+}\pi_{+j} \quad \text{for } i = 1, \ldots, I \text{ and } j = 1, \ldots, J$$

That is, the probability that X falls in row i and Y falls in column j is the product of the probability that X falls in row i with the probability that Y falls in column j.

2.1.5 Binomial and Multinomial Sampling

Section 1.2 introduced the *binomial* and *multinomial* distributions. With random sampling or randomized experiments, it is often sensible to assume that cell counts in contingency tables have one of these distributions.

When the rows of a contingency table refer to different groups, the sample sizes for those groups are often fixed by the sampling design. An example is a randomized experiment in which half the sample is randomly allocated to each of two treatments. When the marginal totals for the levels of X are fixed rather than random, a joint distribution for X and Y is not meaningful, but conditional distributions for Y at each level of X are. When there are two outcome categories for Y, the binomial distribution applies for each conditional distribution. We assume a binomial distribution for the sample in each row, with number of trials equal to the fixed row total. When there are more than two outcome categories for Y, such as (always, sometimes, never), the multinomial distribution applies for each conditional distribution.

Likewise, when the columns are a response variable and the rows are an explanatory variable, it is sensible to divide the cell counts by the row totals to form conditional distributions on the response. In doing so, we inherently treat the row totals as fixed and analyze the data the same way as if the two rows formed separate samples. For example, Table 2.1 cross classifies a random sample of 1127 subjects according to gender and belief in afterlife. Since belief in afterlife is the response variable, we might treat the results for females as a binomial sample with outcome categories "yes" and "no or undecided" for belief in an afterlife, and the results for males as a separate binomial sample on that response. For a multicategory response variable, we treat the samples as separate multinomial samples.

When the total sample size n is fixed and we cross classify the sample on two categorical response variables, the multinomial distribution is the actual joint distribution over the cells. The cells of the contingency table are the possible outcomes, and the cell probabilities are the multinomial parameters. In Table 2.1, for example, the four cell counts are sample values from a multinomial distribution having four categories.

2.2 COMPARING PROPORTIONS IN TWO-BY-TWO TABLES

Response variables having two categories are called *binary variables*. For instance, belief in afterlife is binary when measured with categories (yes, no). Many studies compare two groups on a binary response, Y. The data can be displayed in a 2×2 contingency table, in which the rows are the two groups and the columns are the response levels of Y. This section presents measures for comparing groups on binary responses.

2.2.1 Difference of Proportions

As in the discussion of the binomial distribution in Section 1.2, we use the generic terms *success* and *failure* for the outcome categories. For subjects in row 1, let π_1 denote the probability of a success, so $1 - \pi_1$ is the probability of a failure. For subjects in row 2, let π_2 denote the probability of success. These are conditional probabilities.

The *difference of proportions* $\pi_1 - \pi_2$ compares the success probabilities in the two rows. This difference falls between -1 and $+1$. It equals zero when $\pi_1 = \pi_2$, that is, when the response is independent of the group classification. Let p_1 and p_2 denote the *sample* proportions of successes. The sample difference $p_1 - p_2$ estimates $\pi_1 - \pi_2$.

For simplicity, we denote the sample sizes for the two groups (that is, the row totals n_{1+} and n_{2+}) by n_1 and n_2. When the counts in the two rows are independent binomial samples, the estimated standard error of $p_1 - p_2$ is

$$SE = \sqrt{\frac{p_1(1-p_1)}{n_1} + \frac{p_2(1-p_2)}{n_2}} \tag{2.1}$$

The standard error decreases, and hence the estimate of $\pi_1 - \pi_2$ improves, as the sample sizes increase.

A large-sample $100(1 - \alpha)\%$ (Wald) confidence interval for $\pi_1 - \pi_2$ is

$$(p_1 - p_2) \pm z_{\alpha/2}(SE)$$

For small samples the actual coverage probability is closer to the nominal confidence level if you add 1.0 to every cell of the 2×2 table before applying this formula.[1] For a significance test of H_0: $\pi_1 = \pi_2$, a z test statistic divides $(p_1 - p_2)$ by a pooled SE that applies under H_0. Because z^2 is the Pearson chi-squared statistic presented in Section 2.4.3, we will not discuss this test here.

2.2.2 Example: Aspirin and Heart Attacks

Table 2.3 is from a report on the relationship between aspirin use and myocardial infarction (heart attacks) by the Physicians' Health Study Research Group at Harvard

[1]A. Agresti and B. Caffo, *Am. Statist.*, **54**: 280–288, 2000.

Table 2.3. Cross Classification of Aspirin Use and Myocardial Infarction

	Myocardial Infarction		
Group	Yes	No	Total
Placebo	189	10,845	11,034
Aspirin	104	10,933	11,037

Source: Preliminary Report: Findings from the Aspirin Component of the Ongoing Physicians' Health Study. *New Engl. J. Med.*, **318**: 262–264, 1988.

Medical School. The Physicians' Health Study was a five-year randomized study testing whether regular intake of aspirin reduces mortality from cardiovascular disease. Every other day, the male physicians participating in the study took either one aspirin tablet or a placebo. The study was "blind" – the physicians in the study did not know which type of pill they were taking.

We treat the two rows in Table 2.3 as independent binomial samples. Of the $n_1 = 11{,}034$ physicians taking placebo, 189 suffered myocardial infarction (MI) during the study, a proportion of $p_1 = 189/11{,}034 = 0.0171$. Of the $n_2 = 11{,}037$ physicians taking aspirin, 104 suffered MI, a proportion of $p_2 = 0.0094$. The sample difference of proportions is $0.0171 - 0.0094 = 0.0077$. From equation (2.1), this difference has an estimated standard error of

$$SE = \sqrt{\frac{(0.0171)(0.9829)}{11,034} + \frac{(0.0094)(0.9906)}{11,037}} = 0.0015$$

A 95% confidence interval for the true difference $\pi_1 - \pi_2$ is $0.0077 \pm 1.96(0.0015)$, which is 0.008 ± 0.003, or $(0.005, 0.011)$. Since this interval contains only positive values, we conclude that $\pi_1 - \pi_2 > 0$, that is, $\pi_1 > \pi_2$. For males, taking aspirin appears to result in a diminished risk of heart attack.

2.2.3 Relative Risk

A difference between two proportions of a certain fixed size usually is more important when both proportions are near 0 or 1 than when they are near the middle of the range. Consider a comparison of two drugs on the proportion of subjects who had adverse reactions when using the drug. The difference between 0.010 and 0.001 is the same as the difference between 0.410 and 0.401, namely 0.009. The first difference is more striking, since 10 times as many subjects had adverse reactions with one drug as the other. In such cases, the ratio of proportions is a more relevant descriptive measure.

For 2×2 tables, the *relative risk* is the ratio

$$\text{relative risk} = \frac{\pi_1}{\pi_2} \tag{2.2}$$

It can be any nonnegative real number. The proportions 0.010 and 0.001 have a relative risk of $0.010/0.001 = 10.0$, whereas the proportions 0.410 and 0.401 have a relative risk of $0.410/0.401 = 1.02$. A relative risk of 1.00 occurs when $\pi_1 = \pi_2$, that is, when the response is independent of the group.

Two groups with *sample* proportions p_1 and p_2 have a sample relative risk of p_1/p_2. For Table 2.3, the sample relative risk is $p_1/p_2 = 0.0171/0.0094 = 1.82$. The sample proportion of MI cases was 82% higher for the group taking placebo. The sample difference of proportions of 0.008 makes it seem as if the two groups differ by a trivial amount. By contrast, the relative risk shows that the difference may have important public health implications. Using the difference of proportions alone to compare two groups can be misleading when the proportions are both close to zero.

The sampling distribution of the sample relative risk is highly skewed unless the sample sizes are quite large. Because of this, its confidence interval formula is rather complex (Exercise 2.15). For Table 2.3, software (e.g., SAS – PROC FREQ) reports a 95% confidence interval for the true relative risk of (1.43, 2.30). We can be 95% confident that, after 5 years, the proportion of MI cases for male physicians taking placebo is between 1.43 and 2.30 times the proportion of MI cases for male physicians taking aspirin. This indicates that the risk of MI is at least 43% higher for the placebo group.

The ratio of failure probabilities, $(1 - \pi_1)/(1 - \pi_2)$, takes a different value than the ratio of the success probabilities. When one of the two outcomes has small probability, normally one computes the ratio of the probabilities for that outcome.

2.3 THE ODDS RATIO

We will next study the *odds ratio*, another measure of association for 2×2 contingency tables. It occurs as a parameter in the most important type of model for categorical data.

For a probability of success π, the *odds* of success are defined to be

$$\text{odds} = \pi/(1 - \pi)$$

For instance, if $\pi = 0.75$, then the odds of success equal $0.75/0.25 = 3$.

The odds are nonnegative, with value greater than 1.0 when a success is more likely than a failure. When odds = 4.0, a success is four times as likely as a failure. The probability of success is 0.8, the probability of failure is 0.2, and the odds equal $0.8/0.2 = 4.0$. We then expect to observe four successes for every one failure. When odds $= 1/4$, a failure is four times as likely as a success. We then expect to observe one success for every four failures.

The success probability itself is the function of the odds,

$$\pi = \text{odds}/(\text{odds} + 1)$$

For instance, when odds $= 4$, then $\pi = 4/(4 + 1) = 0.8$.

In 2×2 tables, within row 1 the *odds* of success are $\text{odds}_1 = \pi_1/(1 - \pi_1)$, and within row 2 the odds of success equal $\text{odds}_2 = \pi_2/(1 - \pi_2)$. The ratio of the odds from the two rows,

$$\theta = \frac{\text{odds}_1}{\text{odds}_2} = \frac{\pi_1/(1 - \pi_1)}{\pi_2/(1 - \pi_2)} \tag{2.3}$$

is the *odds ratio*. Whereas the relative risk is a ratio of two probabilities, the odds ratio θ is a ratio of two odds.

2.3.1 Properties of the Odds Ratio

The odds ratio can equal any nonnegative number. When X and Y are independent, $\pi_1 = \pi_2$, so $\text{odds}_1 = \text{odds}_2$ and $\theta = \text{odds}_1/\text{odds}_2 = 1$. The independence value $\theta = 1$ is a baseline for comparison. Odds ratios on each side of 1 reflect certain types of associations. When $\theta > 1$, the odds of success are higher in row 1 than in row 2. For instance, when $\theta = 4$, the odds of success in row 1 are four times the odds of success in row 2. Thus, subjects in row 1 are more likely to have successes than are subjects in row 2; that is, $\pi_1 > \pi_2$. When $\theta < 1$, a success is less likely in row 1 than in row 2; that is, $\pi_1 < \pi_2$.

Values of θ farther from 1.0 in a given direction represent stronger association. An odds ratio of 4 is farther from independence than an odds ratio of 2, and an odds ratio of 0.25 is farther from independence than an odds ratio of 0.50.

Two values for θ represent the same strength of association, but in opposite directions, when one value is the inverse of the other. When $\theta = 0.25$, for example, the odds of success in row 1 are 0.25 times the odds of success in row 2, or equivalently $1/0.25 = 4.0$ times as high in row 2 as in row 1. When the order of the rows is reversed or the order of the columns is reversed, the new value of θ is the inverse of the original value. This ordering is usually arbitrary, so whether we get 4.0 or 0.25 for the odds ratio is merely a matter of how we label the rows and columns.

The odds ratio does not change value when the table orientation reverses so that the rows become the columns and the columns become the rows. The same value occurs when we treat the columns as the response variable and the rows as the explanatory variable, or the rows as the response variable and the columns as the explanatory variable. Thus, it is unnecessary to identify one classification as a response variable in order to estimate θ. By contrast, the relative risk requires this, and its value also depends on whether it is applied to the first or to the second outcome category.

When both variables are response variables, the odds ratio can be defined using joint probabilities as

$$\theta = \frac{\pi_{11}/\pi_{12}}{\pi_{21}/\pi_{22}} = \frac{\pi_{11}\pi_{22}}{\pi_{12}\pi_{21}}$$

The odds ratio is also called the *cross-product ratio*, because it equals the ratio of the products $\pi_{11}\pi_{22}$ and $\pi_{12}\pi_{21}$ of cell probabilities from diagonally opposite cells.

The sample odds ratio equals the ratio of the sample odds in the two rows,

$$\hat{\theta} = \frac{p_1/(1 - p_1)}{p_2/(1 - p_2)} = \frac{n_{11}/n_{12}}{n_{21}/n_{22}} = \frac{n_{11}n_{22}}{n_{12}n_{21}} \tag{2.4}$$

For a multinomial distribution over the four cells or for independent binomial distributions for the two rows, this is the ML estimator of θ.

2.3.2 Example: Odds Ratio for Aspirin Use and Heart Attacks

Let us revisit Table 2.3 from Section 2.2.2 on aspirin use and myocardial infarction. For the physicians taking placebo, the estimated odds of MI equal $n_{11}/n_{12} = 189/10{,}845 = 0.0174$. Since $0.0174 = 1.74/100$, the value 0.0174 means there were 1.74 "yes" outcomes for every 100 "no" outcomes. The estimated odds equal $104/10{,}933 = 0.0095$ for those taking aspirin, or 0.95 "yes" outcomes per every 100 "no" outcomes.

The sample odds ratio equals $\hat{\theta} = 0.0174/0.0095 = 1.832$. This also equals the cross-product ratio $(189 \times 10{,}933)/(10{,}845 \times 104)$. The estimated odds of MI for male physicians taking placebo equal 1.83 times the estimated odds for male physicians taking aspirin. The estimated odds were 83% higher for the placebo group.

2.3.3 Inference for Odds Ratios and Log Odds Ratios

Unless the sample size is extremely large, the sampling distribution of the odds ratio is highly skewed. When $\theta = 1$, for example, $\hat{\theta}$ cannot be much smaller than θ (since $\hat{\theta} \geq 0$), but it could be much larger with nonnegligible probability.

Because of this skewness, statistical inference for the odds ratio uses an alternative but equivalent measure – its natural logarithm, $\log(\theta)$. Independence corresponds to $\log(\theta) = 0$. That is, an odds ratio of 1.0 is equivalent to a log odds ratio of 0.0. An odds ratio of 2.0 has a log odds ratio of 0.7. The log odds ratio is symmetric about zero, in the sense that reversing rows or reversing columns changes its sign. Two values for $\log(\theta)$ that are the same except for sign, such as $\log(2.0) = 0.7$ and $\log(0.5) = -0.7$, represent the same strength of association. Doubling a log odds ratio corresponds to squaring an odds ratio. For instance, log odds ratios of $2(0.7) = 1.4$ and $2(-0.7) = -1.4$ correspond to odds ratios of $2^2 = 4$ and $0.5^2 = 0.25$.

The sample log odds ratio, $\log \hat{\theta}$, has a less skewed sampling distribution that is bell-shaped. Its approximating normal distribution has a mean of $\log \theta$ and a standard error of

$$SE = \sqrt{\frac{1}{n_{11}} + \frac{1}{n_{12}} + \frac{1}{n_{21}} + \frac{1}{n_{22}}} \tag{2.5}$$

The *SE* decreases as the cell counts increase.

Because the sampling distribution is closer to normality for $\log \hat{\theta}$ than $\hat{\theta}$, it is better to construct confidence intervals for $\log \theta$. Transform back (that is, take antilogs, using the *exponential function*, discussed below) to form a confidence interval for θ. A large-sample confidence interval for $\log \theta$ is

$$\log \hat{\theta} \pm z_{\alpha/2}(SE)$$

Exponentiating endpoints of this confidence interval yields one for θ.

For Table 2.3, the natural log of $\hat{\theta}$ equals $\log(1.832) = 0.605$. From (2.5), the *SE* of $\log \hat{\theta}$ equals

$$SE = \sqrt{\frac{1}{189} + \frac{1}{10{,}933} + \frac{1}{104} + \frac{1}{10{,}845}} = 0.123$$

For the population, a 95% confidence interval for $\log \theta$ equals $0.605 \pm 1.96(0.123)$, or $(0.365, 0.846)$. The corresponding confidence interval for θ is

$$[\exp(0.365), \exp(0.846)] = (\mathrm{e}^{0.365}, \mathrm{e}^{0.846}) = (1.44, 2.33)$$

[The symbol e^x, also expressed as $\exp(x)$, denotes the *exponential function* evaluated at x. The exponential function is the antilog for the logarithm using the natural log scale.[2] This means that $\mathrm{e}^x = c$ is equivalent to $\log(c) = x$. For instance, $\mathrm{e}^0 = \exp(0) = 1$ corresponds to $\log(1) = 0$; similarly, $\mathrm{e}^{0.7} = \exp(0.7) = 2.0$ corresponds to $\log(2) = 0.7$.]

Since the confidence interval (1.44, 2.33) for θ does not contain 1.0, the true odds of MI seem different for the two groups. We estimate that the odds of MI are at least 44% higher for subjects taking placebo than for subjects taking aspirin. The endpoints of the interval are not equally distant from $\hat{\theta} = 1.83$, because the sampling distribution of $\hat{\theta}$ is skewed to the right.

The sample odds ratio $\hat{\theta}$ equals 0 or ∞ if any $n_{ij} = 0$, and it is undefined if both entries in a row or column are zero. The slightly amended estimator

$$\tilde{\theta} = \frac{(n_{11} + 0.5)(n_{22} + 0.5)}{(n_{12} + 0.5)(n_{21} + 0.5)}$$

corresponding to adding 1/2 to each cell count, is preferred when any cell counts are very small. In that case, the SE formula (2.5) replaces $\{n_{ij}\}$ by $\{n_{ij} + 0.5\}$.

[2]All logarithms in this text use this natural log scale, which has $\mathrm{e} = \mathrm{e}^1 = 2.718\ldots$ as the base. To find e^x on pocket calculators, enter the value for x and press the e^x key.

For Table 2.3, $\tilde{\theta} = (189.5 \times 10{,}933.5)/(10{,}845.5 \times 104.5) = 1.828$ is close to $\hat{\theta} = 1.832$, since no cell count is especially small.

2.3.4 Relationship Between Odds Ratio and Relative Risk

A sample odds ratio of 1.83 does *not* mean that p_1 is 1.83 times p_2. That's the interpretation of a *relative risk* of 1.83, since that measure is a ratio of proportions rather than odds. Instead, $\hat{\theta} = 1.83$ means that the *odds* value $p_1/(1 - p_1)$ is 1.83 times the odds value $p_2/(1 - p_2)$.

From equation (2.4) and from the sample analog of definition (2.2),

$$\text{Odds ratio} = \frac{p_1/(1 - p_1)}{p_2/(1 - p_2)} = \text{Relative risk} \times \left(\frac{1 - p_2}{1 - p_1}\right)$$

When p_1 and p_2 are both close to zero, the fraction in the last term of this expression equals approximately 1.0. The odds ratio and relative risk then take similar values. Table 2.3 illustrates this similarity. For each group, the sample proportion of MI cases is close to zero. Thus, the sample odds ratio of 1.83 is similar to the sample relative risk of 1.82 that Section 2.2.3 reported. In such a case, an odds ratio of 1.83 *does* mean that p_1 is *approximately* 1.83 times p_2.

This relationship between the odds ratio and the relative risk is useful. For some data sets direct estimation of the relative risk is not possible, yet one can estimate the odds ratio and use it to approximate the relative risk, as the next example illustrates.

2.3.5 The Odds Ratio Applies in Case–Control Studies

Table 2.4 refers to a study that investigated the relationship between smoking and myocardial infarction. The first column refers to 262 young and middle-aged women (age < 69) admitted to 30 coronary care units in northern Italy with acute MI during a 5-year period. Each case was matched with two control patients admitted to the same hospitals with other acute disorders. The controls fall in the second column of the table. All subjects were classified according to whether they had ever been smokers. The "yes" group consists of women who were current smokers or ex-smokers, whereas

Table 2.4. Cross Classification of Smoking Status and Myocardial Infarction

Ever Smoker	MI Cases	Controls
Yes	172	173
No	90	346

Source: A. Gramenzi et al., *J. Epidemiol. Community Health*, **43**: 214–217, 1989. Reprinted with permission by BMJ Publishing Group.

the "no" group consists of women who never were smokers. We refer to this variable as *smoking status*.

We would normally regard MI as a response variable and smoking status as an explanatory variable. In this study, however, the marginal distribution of MI is fixed by the sampling design, there being two controls for each case. The outcome measured for each subject is whether she ever was a smoker. The study, which uses a *retrospective* design to look into the past, is called a *case–control study*. Such studies are common in health-related applications, for instance to ensure a sufficiently large sample of subjects having the disease studied.

We might wish to compare ever-smokers with nonsmokers in terms of the proportion who suffered MI. These proportions refer to the conditional distribution of MI, given smoking status. We cannot estimate such proportions for this data set. For instance, about a third of the sample suffered MI. This is because the study matched each MI case with two controls, and it does not make sense to use $1/3$ as an estimate of the probability of MI. We *can* estimate proportions in the reverse direction, for the conditional distribution of smoking status, given myocardial infarction status. For women suffering MI, the proportion who ever were smokers was $172/262 = 0.656$, while it was $173/519 = 0.333$ for women who had not suffered MI.

When the sampling design is retrospective, we can construct conditional distributions for the explanatory variable, within levels of the fixed response. It is not possible to estimate the probability of the response outcome of interest, or to compute the difference of proportions or relative risk for that outcome. Using Table 2.4, for instance, we cannot estimate the difference between nonsmokers and ever smokers in the probability of suffering MI. We can compute the odds ratio, however. This is because the odds ratio takes the same value when it is defined using the conditional distribution of X given Y as it does when defined [as in equation (2.3)] using the distribution of Y given X; that is, it treats the variables symmetrically. The odds ratio is determined by the conditional distributions in *either* direction. It can be calculated even if we have a study design that measures a response on X within each level of Y.

In Table 2.4, the sample odds ratio is $[0.656/(1 - 0.656)]/[0.333/(1 - 0.333)] = (172 \times 346)/(173 \times 90) = 3.8$. The estimated odds of ever being a smoker were about 2 for the MI cases (i.e., $0.656/0.344$) and about $1/2$ for the controls (i.e., $0.333/0.667$), yielding an odds ratio of about $2/(1/2) = 4$.

We noted that, when $P(Y = 1)$ is small for each value of X, the odds ratio and relative risk take similar values. Even if we can estimate only conditional probabilities of X given Y, if we expect $P(Y = 1 \mid X)$ to be small, then the sample odds ratio is a rough indication of the relative risk. For Table 2.4, we cannot estimate the relative risk of MI or the difference of proportions suffering MI. Since the probability of young or middle-aged women suffering MI is probably small regardless of smoking status, however, the odds ratio value of 3.8 is also a rough estimate of the relative risk. We estimate that women who had ever smoked were about four times as likely to suffer MI as women who had never smoked.

In Table 2.4, it makes sense to treat each column, rather than each row, as a binomial sample. Because of the matching that occurs in case–control studies, however, the binomial samples in the two columns are *dependent* rather than independent. Each observation in column 1 is naturally paired with two of the observations in column 2. Chapters 8–10 present specialized methods for analyzing correlated binomial responses.

2.3.6 Types of Observational Studies

By contrast to the study summarized by Table 2.4, imagine a study that follows a sample of women for the next 20 years, observing the rates of MI for smokers and nonsmokers. Such a sampling design is *prospective*.

There are two types of prospective studies. In *cohort studies*, the subjects make their own choice about which group to join (e.g., whether to be a smoker), and we simply observe in future time who suffers MI. In *clinical trials*, we randomly allocate subjects to the two groups of interest, such as in the aspirin study described in Section 2.2.2, again observing in future time who suffers MI.

Yet another approach, a *cross-sectional design*, samples women and classifies them simultaneously on the group classification and their current response. As in a case–control study, we can then gather the data at once, rather than waiting for future events.

Case–control, cohort, and cross-sectional studies are *observational studies*. We observe who chooses each group and who has the outcome of interest. By contrast, a clinical trial is an *experimental study*, the investigator having control over which subjects enter each group, for instance, which subjects take aspirin and which take placebo. Experimental studies have fewer potential pitfalls for comparing groups, because the randomization tends to balance the groups on lurking variables that could be associated both with the response and the group identification. However, observational studies are often more practical for biomedical and social science research.

2.4 CHI-SQUARED TESTS OF INDEPENDENCE

Consider the null hypothesis (H_0) that cell probabilities equal certain fixed values $\{\pi_{ij}\}$. For a sample of size n with cell counts $\{n_{ij}\}$, the values $\{\mu_{ij} = n\pi_{ij}\}$ are *expected frequencies*. They represent the values of the expectations $\{E(n_{ij})\}$ when H_0 is true.

This notation refers to two-way tables, but similar notions apply to a set of counts for a single categorical variable or to multiway tables. To illustrate, for each of n observations of a binary variable, let π denote the probability of success. For the null hypothesis that $\pi = 0.50$, the expected frequency of successes equals $\mu = n\pi = n/2$, which also equals the expected frequency of failures. If H_0 is true, we expect about half the sample to be of each type.

To judge whether the data contradict H_0, we compare $\{n_{ij}\}$ to $\{\mu_{ij}\}$. If H_0 is true, n_{ij} should be close to μ_{ij} in each cell. The larger the differences $\{n_{ij} - \mu_{ij}\}$, the stronger the evidence against H_0. The test statistics used to make such comparisons have large-sample chi-squared distributions.

2.4.1 Pearson Statistic and the Chi-Squared Distribution

The *Pearson chi-squared statistic* for testing H_0 is

$$X^2 = \sum \frac{(n_{ij} - \mu_{ij})^2}{\mu_{ij}} \tag{2.6}$$

It was proposed in 1900 by Karl Pearson, the British statistician known also for the Pearson product–moment correlation estimate, among many contributions. This statistic takes its minimum value of zero when all $n_{ij} = \mu_{ij}$. For a fixed sample size, greater differences $\{n_{ij} - \mu_{ij}\}$ produce larger X^2 values and stronger evidence against H_0.

Since larger X^2 values are more contradictory to H_0, the P-value is the null probability that X^2 is at least as large as the observed value. The X^2 statistic has approximately a chi-squared distribution, for large n. The P-value is the chi-squared right-tail probability above the observed X^2 value. The chi-squared approximation improves as $\{\mu_{ij}\}$ increase, and $\{\mu_{ij} \geq 5\}$ is usually sufficient for a decent approximation.

The chi-squared distribution is concentrated over nonnegative values. It has mean equal to its degrees of freedom (*df*), and its standard deviation equals $\sqrt{(2df)}$. As *df* increases, the distribution concentrates around larger values and is more spread out. The distribution is skewed to the right, but it becomes more bell-shaped (normal) as *df* increases. Figure 2.2 displays chi-squared densities having *df* = 1, 5, 10, and 20.

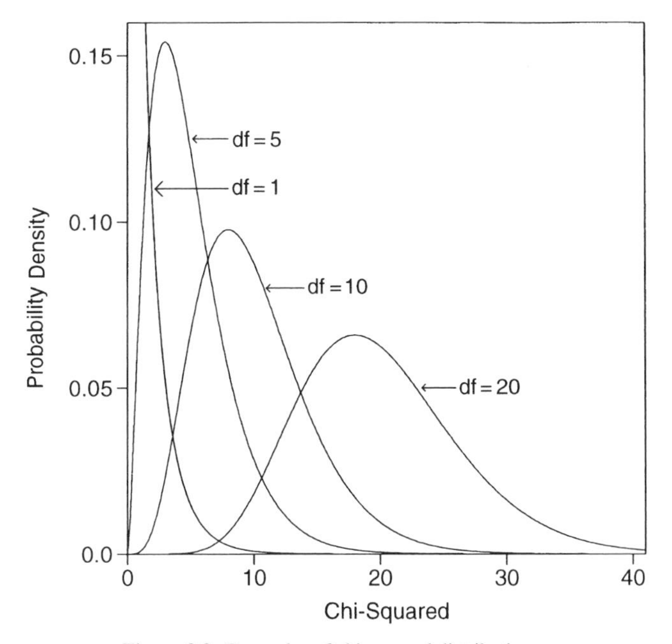

Figure 2.2. Examples of chi-squared distributions.

The *df* value equals the difference between the number of parameters in the alternative hypothesis and in the null hypothesis, as explained later in this section.

2.4.2 Likelihood-Ratio Statistic

Of the types of statistics Section 1.4.1 summarized, the Pearson statistic X^2 is a *score statistic*. (This means that X^2 is based on a covariance matrix for the counts that is estimated under H_0.) An alternative statistic presented in Section 1.4.1 results from the *likelihood-ratio* method for significance tests.

Recall that the likelihood function is the probability of the data, viewed as a function of the parameter once the data are observed. The likelihood-ratio test determines the parameter values that maximize the likelihood function (a) under the assumption that H_0 is true, (b) under the more general condition that H_0 may or may not be true. As Section 1.4.1 explained, the test statistic uses the ratio of the maximized likelihoods, through

$$-2\log\left(\frac{\text{maximum likelihood when parameters satisfy } H_0}{\text{maximum likelihood when parameters are unrestricted}}\right)$$

The test statistic value is nonnegative. When H_0 is false, the ratio of maximized likelihoods tends to be far below 1, for which the logarithm is negative; then, -2 times the log ratio tends to be a large positive number, more so as the sample size increases.

For two-way contingency tables with likelihood function based on the multinomial distribution, the likelihood-ratio statistic simplifies to

$$G^2 = 2\sum n_{ij}\log\left(\frac{n_{ij}}{\mu_{ij}}\right) \tag{2.7}$$

This statistic is called the *likelihood-ratio chi-squared statistic*. Like the Pearson statistic, G^2 takes its minimum value of 0 when all $n_{ij} = \mu_{ij}$, and larger values provide stronger evidence against H_0.

The Pearson X^2 and likelihood-ratio G^2 provide separate test statistics, but they share many properties and usually provide the same conclusions. When H_0 is true and the expected frequencies are large, the two statistics have the same chi-squared distribution, and their numerical values are similar.

2.4.3 Tests of Independence

In two-way contingency tables with joint probabilities $\{\pi_{ij}\}$ for two response variables, the null hypothesis of statistical independence is

$$H_0\colon \pi_{ij} = \pi_{i+}\pi_{+j} \quad \text{for all } i \text{ and } j$$

The marginal probabilities then determine the joint probabilities. To test H_0, we identify $\mu_{ij} = n\pi_{ij} = n\pi_{i+}\pi_{+j}$ as the expected frequency. Here, μ_{ij} is the expected value of n_{ij} assuming independence. Usually, $\{\pi_{i+}\}$ and $\{\pi_{+j}\}$ are unknown, as is this expected value.

To estimate the expected frequencies, substitute sample proportions for the unknown marginal probabilities, giving

$$\hat{\mu}_{ij} = np_{i+}p_{+j} = n\left(\frac{n_{i+}}{n}\right)\left(\frac{n_{+j}}{n}\right) = \frac{n_{i+}n_{+j}}{n}$$

This is the row total for the cell multiplied by the column total for the cell, divided by the overall sample size. The $\{\hat{\mu}_{ij}\}$ are called *estimated expected frequencies*. They have the same row and column totals as the observed counts, but they display the pattern of independence.

For testing independence in $I \times J$ contingency tables, the Pearson and likelihood-ratio statistics equal

$$X^2 = \sum \frac{(n_{ij} - \hat{\mu}_{ij})^2}{\hat{\mu}_{ij}}, \quad G^2 = 2\sum n_{ij} \log\left(\frac{n_{ij}}{\hat{\mu}_{ij}}\right) \tag{2.8}$$

Their large-sample chi-squared distributions have $df = (I - 1)(J - 1)$.

The df value means the following: under H_0, $\{\pi_{i+}\}$ and $\{\pi_{+j}\}$ determine the cell probabilities. There are $I - 1$ nonredundant row probabilities. Because they sum to 1, the first $I - 1$ determine the last one through $\pi_{I+} = 1 - (\pi_{1+} + \cdots + \pi_{I-1,+})$. Similarly, there are $J - 1$ nonredundant column probabilities. So, under H_0, there are $(I - 1) + (J - 1)$ parameters. The alternative hypothesis H_a merely states that there is not independence. It does not specify a pattern for the IJ cell probabilities. The probabilities are then solely constrained to sum to 1, so there are $IJ - 1$ nonredundant parameters. The value for df is the difference between the number of parameters under H_a and H_0, or

$$df = (IJ - 1) - [(I - 1) + (J - 1)] = IJ - I - J + 1 = (I - 1)(J - 1)$$

2.4.4 Example: Gender Gap in Political Affiliation

Table 2.5, from the 2000 General Social Survey, cross classifies gender and political party identification. Subjects indicated whether they identified more strongly with the Democratic or Republican party or as Independents. Table 2.5 also contains estimated expected frequencies for H_0: independence. For instance, the first cell has $\hat{\mu}_{11} = n_{1+}n_{+1}/n = (1557 \times 1246)/2757 = 703.7$.

The chi-squared test statistics are $X^2 = 30.1$ and $G^2 = 30.0$, with $df = (I - 1)(J - 1) = (2 - 1)(3 - 1) = 2$. This chi-squared distribution has a mean of $df = 2$ and a standard deviation of $\sqrt{(2df)} = \sqrt{4} = 2$. So, a value of 30 is far out in the right-hand tail. Each statistic has a P-value < 0.0001. This evidence of association

Table 2.5. Cross Classification of Party Identification by Gender

	Party Identification			
Gender	Democrat	Independent	Republican	Total
Females	762 (703.7)	327 (319.6)	468 (533.7)	1557
Males	484 (542.3)	239 (246.4)	477 (411.3)	1200
Total	1246	566	945	2757

Note: Estimated expected frequencies for hypothesis of independence in parentheses. Data from 2000 General Social Survey.

would be rather unusual if the variables were truly independent. Both test statistics suggest that political party identification and gender are associated.

2.4.5 Residuals for Cells in a Contingency Table

A test statistic and its P-value describe the evidence against the null hypothesis. A cell-by-cell comparison of observed and estimated expected frequencies helps us better understand the nature of the evidence. Larger differences between n_{ij} and $\hat{\mu}_{ij}$ tend to occur for cells that have larger expected frequencies, so the raw difference $n_{ij} - \hat{\mu}_{ij}$ is insufficient. For the test of independence, a useful cell residual is

$$\frac{n_{ij} - \hat{\mu}_{ij}}{\sqrt{\hat{\mu}_{ij}(1 - p_{i+})(1 - p_{+j})}} \tag{2.9}$$

The denominator is the estimated standard error of $n_{ij} - \hat{\mu}_{ij}$, under H_0. The ratio (2.9) is called a *standardized residual*, because it divides $n_{ij} - \hat{\mu}_{ij}$ by its SE.

When H_0 is true, each standardized residual has a large-sample standard normal distribution. A standardized residual having absolute value that exceeds about 2 when there are few cells or about 3 when there are many cells indicates lack of fit of H_0 in that cell. (Under H_0, we expect about 5% of the standardized residuals to be farther from 0 than ± 2 by chance alone.)

Table 2.6 shows the standardized residuals for testing independence in Table 2.5. For the first cell, for instance, $n_{11} = 762$ and $\hat{\mu}_{11} = 703.7$. The first row and first column marginal proportions equal $p_{1+} = 1557/2757 = 0.565$ and $p_{+1} = 1246/2757 = 0.452$. Substituting into (2.9), the standardized residual for this cell equals

$$(762 - 703.7)/\sqrt{703.7(1 - 0.565)(1 - 0.452)} = 4.50$$

This cell shows a greater discrepancy between n_{11} and $\hat{\mu}_{11}$ than we would expect if the variables were truly independent.

Table 2.6. Standardized Residuals (in Parentheses) for Testing Independence in Table 2.5

	Party Identification		
Gender	Democrat	Independent	Republican
Females	762 (4.50)	327 (0.70)	468 (−5.32)
Males	484 (−4.50)	239 (−0.70)	477 (5.32)

Table 2.6 has large *positive* residuals for female Democrats and male Republicans. Thus, there were more female Democrats and male Republicans than the hypothesis of independence predicts. The table has large *negative* residuals for female Republicans and male Democrats. Thus, there were fewer female Republicans and male Democrats than the hypothesis of independence predicts. An odds ratio describes this evidence of a gender gap. The 2×2 table of Democrat and Republican identifiers has a sample odds ratio of $(762 \times 477)/(468 \times 484) = 1.60$. Of those subjects identifying with one of the two parties, the estimated odds of identifying with the Democrats rather than the Republicans were 60% higher for females than males.

For each political party, Table 2.6 shows that the residual for females is the negative of the one for males. This is because the observed counts and the estimated expected frequencies have the same row and column totals. Thus, in a given column, if $n_{ij} > \hat{\mu}_{ij}$ in one cell, the reverse must happen in the other cell. The differences $n_{1j} - \hat{\mu}_{1j}$ and $n_{2j} - \hat{\mu}_{2j}$ have the same magnitude but different signs, implying the same pattern for their standardized residuals.

2.4.6 Partitioning Chi-Squared

Chi-squared statistics sum and break up into other chi-squared statistics. If one chi-squared statistic has $df = df_1$ and a separate, independent, chi-squared statistic has $df = df_2$, then their sum has a chi-squared distribution with $df = df_1 + df_2$. For example, suppose we cross classify gender and political party ID with a 2×3 table for college-educated subjects and a separate 2×3 table for subjects not having a college education. Then, the sum of the X^2 values or the G^2 values from the two tables is a chi-squared statistic with $df = 2 + 2 = 4$.

Likewise, chi-squared statistics having $df > 1$ can be broken into components with fewer degrees of freedom. Another supplement to a test of independence partitions its chi-squared test statistic so that the components represent certain aspects of the association. A partitioning may show that an association primarily reflects differences between certain categories or groupings of categories.

For testing independence in $2 \times J$ tables, $df = (J - 1)$ and a chi-squared statistic can partition into $J - 1$ components. For example, G^2 equals the sum of a G^2 statistic that compares the first two columns, plus a G^2 statistic for the 2×2 table that combines the first two columns and compares them to the third column, and so on, up to a G^2 statistic for the 2×2 table that combines the first $J - 1$ columns and compares them to the last column. Each component G^2 statistic has $df = 1$.

Consider again Table 2.5. The first two columns of this table form a 2×2 table with cell counts, by row, of (762, 327/484, 239). For this component table, $G^2 = 1.8$, with $df = 1$. Of those subjects who identify either as Democrats or Independents, there is not much evidence (P-value $= 0.17$) of a difference between females and males in the relative numbers in the two categories. The second 2×2 table combines these columns and compares them to the Republican column, giving the table with rows $(762 + 327, 468/484 + 239, 477) = (1089, 468/723, 477)$. This table has $G^2 = 28.2$, based on $df = 1$. There is strong evidence of a difference between females and males in the relative numbers identifying as Republican instead of Democrat or Independent. Note that $1.8 + 28.2 = 30.0$; that is, the sum of these G^2 components equals G^2 for the test of independence for the complete 2×3 table. This overall statistic primarily reflects differences between genders in choosing between Republican and the other two categories.

It might seem more natural to compute G^2 for separate 2×2 tables that pair each column with a particular one, say the last. This is a reasonable way to investigate association in many data sets. However, these component statistics are not independent and do not sum to G^2 for the complete table. Certain rules determine ways of forming tables so that chi-squared partitions, but they are beyond the scope of this text. A necessary condition is that the G^2 values for the component tables sum to G^2 for the original table.

The G^2 statistic has exact partitionings. The Pearson X^2 does not equal the sum of X^2 values for the separate tables in a partition. However, it is valid to use the X^2 statistics for the separate tables in the partition. They simply do not provide an exact algebraic partitioning of X^2 for the overall table.

2.4.7 Comments About Chi-Squared Tests

Chi-squared tests of independence, like any significance test, have limitations. They merely indicate the degree of evidence for an association. They are rarely adequate for answering all questions we have about a data set. Rather than relying solely on these tests, study the nature of the association. It is sensible to study residuals and estimate parameters such as odds ratios that describe the strength of association.

The X^2 and G^2 chi-squared tests also have limitations in the types of data sets for which they are applicable. For instance, they require large samples. The sampling distributions of X^2 and G^2 get closer to chi-squared as the sample size n increases, relative to the number of cells IJ. The convergence is quicker for X^2 than G^2. The chi-squared approximation is often poor for G^2 when some expected frequencies are less than about 5. When I or J is large, it can be decent for X^2 when some expected frequencies are as small as 1. To play safe, you can instead use a small-sample procedure whenever at least one expected frequency is less than 5. Section 2.6 discusses small-sample methods.

The $\{\hat{\mu}_{ij} = n_{i+}n_{+j}/n\}$ used in X^2 and G^2 depend on the row and column marginal totals, but not on the order in which the rows and columns are listed. Thus, X^2 and G^2

do not change value with reorderings of rows or of columns. This means that these tests treat both classifications as nominal. When at least one variable is ordinal, more powerful tests of independence usually exist. The next section presents such a test.

2.5 TESTING INDEPENDENCE FOR ORDINAL DATA

When the rows and/or the columns are ordinal, the chi-squared test of independence using test statistic X^2 or G^2 ignores the ordering information. Test statistics that use the ordinality by treating ordinal variables as quantitative rather than qualitative (nominal scale) are usually more appropriate and provide greater power.

2.5.1 Linear Trend Alternative to Independence

When the variables are ordinal, a trend association is common. As the level of X increases, responses on Y tend to increase toward higher levels, or responses on Y tend to decrease toward lower levels.

To detect a trend association, a simple analysis assigns scores to categories and measures the degree of *linear trend*. The test statistic, which is sensitive to positive or negative linear trends, utilizes correlation information in the data. Let $u_1 \leq u_2 \leq \cdots \leq u_I$ denote scores for the rows, and let $v_1 \leq v_2 \leq \cdots \leq v_J$ denote scores for the columns. The scores have the same ordering as the category levels. You should choose the scores to reflect distances between categories, with greater distances between categories regarded as farther apart.

Let $\bar{u} = \sum_i u_i p_{i+}$ denote the sample mean of the row scores, and let $\bar{v} = \sum_j v_j p_{+j}$ denote the sample mean of the column scores. The sum $\sum_{i,j}(u_i - \bar{u})(v_j - \bar{v})p_{ij}$ weights cross-products of deviation scores by their relative frequency. This is the sample *covariance* of X and Y. The correlation r between X and Y equals the covariance divided by the product of the sample standard deviations of X and Y. That is,

$$r = \frac{\sum_{i,j}(u_i - \bar{u})(v_j - \bar{v})p_{ij}}{\sqrt{\left[\sum_i (u_i - \bar{u})^2 p_{i+}\right]\left[\sum_j (v_j - \bar{v})^2 p_{+j}\right]}}$$

It is simple to compute r using software, entering for each subject their score on each classification. The correlation falls between -1 and $+1$. Independence between the variables implies that its population value ρ equals zero. The larger the correlation is in absolute value, the farther the data fall from independence in the linear dimension.

For testing H_0: independence against the two-sided H_a: $\rho \neq 0$, a test statistic is

$$M^2 = (n - 1)r^2 \tag{2.10}$$

This test statistic increases as r increases in magnitude and as the sample size n grows. For large n, M^2 has approximately a chi-squared distribution with $df = 1$.

Large values contradict independence, so, as with X^2 and G^2, the P-value is the right-tail probability above the observed value. The square root, $M = \sqrt{(n-1)}r$, has approximately a standard normal null distribution. It applies to one-sided alternative hypotheses, such as H_a: $\rho > 0$.

Like X^2 and G^2, M^2 does not distinguish between response and explanatory variables. We get the same value regardless of which is the row variable and which is the column variable.

2.5.2 Example: Alcohol Use and Infant Malformation

Table 2.7 refers to a prospective study of maternal drinking and congenital malformations. After the first 3 months of pregnancy, the women in the sample completed a questionnaire about alcohol consumption. Following childbirth, observations were recorded on the presence or absence of congenital sex organ malformations. Alcohol consumption, measured as average number of drinks per day, is an explanatory variable with ordered categories. Malformation, the response variable, is nominal.

When a variable is nominal but has only two categories, statistics (such as M^2) that treat the variable as ordinal are still valid. For instance, we could artificially regard malformation as ordinal, treating "absent" as "low" and "present" as "high." Any choice of two scores, such as 0 for "absent" and 1 for "present," yields the same value of M^2.

Table 2.7 has a mixture of very small, moderate, and extremely large counts. Even though the sample size is large ($n = 32{,}574$), in such cases the actual sampling distributions of X^2 or G^2 may not be close to chi-squared. For these data, having $df = 4$, $G^2 = 6.2 (P = 0.19)$ and $X^2 = 12.1 (P = 0.02)$ provide mixed signals. In any case, they ignore the ordinality of alcohol consumption.

From Table 2.7, the percentage of malformation cases has roughly an increasing trend across the levels of alcohol consumption. The first two are similar and the next two are also similar, however, and any of the last three percentages changes dramatically if we do a sensitivity analysis by deleting one malformation case. Table 2.7 also reports standardized residuals for the "present" category. They are negative at low levels of alcohol consumption and positive at high levels of consumption, although most are small and they also change substantially with slight changes in the

Table 2.7. Infant Malformation and Mother's Alcohol Consumption

Alcohol Consumption	Malformation: Absent	Malformation: Present	Total	Percentage Present	Standardized Residual
0	17,066	48	17,114	0.28	−0.18
<1	14,464	38	14,502	0.26	−0.71
1–2	788	5	793	0.63	1.84
3–5	126	1	127	0.79	1.06
≥6	37	1	38	2.63	2.71

Source: B. I. Graubard and E. L. Korn, *Biometrics*, **43**: 471–476, 1987. Reprinted with permission from the Biometric Society.

data. The sample percentages and the standardized residuals both suggest a possible tendency for malformations to be more likely at higher levels of alcohol consumption.

To use the ordinal test statistic M^2, we assign scores to alcohol consumption that are midpoints of the categories; that is, $v_1 = 0$, $v_2 = 0.5$, $v_3 = 1.5$, $v_4 = 4.0$, $v_5 = 7.0$, the last score being somewhat arbitrary. From PROC FREQ in SAS, the sample correlation between alcohol consumption and malformation is $r = 0.0142$. The test statistic $M^2 = (32{,}573)(0.0142)^2 = 6.6$ has P-value $= 0.01$, suggesting strong evidence of a nonzero correlation. The standard normal statistic $M = 2.56$ has $P = 0.005$ for H_a: $\rho > 0$.

For the chosen scores, $r = 0.014$ seems weak. However, for tables such as this one that are highly discrete and unbalanced, it is not possible to obtain a large value for r, and r is not very useful for describing association. Future chapters present tests such as M^2 as part of a model-based analysis. Model-based approaches yield estimates of the effect size as well as smoothed estimates of cell probabilities. These estimates are more informative than mere significance tests.

2.5.3 Extra Power with Ordinal Tests

For testing H_0: independence, X^2 and G^2 refer to the most general H_a possible, whereby cell probabilities exhibit *any* type of statistical dependence. Their df value of $(I-1)(J-1)$ reflects that H_a has $(I-1)(J-1)$ more parameters than H_0 (recall the discussion at the end of Section 2.4.3). These statistics are designed to detect *any* type of pattern for the additional parameters. In achieving this generality, they sacrifice sensitivity for detecting particular patterns.

When the row and column variables are ordinal, one can attempt to describe the association using a single extra parameter. For instance, the test statistic M^2 is based on a correlation measure of linear trend. When a chi-squared test statistic refers to a single parameter, it has $df = 1$.

When the association truly has a positive or negative trend, the ordinal test using M^2 has a power advantage over the tests based on X^2 or G^2. Since df equals the mean of the chi-squared distribution, a relatively large M^2 value based on $df = 1$ falls farther out in its right-hand tail than a comparable value of X^2 or G^2 based on $df = (I-1)(J-1)$. Falling farther out in the tail produces a smaller P-value. When there truly is a linear trend, M^2 often has similar size to X^2 or G^2, so it tends to provide smaller P-values.

Another advantage of chi-squared tests having small df values relates to the accuracy of chi-squared approximations. For small to moderate sample sizes, the true sampling distributions tend to be closer to chi-squared when df is smaller. When several cell counts are small, the chi-squared approximation is usually worse for X^2 or G^2 than it is for M^2.

2.5.4 Choice of Scores

For most data sets, the choice of scores has little effect on the results. Different choices of ordered scores usually give similar results. This may not happen, however, when the data are very unbalanced, such as when some categories have many more observations than other categories. Table 2.7 illustrates this. For the equally spaced row scores (1, 2, 3, 4, 5), $M^2 = 1.83$, giving a much weaker conclusion ($P = 0.18$). The magnitudes of r and M^2 do not change with transformations of the scores that maintain the same relative spacings between the categories. For example, scores (1, 2, 3, 4, 5) yield the same correlation as scores (0, 1, 2, 3, 4) or (2, 4, 6, 8, 10) or (10, 20, 30, 40, 50).

An alternative approach assigns ranks to the subjects and uses them as the category scores. For all subjects in a category, one assigns the average of the ranks that would apply for a complete ranking of the sample from 1 to n. These are called *midranks*. For example, in Table 2.7 the 17,114 subjects at level 0 for alcohol consumption share ranks 1 through 17,114. We assign to each of them the average of these ranks, which is the midrank $(1 + 17{,}114)/2 = 8557.5$. The 14,502 subjects at level <1 for alcohol consumption share ranks 17,115 through $17{,}114 + 14{,}502 = 31{,}616$, for a midrank of $(17{,}115 + 31{,}616)/2 = 24{,}365.5$. Similarly the midranks for the last three categories are 32,013, 32,473, and 32,555.5. These scores yield $M^2 = 0.35$ and a weaker conclusion yet ($P = 0.55$).

Why does this happen? Adjacent categories having relatively few observations necessarily have similar midranks. The midranks (8557.5, 24,365.5, 32,013, 32,473, 32,555.5) for Table 2.7 are similar for the final three categories, since those categories have considerably fewer observations than the first two categories. A consequence is that this scoring scheme treats alcohol consumption level 1–2 (category 3) as much closer to consumption level ≥ 6 (category 5) than to consumption level 0 (category 1). This seems inappropriate. It is better to use your judgment by selecting scores that reflect well the distances between categories. When uncertain, perform a sensitivity analysis. Select two or three sensible choices, and check that the results are similar for each. Equally spaced scores often are a reasonable compromise when the category labels do not suggest any obvious choices, such as the categories (liberal, moderate, conservative) for political philosophy.

The M^2 statistic using midrank scores for each variable is sensitive to detecting nonzero values of a rank correlation called *Spearman's rho*. Alternative ordinal tests for $I \times J$ tables utilize versions of other ordinal association measures. For instance, *gamma* and *Kendall's tau-b* are contingency-table generalizations of the ordinal measure *Kendall's tau*. The sample value of any such measure divided by its standard error has a large-sample standard normal distribution for testing independence. Like the test based on M^2, these tests share the potential power advantage that results from using a single parameter to describe the association.

2.5.5 Trend Tests for $I \times 2$ and $2 \times J$ Tables

When X is binary, the table has size $2 \times J$. Such tables occur in comparisons of two groups, such as when the rows represent two treatments. The M^2 statistic then detects differences between the two row means of the scores $\{v_j\}$ on Y. Small P-values suggest that the true difference in row means is nonzero. With midrank scores for Y, the test is sensitive to differences in mean ranks for the two rows. This test is called the *Wilcoxon* or *Mann–Whitney test*. Most nonparametric statistics texts present this test for fully ranked response data, whereas for a $2 \times J$ table sets of subjects at the same level of Y are tied and use midranks. The large-sample version of that nonparametric test uses a standard normal z statistic. The square of the z statistic is equivalent to M^2.

Tables of size $I \times 2$, such as Table 2.7, have a binary response variable Y. We then focus on how the proportion of "successes" varies across the levels of X. For the chosen row scores, M^2 detects a linear trend in this proportion and relates to models presented in Section 3.2.1. Small P-values suggest that the population slope for this linear trend is nonzero. This version of the ordinal test is called the *Cochran–Armitage trend test*.

2.5.6 Nominal–Ordinal Tables

The M^2 test statistic treats both classifications as ordinal. When one variable (say X) is nominal but has only two categories, we can still use it. When X is nominal with more than two categories, it is inappropriate. One possible test statistic finds the mean response on the ordinal variable (for the chosen scores) in each row and summarizes the variation among the row means. The statistic, which has a large-sample chi-squared distribution with $df = (I - 1)$, is rather complex computationally. We defer discussion of this case to Section 6.4.3. When $I = 2$, it is identical to M^2.

2.6 EXACT INFERENCE FOR SMALL SAMPLES

The confidence intervals and tests presented so far in this chapter are large-sample methods. As the sample size n grows, "chi-squared" statistics such as X^2, G^2, and M^2 have distributions that are more nearly chi-squared. When n is small, one can perform inference using *exact* distributions rather than large-sample approximations.

2.6.1 Fisher's Exact Test for 2 × 2 Tables

For 2×2 tables, independence corresponds to an odds ratio of $\theta = 1$. Suppose the cell counts $\{n_{ij}\}$ result from two independent binomial samples or from a single multinomial sample over the four cells. A small-sample null probability distribution for the cell counts that does not depend on any unknown parameters results from considering the set of tables having the same row and column totals as the observed data. Once we condition on this restricted set of tables, the cell counts have the *hypergeometric* distribution.

For given row and column marginal totals, n_{11} determines the other three cell counts. Thus, the hypergeometric formula expresses probabilities for the four cell counts in terms of n_{11} alone. When $\theta = 1$, the probability of a particular value

n_{11} equals

$$P(n_{11}) = \frac{\binom{n_{1+}}{n_{11}}\binom{n_{2+}}{n_{+1} - n_{11}}}{\binom{n}{n_{+1}}} \tag{2.11}$$

The binomial coefficients equal $\binom{a}{b} = a!/b!(a-b)!$.

To test H_0: independence, the P-value is the sum of hypergeometric probabilities for outcomes at least as favorable to H_a as the observed outcome. We illustrate for H_a: $\theta > 1$. Given the marginal totals, tables having larger n_{11} values also have larger sample odds ratios $\hat{\theta} = (n_{11}n_{22})/(n_{12}n_{21})$; hence, they provide stronger evidence in favor of this alternative. The P-value equals the right-tail hypergeometric probability that n_{11} is at least as large as the observed value. This test, proposed by the eminent British statistician R. A. Fisher in 1934, is called *Fisher's exact test.*

2.6.2 Example: Fisher's Tea Taster

To illustrate this test in his 1935 book, *The Design of Experiments*, Fisher described the following experiment: When drinking tea, a colleague of Fisher's at Rothamsted Experiment Station near London claimed she could distinguish whether milk or tea was added to the cup first. To test her claim, Fisher designed an experiment in which she tasted eight cups of tea. Four cups had milk added first, and the other four had tea added first. She was told there were four cups of each type and she should try to select the four that had milk added first. The cups were presented to her in random order.

Table 2.8 shows a potential result of the experiment. The null hypothesis H_0: $\theta = 1$ for Fisher's exact test states that her guess was independent of the actual order of pouring. The alternative hypothesis that reflects her claim, predicting a positive association between true order of pouring and her guess, is H_a: $\theta > 1$. For this experimental design, the column margins are identical to the row margins (4, 4), because she knew that four cups had milk added first. Both marginal distributions are naturally fixed.

Table 2.8. Fisher's Tea Tasting Experiment

	Guess Poured First		
Poured First	Milk	Tea	Total
Milk	3	1	4
Tea	1	3	4
Total	4	4	

The null distribution of n_{11} is the hypergeometric distribution defined for all 2×2 tables having row and column margins (4, 4). The potential values for n_{11} are (0, 1,

2, 3, 4). The observed table, three correct guesses of the four cups having milk added first, has probability

$$P(3) = \frac{\binom{4}{3}\binom{4}{1}}{\binom{8}{4}} = \frac{[4!/(3!)(1!)][4!/(1!)(3!)]}{[8!/(4!)(4!)]} = \frac{16}{70} = 0.229$$

For H_a: $\theta > 1$, the only table that is more extreme consists of four correct guesses. It has $n_{11} = n_{22} = 4$ and $n_{12} = n_{21} = 0$, and a probability of

$$P(4) = \binom{4}{4}\binom{4}{0}\bigg/\binom{8}{4} = 1/70 = 0.014$$

Table 2.9 summarizes the possible values of n_{11} and their probabilities.

The P-value for H_a: $\theta > 1$ equals the right-tail probability that n_{11} is at least as large as observed; that is, $P = P(3) + P(4) = 0.243$. This is not much evidence against H_0: independence. The experiment did not establish an association between the actual order of pouring and the guess, but it is difficult to show effects with such a small sample.

For the potential n_{11} values, Table 2.9 shows P-values for H_a: $\theta > 1$. If the tea taster had guessed all cups correctly (i.e., $n_{11} = 4$), the observed result would have been the most extreme possible in the right tail of the hypergeometric distribution. Then, $P = P(4) = 0.014$, giving more reason to believe her claim.

2.6.3 P-values and Conservatism for Actual P(Type I Error)

The two-sided alternative H_a: $\theta \neq 1$ is the general alternative of statistical dependence, as in chi-squared tests. Its exact P-value is usually defined as the two-tailed sum of the probabilities of tables no more likely than the observed table. For Table 2.8, summing all probabilities that are no greater than the probability $P(3) = 0.229$ of

Table 2.9. Hypergeometric Distribution for Tables with Margins of Table 2.8

n_{11}	Probability	P-value	X^2
0	0.014	1.000	8.0
1	0.229	0.986	2.0
2	0.514	0.757	0.0
3	0.229	0.243	2.0
4	0.014	0.014	8.0

Note: P-value refers to right-tail hypergeometric probability for one-sided alternative.

the observed table gives $P = P(0) + P(1) + P(3) + P(4) = 0.486$. When the row or column marginal totals are equal, the hypergeometric distribution is unimodal and symmetric, and the two-sided P-value doubles the one-sided one.

For small samples, the exact distribution (2.11) has relatively few possible values for n_{11}. The P-value also has relatively few possible values. For Table 2.8, it can assume five values for the one-sided test and three values for the two-sided test. As Section 1.4.4 explained, discreteness affects error rates. Suppose, like many methodologists, you will reject H_0 if the P-value is less than or equal to 0.05. Because of the test's discreteness, the actual probability of type I error may be much less than 0.05. For the one-sided alternative, the tea-tasting experiment yields a P-value below 0.05 only when $n_{11} = 4$, in which case $P = 0.014$. When H_0 is true, the probability of this outcome is 0.014. So, $P(\text{type I error}) = 0.014$, not 0.05. The test is *conservative*, because the actual error rate is smaller than the intended one.

To diminish the conservativeness, we recommend using the *mid P-value*. Section 1.4.5 defined this as *half* the probability of the observed result plus the probability of more extreme results. For the tea-tasting data, with $n_{11} = 3$, the one-sided mid P-value equals $P(3)/2 + P(4) = 0.229/2 + 0.014 = 0.129$, compared with 0.243 for the ordinary P-value.

In Table 2.8, both margins are naturally fixed. It is more common that only one set is fixed, such as when rows totals are fixed with independent binomial samples. Then, alternative exact tests are *unconditional*, not conditioning on the other margin. They are less conservative than Fisher's exact test. Such tests are computationally intensive and beyond the scope of this text, but are available in some software (e.g., StatXact).

Exact tests of independence for tables of size larger than 2×2 use a multivariate version of the hypergeometric distribution. Such tests are not practical to compute by hand or calculator but are feasible with software (e.g., StatXact, PROC FREQ in SAS).

2.6.4 Small-Sample Confidence Interval for Odds Ratio*

It is also possible to construct small-sample confidence intervals for the odds ratio. They correspond to a generalization of Fisher's exact test that tests an arbitrary value, H_0: $\theta = \theta_0$. A 95% confidence interval contains all θ_0 values for which the exact test of H_0: $\theta = \theta_0$ has $P > 0.05$. This is available in some software (e.g., StatXact, PROC FREQ in SAS).

As happens with exact tests, discreteness makes these confidence intervals conservative. The true confidence level may actually be considerably larger than the selected one. Moreover, the true level is unknown. To reduce the conservativeness, we recommend constructing the confidence interval that corresponds to the test using a mid P-value. This interval is shorter.

For the tea-tasting data (Table 2.8), the exact 95% confidence interval for the true odds ratio equals (0.21, 626.17). The confidence interval based on the test using the mid-P value equals (0.31, 308.55). Both intervals are very wide, because the sample size is so small.

2.7 ASSOCIATION IN THREE-WAY TABLES

An important part of most research studies is the choice of control variables. In studying the effect of an explanatory variable X on a response variable Y, we should adjust for confounding variables that can influence that relationship because they are associated both with X and with Y. Otherwise, an observed XY association may merely reflect effects of those variables on X and Y. This is especially vital for observational studies, where one cannot remove effects of such variables by randomly assigning subjects to different treatments.

Consider a study of the effects of passive smoking; that is, the effects on a nonsmoker of living with a smoker. To analyze whether passive smoking is associated with lung cancer, a cross-sectional study might compare lung cancer rates between nonsmokers whose spouses smoke and nonsmokers whose spouses do not smoke. In doing so, the study should attempt to control for age, socioeconomic status, or other factors that might relate both to whether one's spouse smokes and to whether one has lung cancer. A statistical control would hold such variables constant while studying the association. Without such controls, results will have limited usefulness. Suppose that spouses of nonsmokers tend to be younger than spouses of smokers and that younger people are less likely to have lung cancer. Then, a lower proportion of lung cancer cases among nonsmoker spouses may merely reflect their lower average age and not an effect of passive smoking.

Including control variables in an analysis requires a multivariate rather than a bivariate analysis. We illustrate basic concepts for a single control variable Z, which is categorical. A three-way contingency table displays counts for the three variables.

2.7.1 Partial Tables

Two-way cross-sectional slices of the three-way table cross classify X and Y at separate levels of Z. These cross sections are called *partial tables*. They display the XY relationship at fixed levels of Z, hence showing the effect of X on Y while controlling for Z. The partial tables remove the effect of Z by holding its value constant.

The two-way contingency table that results from combining the partial tables is called the *XY marginal table*. Each cell count in it is a sum of counts from the same cell location in the partial tables. The marginal table contains no information about Z, so rather than controlling Z, it ignores Z. It is simply a two-way table relating X and Y. Methods for two-way tables do not take into account effects of other variables.

The associations in partial tables are called *conditional associations*, because they refer to the effect of X on Y conditional on fixing Z at some level. Conditional associations in partial tables can be quite different from associations in marginal tables, as the next example shows.

2.7.2 Conditional Versus Marginal Associations: Death Penalty Example

Table 2.10 is a $2 \times 2 \times 2$ contingency table – two rows, two columns, and two layers – from an article that studied effects of racial characteristics on whether subjects

Table 2.10. Death Penalty Verdict by Defendant's Race and Victims' Race

Victims' Race	Defendant's Race	Death Penalty Yes	Death Penalty No	Percentage Yes
White	White	53	414	11.3
	Black	11	37	22.9
Black	White	0	16	0.0
	Black	4	139	2.8
Total	White	53	430	11.0
	Black	15	176	7.9

Source: M. L. Radelet and G. L. Pierce, *Florida Law Rev.*, **43**: 1–34, 1991. Reprinted with permission of the *Florida Law Review*.

convicted of homicide receive the death penalty. The 674 subjects were the defendants in indictments involving cases with multiple murders, in Florida between 1976 and 1987. The variables are Y = death penalty verdict, having categories (yes, no), and X = race of defendant and Z = race of victims, each having categories (white, black). We study the effect of defendant's race on the death penalty verdict, treating victims' race as a control variable. Table 2.10 has a 2×2 partial table relating defendant's race and the death penalty verdict at each level of victims' race.

For each combination of defendant's race and victims' race, Table 2.10 lists and Figure 2.3 displays the percentage of defendants who received the death penalty.

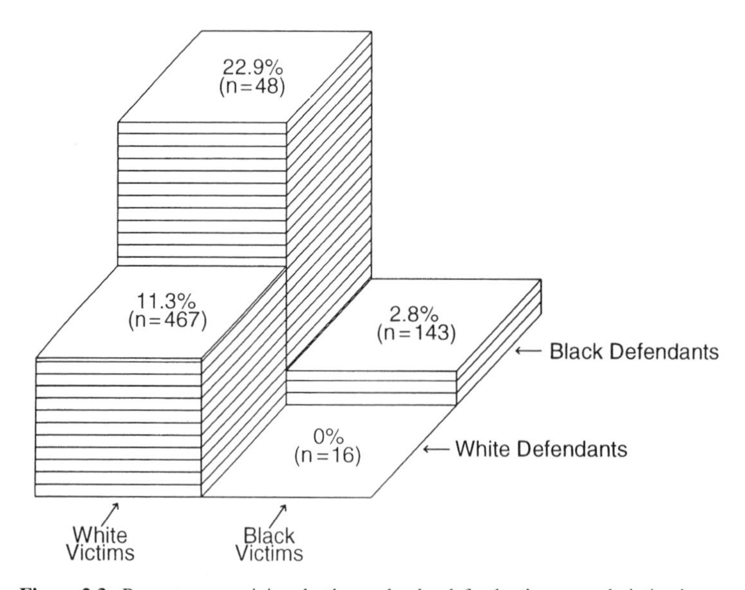

Figure 2.3. Percentage receiving death penalty, by defendant's race and victims' race.

We use these to describe the conditional associations between defendant's race and the death penalty verdict, controlling for victims' race. When the victims were white, the death penalty was imposed $22.9 - 11.3\% = 11.6\%$ more often for black defendants than for white defendants. When the victim was black, the death penalty was imposed $2.8 - 0.0\% = 2.8\%$ more often for black defendants than for white defendants. Thus, *controlling* for victims' race by keeping it fixed, the percentage of "yes" death penalty verdicts was higher for black defendants than for white defendants.

The bottom portion of Table 2.10 displays the marginal table for defendant's race and the death penalty verdict. We obtain it by summing the cell counts in Table 2.10 over the two levels of victims' race, thus combining the two partial tables (e.g., $11 + 4 = 15$). We see that, overall, 11.0% of white defendants and 7.9% of black defendants received the death penalty. *Ignoring* victims' race, the percentage of "yes" death penalty verdicts was lower for black defendants than for white defendants. The association reverses direction compared with the partial tables.

2.7.3 Simpson's Paradox

The result that a marginal association can have different direction from the conditional associations is called *Simpson's paradox*. This result applies to quantitative as well as categorical variables.

In the death penalty example, why does the association between death penalty verdict and defendant's race differ so much when we ignore vs control victims' race? This relates to the nature of the association between the control variable, victims' race, and the other variables. First, the association between victims' race and defendant's race is extremely strong. The marginal table relating these variables has odds ratio $(467 \times 143)/(48 \times 16) = 87.0$. The odds that a white defendant had white victims are estimated to be 87.0 times the odds that a black defendant had white victims. Second, Table 2.10 shows that, regardless of defendant's race, the death penalty was considerably more likely when the victims were white than when the victims were black. So, whites are tending to kill whites, and killing whites is more likely to result in the death penalty. This suggests that the marginal association should show a greater tendency for white defendants to receive the death penalty than do the conditional associations. In fact, Table 2.10 shows this pattern.

Figure 2.4 may clarify why Simpson's paradox happens. For each defendant's race, the figure plots the proportion receiving the death penalty at each level of victims' race. Each proportion is labeled by a letter symbol giving the level of victims' race. Surrounding each observation is a circle having area proportional to the number of observations at that combination of defendant's race and victims' race. For instance, the W in the largest circle represents a proportion of 0.113 receiving the death penalty for cases with white defendants and white victims. That circle is largest, because the number of cases at that combination $(53 + 414 = 467)$ is larger than at the other three combinations. The next largest circle relates to cases in which blacks kill blacks.

To control for victims' race, we compare circles having the same victims' race letter at their centers. The line connecting the two W circles has a positive slope, as does the line connecting the two B circles. Controlling for victims' race, this reflects

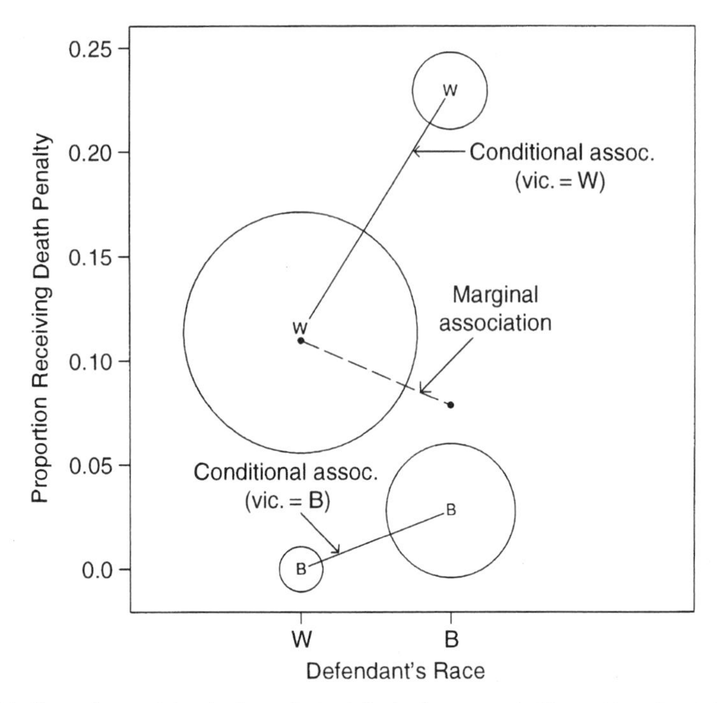

Figure 2.4. Proportion receiving death penalty by defendant's race, controlling and ignoring victims' race.

a higher chance of the death penalty for black defendants than white defendants. When we add results across victims' race to get a summary result for the marginal effect of defendant's race on the death penalty verdict, the larger circles having the greater number of cases have greater influence. Thus, the summary proportions for each defendant's race, marked on the figure by periods, fall closer to the center of the larger circles than the smaller circles. A line connecting the summary marginal proportions has negative slope. This indicates that white defendants are more likely than black defendants to receive the death penalty.

2.7.4 Conditional and Marginal Odds Ratios

Conditional associations, like marginal associations, can be described using odds ratios. We refer to odds ratios for partial tables as *conditional odds ratios*. For binary X and Y, within a fixed level k of Z, let $\theta_{XY(k)}$ denote the odds ratio between X and Y computed for the true probabilities.

Consider the conditional association between defendant's race and the death penalty. From Table 2.10, the estimated odds ratio in the first partial table, for which victims' race is white, equals

$$\hat{\theta}_{XY(1)} = \frac{53 \times 37}{414 \times 11} = 0.43$$

The sample odds for white defendants receiving the death penalty were 43% of the sample odds for black defendants. In the second partial table, for which victim's race is black, $\hat{\theta}_{XY(2)} = (0 \times 139)/(16 \times 4) = 0.0$, because the death penalty was never given to white defendants having black victims.

The conditional odds ratios can be quite different from the marginal odds ratio, for which the third variable is ignored rather than controlled. The marginal odds ratio for defendant's race and the death penalty uses the 2×2 marginal table in Table 2.10, collapsing over victims' race. The estimate equals $(53 \times 176)/(430 \times 15) = 1.45$. The sample odds of the death penalty were 45% higher for white defendants than for black defendants. Yet, we just observed that those odds were smaller for a white defendant than for a black defendant, within each level of victims' race. This reversal in the association when we control for victims' race illustrates Simpson's paradox.

2.7.5 Conditional Independence Versus Marginal Independence

If X and Y are independent in each partial table, then X and Y are said to be *conditionally independent, given Z*. All conditional odds ratios between X and Y then equal 1. Conditional independence of X and Y, given Z, does not imply marginal independence of X and Y. That is, when odds ratios between X and Y equal 1 at each level of Z, the marginal odds ratio may differ from 1.

The expected frequencies in Table 2.11 show a hypothetical relationship among three variables: Y = response (success, failure), X = drug treatment (A, B), and Z = clinic$(1, 2)$. The conditional odds ratios between X and Y at the two levels of Z are

$$\theta_{XY(1)} = \frac{18 \times 8}{12 \times 12} = 1.0, \quad \theta_{XY(2)} = \frac{2 \times 32}{8 \times 8} = 1.0$$

Given clinic, response and treatment are conditionally independent. The marginal table adds together the tables for the two clinics. The odds ratio for that

Table 2.11. Conditional Independence Does Not Imply Marginal Independence

		Response	
Clinic	Treatment	Success	Failure
1	A	18	12
	B	12	8
2	A	2	8
	B	8	32
Total	A	20	20
	B	20	40

marginal table equals $(20 \times 40)/(20 \times 20) = 2.0$, so the variables are not marginally independent.

Why are the odds of a success twice as high for treatment A as treatment B, when we ignore clinic? The conditional XZ and YZ odds ratios give a clue. The odds ratio between Z and either X or Y, at each fixed level of the other variable, equals 6.0. For instance, the XZ odds ratio at the first level of Y equals $(18 \times 8)/(12 \times 2) = 6.0$. The conditional odds of receiving treatment A are six times higher at clinic 1 than clinic 2, and the conditional odds of success are six times higher at clinic 1 than at clinic 2. Clinic 1 tends to use treatment A more often, and clinic 1 also tends to have more successes. For instance, if subjects who attend clinic 1 tend to be in better health or tend to be younger than those who go to clinic 2, perhaps they have a better success rate than subjects in clinic 2 regardless of the treatment received.

It is misleading to study only the marginal table, concluding that successes are more likely with treatment A than with treatment B. Subjects within a particular clinic are likely to be more homogeneous than the overall sample, and response is independent of treatment in each clinic.

2.7.6 Homogeneous Association

Let K denote the number of categories for Z. When X and Y are binary, there is *homogeneous XY association* when

$$\theta_{XY(1)} = \theta_{XY(2)} = \cdots = \theta_{XY(K)}$$

Conditional independence of X and Y is the special case in which each conditional odds ratio equals 1.0.

In an $I \times J \times K$ table, homogeneous XY association means that any conditional odds ratio formed using two levels of X and two levels of Y is the same at each level of Z. When there is homogeneous XY association, there is also homogeneous XZ association and homogeneous YZ association. Homogeneous association is a symmetric property, applying to any pair of the variables viewed across the levels of the third. When it occurs, there is said to be *no interaction* between two variables in their effects on the third variable.

When there is not homogeneous association, the conditional odds ratio for any pair of variables changes across levels of the third variable. For X = smoking (yes, no), Y = lung cancer (yes, no), and Z = age (<45, 45–65, >65), suppose $\theta_{XY(1)} = 1.2$, $\theta_{XY(2)} = 2.8$, and $\theta_{XY(3)} = 6.2$. Then, smoking has a weak effect on lung cancer for young people, but the effect strengthens considerably with age.

Inference about associations in multi-way contingency tables is best handled in the context of models. Section 4.3 introduces a model that has the property of homogeneous association. We will see there and in Section 5.2 how to judge whether conditional independence or homogeneous association are plausible.

PROBLEMS

2.1 An article in the *New York Times* (February 17, 1999) about the PSA blood test for detecting prostate cancer stated that, of men who had this disease, the test fails to detect prostate cancer in 1 in 4 (so called false-negative results), and of men who did not have it, as many as two-thirds receive false-positive results. Let C ($\bar{C}$) denote the event of having (not having) prostate cancer and let + (−) denote a positive (negative) test result.

a. Which is true: $P(-|C) = 1/4$ or $P(C|-) = 1/4$? $P(\bar{C}|+) = 2/3$ or $P(+|\bar{C}) = 2/3$?

b. What is the sensitivity of this test?

c. Of men who take the PSA test, suppose $P(C) = 0.01$. Find the cell probabilities in the 2×2 table for the joint distribution that cross classifies Y = diagnosis (+, −) with X = true disease status ($C, \bar{C}$).

d. Using (c), find the marginal distribution for the diagnosis.

e. Using (c) and (d), find $P(C|+)$, and interpret.

2.2 For diagnostic testing, let X = true status (1 = disease, 2 = no disease) and Y = diagnosis (1 = positive, 2 = negative). Let $\pi_i = P(Y = 1|X = i)$, $i = 1, 2$.

a. Explain why sensitivity $= \pi_1$ and specificity $= 1 - \pi_2$.

b. Let γ denote the probability that a subject has the disease. Given that the diagnosis is positive, use Bayes's theorem to show that the probability a subject truly has the disease is

$$\pi_1\gamma/[\pi_1\gamma + \pi_2(1 - \gamma)]$$

c. For mammograms for detecting breast cancer, suppose $\gamma = 0.01$, sensitivity = 0.86, and specificity = 0.88. Given a positive test result, find the probability that the woman truly has breast cancer.

d. To better understand the answer in (c), find the joint probabilities for the 2×2 cross classification of X and Y. Discuss their relative sizes in the two cells that refer to a positive test result.

2.3 According to recent UN figures, the annual gun homicide rate is 62.4 per one million residents in the United States and 1.3 per one million residents in the UK.

a. Compare the proportion of residents killed annually by guns using the (i) difference of proportions, (ii) relative risk.

b. When both proportions are very close to 0, as here, which measure is more useful for describing the strength of association? Why?

2.4 A newspaper article preceding the 1994 World Cup semifinal match between Italy and Bulgaria stated that "Italy is favored 10–11 to beat Bulgaria, which is rated at 10–3 to reach the final." Suppose this means that the odds that Italy wins are $11/10$ and the odds that Bulgaria wins are $3/10$. Find the probability that each team wins, and comment.

2.5 Consider the following two studies reported in the *New York Times*:

a. A British study reported (December 3, 1998) that, of smokers who get lung cancer, "women were 1.7 times more vulnerable than men to get small-cell lung cancer." Is 1.7 an odds ratio, or a relative risk?

b. A National Cancer Institute study about tamoxifen and breast cancer reported (April 7, 1998) that the women taking the drug were 45% less likely to experience invasive breast cancer compared with the women taking placebo. Find the relative risk for (i) those taking the drug compared to those taking placebo, (ii) those taking placebo compared to those taking the drug.

2.6 In the United States, the estimated annual probability that a woman over the age of 35 dies of lung cancer equals 0.001304 for current smokers and 0.000121 for nonsmokers [M. Pagano and K. Gauvreau, *Principles of Biostatistics*, Belmont, CA: Duxbury Press (1993), p. 134].

a. Calculate and interpret the difference of proportions and the relative risk. Which is more informative for these data? Why?

b. Calculate and interpret the odds ratio. Explain why the relative risk and odds ratio take similar values.

2.7 For adults who sailed on the Titanic on its fateful voyage, the odds ratio between gender (female, male) and survival (yes, no) was 11.4. (For data, see R. Dawson, *J. Statist. Educ.* **3**, no. 3, 1995.)

a. What is wrong with the interpretation, "The probability of survival for females was 11.4 times that for males"? Give the correct interpretation.

b. The odds of survival for females equaled 2.9. For each gender, find the proportion who survived.

c. Find the value of R in the interpretation, "The probability of survival for females was R times that for males."

2.8 A research study estimated that under a certain condition, the probability a subject would be referred for heart catheterization was 0.906 for whites and 0.847 for blacks.

a. A press release about the study stated that the odds of referral for cardiac catheterization for blacks are 60% of the odds for whites. Explain how they obtained 60% (more accurately, 57%).

b. An Associated Press story that described the study stated "Doctors were only 60% as likely to order cardiac catheterization for blacks as for whites."

What is wrong with this interpretation? Give the correct percentage for this interpretation. (In stating results to the general public, it is better to use the relative risk than the odds ratio. It is simpler to understand and less likely to be misinterpreted. For details, see *New Engl. J. Med.*, **341**: 279–283, 1999.)

2.9 An estimated odds ratio for adult females between the presence of squamous cell carcinoma (yes, no) and smoking behavior (smoker, nonsmoker) equals 11.7 when the smoker category consists of subjects whose smoking level s is $0 < s < 20$ cigarettes per day; it is 26.1 for smokers with $s \geq 20$ cigarettes per day (R. Brownson et al., *Epidemiology*, **3**: 61–64, 1992). Show that the estimated odds ratio between carcinoma and smoking levels ($s \geq 20, 0 < s < 20$) equals $26.1/11.7 = 2.2$.
Data posted at the FBI website (www.fbi.gov)

2.10 Data posted at the FBI website (www.fbi.gov) stated that of all blacks slain in 2005, 91% were slain by blacks, and of all whites slain in 2005, 83% were slain by whites. Let Y denote race of victim and X denote race of murderer.

a. Which conditional distribution do these statistics refer to, Y given X, or X given Y?

b. Calculate and interpret the odds ratio between X and Y.

c. Given that a murderer was white, can you estimate the probability that the victim was white? What additional information would you need to do this? (Hint: How could you use Bayes's Theorem?)

2.11 A 20-year study of British male physicians (R. Doll and R. Peto, *British Med. J.*, **2**: 1525–1536, 1976) noted that the proportion who died from lung cancer was 0.00140 per year for cigarette smokers and 0.00010 per year for nonsmokers. The proportion who died from heart disease was 0.00669 for smokers and 0.00413 for nonsmokers.

a. Describe the association of smoking with lung cancer and with heart disease, using the difference of proportions, the relative risk, and the odds ratio. Interpret.

b. Which response (lung cancer or heart disease) is more strongly related to cigarette smoking, in terms of the reduction in deaths that could occur with an absence of smoking?

2.12 A statistical analysis that combines information from several studies is called a *meta analysis*. A meta analysis compared aspirin with placebo on incidence of heart attack and of stroke, separately for men and for women (*J. Am. Med. Assoc.*, **295**: 306–313, 2006). For the Women's Health Study, heart attacks were reported for 198 of 19,934 taking aspirin and for 193 of 19,942 taking placebo.

a. Construct the 2×2 table that cross classifies the treatment (aspirin, placebo) with whether a heart attack was reported (yes, no).

b. Estimate the odds ratio. Interpret.

c. Find a 95% confidence interval for the population odds ratio for women. Interpret. (As of 2006, results suggested that for women, aspirin was helpful for reducing risk of stroke but not necessarily risk of heart attack.)

2.13 Refer to Table 2.1 about belief in an afterlife.

a. Construct a 90% confidence interval for the difference of proportions, and interpret.

b. Construct a 90% confidence interval for the odds ratio, and interpret.

c. Conduct a test of statistical independence. Report the P-value and interpret.

2.14 A poll by Louis Harris and Associates of 1249 adult Americans indicated that 36% believe in ghosts and 37% believe in astrology. Can you compare the proportions using inferential methods for independent binomial samples? If yes, do so. If not, explain why not.

2.15 A large-sample confidence interval for the log of the relative risk is

$$\log(p_1/p_2) \pm z_{\alpha/2}\sqrt{\frac{1-p_1}{n_1 p_1} + \frac{1-p_2}{n_2 p_2}}$$

Antilogs of the endpoints yield an interval for the true relative risk. Verify the 95% confidence interval of (1.43, 2.30) reported for the relative risk in Section 2.2.3 for the aspirin and heart attack study.

2.16 Table 2.12 comes from one of the first studies of the link between lung cancer and smoking, by Richard Doll and A. Bradford Hill. In 20 hospitals in London, UK, patients admitted with lung cancer in the previous year were queried about their smoking behavior. For each patient admitted, researchers studied the smoking behavior of a noncancer control patient at the same hospital of the

Table 2.12. Data for Problem 2.16

	Lung Cancer	
Have Smoked	Cases	Controls
Yes	688	650
No	21	59
Total	709	709

Based on data reported in Table IV, R. Doll and A. B. Hill, *Br. Med. J.*, 739–748, September 30, 1950.

same sex and within the same 5-year grouping on age. A smoker was defined as a person who had smoked at least one cigarette a day for at least a year.

a. Identify the response variable and the explanatory variable.

b. Identify the type of study this was.

c. Can you use these data to compare smokers with nonsmokers in terms of the proportion who suffered lung cancer? Why or why not?

d. Summarize the association, and explain how to interpret it.

2.17 Refer to Table 2.3. Find the P-value for testing that the incidence of heart attacks is independent of aspirin intake using (a) X^2, (b) G^2. Interpret results.

2.18 Table 2.13 shows data from the 2002 General Social Survey cross classifying a person's perceived happiness with their family income. The table displays the observed and expected cell counts and the standardized residuals for testing independence.

a. Show how to obtain the estimated expected cell count of 35.8 for the first cell.

b. For testing independence, $X^2 = 73.4$. Report the df value and the P-value, and interpret.

c. Interpret the standardized residuals in the corner cells having counts 21 and 83.

d. Interpret the standardized residuals in the corner cells having counts 110 and 94.

Table 2.13. Data for Problem 2.18, with Estimated Expected Frequencies and Standardized Residuals

	Happiness		
Income	Not Too Happy	Pretty Happy	Very Happy
Above average	21 35.8 −2.973	159 166.1 −0.947	110 88.1 3.144
Average	53 79.7 −4.403	372 370.0 0.224	221 196.4 2.907
Below average	94 52.5 7.368	249 244.0 0.595	83 129.5 −5.907

2.19 Table 2.14 was taken from the 2002 General Social Survey.

a. Test the null hypothesis of independence between party identification and race. Interpret.

Table 2.14. Data for Problem 2.19

	Party Identification		
Race	Democrat	Independent	Republican
White	871	444	873
Black	302	80	43

b. Use standardized residuals to describe the evidence.

c. Partition the chi-squared into two components, and use the components to describe the evidence.

2.20 In an investigation of the relationship between stage of breast cancer at diagnosis (local or advanced) and a woman's living arrangement (D. J. Moritz and W. A. Satariano, *J. Clin. Epidemiol.*, **46**: 443–454, 1993), of 144 women living alone, 41.0% had an advanced case; of 209 living with spouse, 52.2% were advanced; of 89 living with others, 59.6% were advanced. The authors reported the P-value for the relationship as 0.02. Reconstruct the analysis they performed to obtain this P-value.

2.21 Each subject in a sample of 100 men and 100 women is asked to indicate which of the following factors (one or more) are responsible for increases in teenage crime: A, the increasing gap in income between the rich and poor; B, the increase in the percentage of single-parent families; C, insufficient time spent by parents with their children. A cross classification of the responses by gender is

Gender	A	B	C
Men	60	81	75
Women	75	87	86

a. Is it valid to apply the chi-squared test of independence to this 2×3 table? Explain.

b. Explain how this table actually provides information needed to cross-classify gender with each of three variables. Construct the contingency table relating gender to opinion about whether factor A is responsible for increases in teenage crime.

2.22 Table 2.15 classifies a sample of psychiatric patients by their diagnosis and by whether their treatment prescribed drugs.

a. Conduct a test of independence, and interpret the P-value.

b. Obtain standardized residuals, and interpret.

Table 2.15. Data for Problem 2.22

Diagnosis	Drugs	No Drugs
Schizophrenia	105	8
Affective disorder	12	2
Neurosis	18	19
Personality disorder	47	52
Special symptoms	0	13

Source: E. Helmes and G. C. Fekken, *J. Clin. Psychol.*, **42**: 569–576, 1986. Copyright by Clinical Psychology Publishing Co., Inc., Brandon, VT. Reproduced by permission of the publisher.

c. Partition chi-squared into three components to describe differences and similarities among the diagnoses, by comparing (i) the first two rows, (ii) the third and fourth rows, (iii) the last row to the first and second rows combined and the third and fourth rows combined.

2.23 Table 2.16, from a recent General Social Survey, cross-classifies the degree of fundamentalism of subjects' religious beliefs by their highest degree of education. The table also shows standardized residuals. For these data, $X^2 = 69.2$. Write a report of about 200 words, summarizing description and inference for these data.

Table 2.16. Table for Problem 2.23, with Standardized Residuals

	Religious Beliefs		
Highest Degree	Fundamentalist	Moderate	Liberal
Less than high school	178 (4.5)	138 (−2.6)	108 (−1.9)
High school or junior college	570 (2.6)	648 (1.3)	442 (−4.0)
Bachelor or graduate	138 (−6.8)	252 (0.7)	252 (6.3)

2.24 Formula (2.8) has alternative formula $X^2 = n \sum (p_{ij} - p_{i+}p_{+j})^2 / p_{i+}p_{+j}$. Hence, for given $\{p_{ij}\}$, X^2 is large when n is sufficiently large, regardless of whether the association is practically important. Explain why chi-squared tests, like other tests, merely indicate the degree of evidence against a hypothesis and do not give information about the strength of association.

2.25 For tests of H_0: independence, $\{\hat{\mu}_{ij} = n_{i+}n_{+j}/n\}$.

a. Show that $\{\hat{\mu}_{ij}\}$ have the same row and column totals as $\{n_{ij}\}$.

b. For 2×2 tables, show that $\hat{\mu}_{11}\hat{\mu}_{22}/\hat{\mu}_{12}\hat{\mu}_{21} = 1.0$. Hence, $\{\hat{\mu}_{ij}\}$ satisfy H_0.

2.26 A chi-squared variate with degrees of freedom equal to df has representation $Z_1^2 + \cdots + Z_{df}^2$, where $Z_1, \ldots, Z_{df}$ are independent standard normal variates.

a. If Z has a standard normal distribution, what distribution does Z^2 have?

b. Show that, if Y_1 and Y_2 are independent chi-squared variates with degrees of freedom df_1 and df_2, then $Y_1 + Y_2$ has a chi-squared distribution with $df = df_1 + df_2$.

2.27 A study on educational aspirations of high school students (S. Crysdale, *Int. J. Comp. Sociol.*, **16**: 19–36, 1975) measured aspirations using the scale (some high school, high school graduate, some college, college graduate). For students whose family income was low, the counts in these categories were (9, 44, 13, 10); when family income was middle, the counts were (11, 52, 23, 22); when family income was high, the counts were (9, 41, 12, 27).

a. Test independence of aspirations and family income using X^2 or G^2. Interpret, and explain the deficiency of this test for these data.

b. Find the standardized residuals. Do they suggest any association pattern?

c. Conduct a more powerful test. Interpret results.

2.28 By trial and error, find a 3×3 table of counts for which the P-value is greater than 0.05 for the X^2 test but less than 0.05 for the M^2 ordinal test. Explain why this happens.

2.29 A study (B. Kristensen et al., *J. Intern. Med.*, **232**: 237–245, 1992) considered the effect of prednisolone on severe hypercalcaemia in women with metastatic breast cancer. Of 30 patients, 15 were randomly selected to receive prednisolone, and the other 15 formed a control group. Normalization in their level of serum-ionized calcium was achieved by seven of the 15 prednisolone-treated patients and by 0 of the 15 patients in the control group. Use Fisher's exact test to find a P-value for testing whether results were significantly better for treatment than control. Interpret.

2.30 Table 2.17 contains results of a study comparing radiation therapy with surgery in treating cancer of the larynx. Use Fisher's exact test to test $H_0\colon \theta = 1$ against $H_a\colon \theta > 1$. Interpret results.

2.31 Refer to the previous exercise.

a. Obtain and interpret a two-sided exact P-value.

b. Obtain and interpret the one-sided mid P-value. Give advantages of this type of P-value, compared with the ordinary one.

Table 2.17. Data for Problem 2.30

	Cancer Controlled	Cancer Not Controlled
Surgery	21	2
Radiation therapy	15	3

Source: W. Mendenhall et al., *Int. J. Radiat. Oncol. Biol. Phys.*, **10**: 357–363, 1984. Reprinted with permission from Elsevier Science Ltd.

2.32 Of the six candidates for three managerial positions, three are female and three are male. Denote the females by F1, F2, F3 and the males by M1, M2, M3. The result of choosing the managers is (F2, M1, M3).

a. Identify the 20 possible samples that could have been selected, and construct the contingency table for the sample actually obtained.

b. Let p_1 denote the sample proportion of males selected and p_2 the sample proportion of females. For the observed table, $p_1 - p_2 = 1/3$. Of the 20 possible samples, show that 10 have $p_1 - p_2 \geq 1/3$. Thus, if the three managers were randomly selected, the probability would equal $10/20 = 0.50$ of obtaining $p_1 - p_2 \geq 1/3$. This reasoning provides the P-value for Fisher's exact test with H_a: $\pi_1 > \pi_2$.

2.33 In murder trials in 20 Florida counties during 1976 and 1977, the death penalty was given in 19 out of 151 cases in which a white killed a white, in 0 out of 9 cases in which a white killed a black, in 11 out of 63 cases in which a black killed a white, and in 6 out of 103 cases in which a black killed a black (M. Radelet, *Am. Sociol. Rev.*, **46**: 918–927, 1981).

a. Exhibit the data as a three-way contingency table.

b. Construct the partial tables needed to study the conditional association between defendant's race and the death penalty verdict. Find and interpret the sample conditional odds ratios, adding 0.5 to each cell to reduce the impact of the 0 cell count.

c. Compute and interpret the sample marginal odds ratio between defendant's race and the death penalty verdict. Do these data exhibit Simpson's paradox? Explain.

2.34 Smith and Jones are baseball players. Smith had a higher batting average than Jones in 2005 and 2006. Is it possible that, for the combined data for these two years, Jones had the higher batting average? Explain, and illustrate using data.

2.35 At each age level, the death rate is higher in South Carolina than in Maine, but overall the death rate is higher in Maine. Explain how this could be possible. (For data, see H. Wainer, *Chance*, **12**: 44, 1999.)

2.36 Give a "real world" example of three variables X, Y, and Z, for which you expect X and Y to be marginally associated but conditionally independent, controlling for Z.

2.37 Based on murder rates in the United States, the Associated Press reported that the probability a newborn child has of eventually being a murder victim is 0.0263 for nonwhite males, 0.0049 for white males, 0.0072 for nonwhite females, and 0.0023 for white females.

a. Find the conditional odds ratios between race and whether a murder victim, given gender. Interpret.

b. If half the newborns are of each gender, for each race, find the marginal odds ratio between race and whether a murder victim.

2.38 For three-way contingency tables:

a. When any pair of variables is conditionally independent, explain why there is homogeneous association.

b. When there is not homogeneous association, explain why no pair of variables can be conditionally independent.

2.39 True, or false?

a. In 2×2 tables, statistical independence is equivalent to a population odds ratio value of $\theta = 1.0$.

b. We found that a 95% confidence interval for the odds ratio relating having a heart attack (yes, no) to drug (placebo, aspirin) is (1.44, 2.33). If we had formed the table with aspirin in the first row (instead of placebo), then the 95% confidence interval would have been $(1/2.33, 1/1.44) = (0.43, 0.69)$.

c. Using a survey of college students, we study the association between opinion about whether it should be legal to (1) use marijuana, (2) drink alcohol if you are 18 years old. We may get a different value for the odds ratio if we treat opinion about marijuana use as the response variable than if we treat alcohol use as the response variable.

d. Interchanging two rows or interchanging two columns in a contingency table has no effect on the value of the X^2 or G^2 chi-squared statistics. Thus, these tests treat both the rows and the columns of the contingency table as nominal scale, and if either or both variables are ordinal, the test ignores that information.

e. Suppose that income (high, low) and gender are conditionally independent, given type of job (secretarial, construction, service, professional, etc.). Then, income and gender are also independent in the 2×2 marginal table (i.e., ignoring, rather than controlling, type of job).

CHAPTER 3

Generalized Linear Models

Chapter 2 presented methods for analyzing contingency tables. Those methods help us investigate effects of explanatory variables on categorical response variables. The rest of this book uses *models* as the basis of such analyses. In fact, the methods of Chapter 2 also result from analyzing effects in certain models, but models can handle more complicated situations, such as analyzing simultaneously the effects of several explanatory variables.

A good-fitting model has several benefits. The structural form of the model describes the patterns of association and interaction. The sizes of the model parameters determine the strength and importance of the effects. Inferences about the parameters evaluate which explanatory variables affect the response variable Y, while controlling effects of possible confounding variables. Finally, the model's predicted values smooth the data and provide improved estimates of the mean of Y at possible explanatory variable values.

The models this book presents are *generalized linear models*. This broad class of models includes ordinary regression and ANOVA models for continuous responses as well as models for discrete responses. This chapter introduces generalized linear models for categorical and other discrete response data. The acronym *GLM* is shorthand for *g*eneralized *l*inear *m*odel.

Section 3.1 defines GLMs. Section 3.2 introduces GLMs for binary responses. An important special case is the *logistic regression model*, which Chapters 4–6 present in detail. Section 3.3 introduces GLMs for responses for an outcome that is a count. An important special case is the *loglinear model*, the subject of Chapter 7. Section 3.4 discusses inference and model checking for GLMs, and Section 3.5 discusses ML fitting.

An Introduction to Categorical Data Analysis, Second Edition. By Alan Agresti

3.1 COMPONENTS OF A GENERALIZED LINEAR MODEL

All generalized linear models have three components: The *random component* identifies the response variable Y and assumes a probability distribution for it. The *systematic component* specifies the explanatory variables for the model. The *link function* specifies a function of the expected value (mean) of Y, which the GLM relates to the explanatory variables through a prediction equation having linear form.

3.1.1 Random Component

The *random component* of a GLM identifies the response variable Y and selects a probability distribution for it. Denote the observations on Y by $(Y_1, \ldots, Y_n)$. Standard GLMs treat $Y_1, \ldots, Y_n$ as independent.

In many applications, the observations on Y are binary, such as "success" or "failure." More generally, each Y_i might be the number of successes out of a certain fixed number of trials. In either case, we assume a *binomial* distribution for Y. In some applications, each observation is a count. We might then assume a distribution for Y that applies to all the nonnegative integers, such as the *Poisson* or *negative binomial*. If each observation is continuous, such as a subject's weight in a dietary study, we might assume a *normal* distribution for Y.

3.1.2 Systematic Component

The *systematic component* of a GLM specifies the explanatory variables. These enter linearly as predictors on the right-hand side of the model equation. That is, the systematic component specifies the variables that are the $\{x_j\}$ in the formula

$$\alpha + \beta_1 x_1 + \cdots + \beta_k x_k$$

This linear combination of the explanatory variables is called the *linear predictor*.

Some $\{x_j\}$ can be based on others in the model. For example, perhaps $x_3 = x_1 x_2$, to allow interaction between x_1 and x_2 in their effects on Y, or perhaps $x_3 = x_1^2$, to allow a curvilinear effect of x_1. (GLMs use lower case for each x to emphasize that x-values are treated as fixed rather than as a random variable.)

3.1.3 Link Function

Denote the expected value of Y, the mean of its probability distribution, by $\mu = E(Y)$. The third component of a GLM, the *link function*, specifies a function $g(\cdot)$ that relates μ to the linear predictor as

$$g(\mu) = \alpha + \beta_1 x_1 + \cdots + \beta_k x_k$$

The function $g(\cdot)$, the link function, connects the random and systematic components.

The simplest link function is $g(\mu) = \mu$. This models the mean directly and is called the *identity link*. It specifies a linear model for the mean response,

$$\mu = \alpha + \beta_1 x_1 + \cdots + \beta_k x_k \tag{3.1}$$

This is the form of ordinary regression models for continuous responses.

Other link functions permit μ to be nonlinearly related to the predictors. For instance, the link function $g(\mu) = \log(\mu)$ models the log of the mean. The log function applies to positive numbers, so the log link function is appropriate when μ cannot be negative, such as with count data. A GLM that uses the log link is called a *loglinear model*. It has form

$$\log(\mu) = \alpha + \beta_1 x_1 + \cdots + \beta_k x_k$$

The link function $g(\mu) = \log[\mu/(1 - \mu)]$ models the log of an odds. It is appropriate when μ is between 0 and 1, such as a probability. This is called the *logit* link. A GLM that uses the logit link is called a *logistic regression model*.

Each potential probability distribution for Y has one special function of the mean that is called its *natural parameter*. For the normal distribution, it is the mean itself. For the binomial, the natural parameter is the logit of the success probability. The link function that uses the natural parameter as $g(\mu)$ in the GLM is called the *canonical link*. Although other link functions are possible, in practice the canonical links are most common.

3.1.4 Normal GLM

Ordinary regression models for continuous responses are special cases of GLMs. They assume a normal distribution for Y and model its mean directly, using the identity link function, $g(\mu) = \mu$. A GLM generalizes ordinary regression models in two ways: First, it allows Y to have a distribution other than the normal. Second, it allows modeling some function of the mean. Both generalizations are important for categorical data.

Historically, early analyses of nonnormal responses often attempted to transform Y so it is approximately normal, with constant variance. Then, ordinary regression methods using least squares are applicable. In practice, this is difficult to do. With the theory and methodology of GLMs, it is unnecessary to transform data so that methods for normal responses apply. This is because the GLM fitting process uses ML methods for our choice of random component, and we are not restricted to normality for that choice. The GLM choice of link function is separate from the choice of random component. It is not chosen to produce normality or stabilize the variance.

GLMs unify a wide variety of statistical methods. Regression, ANOVA, and models for categorical data are special cases of one super model. In fact, the same algorithm yields ML estimates of parameters for all GLMs. This algorithm is the basis of software for fitting GLMs, such as PROC GENMOD in SAS, the glm function in R and Splus, and the glm command in Stata.

The next two sections illustrate the GLM components by introducing the two most important GLMs for discrete responses – logistic regression models for binary data and loglinear models for count data.

3.2 GENERALIZED LINEAR MODELS FOR BINARY DATA

Many categorical response variables have only two categories: for example, whether you take public transportation today (yes, no), or whether you have had a physical exam in the past year (yes, no). Denote a binary response variable by Y and its two possible outcomes by 1 ("success") and 0 ("failure").

The distribution of Y is specified by probabilities $P(Y = 1) = \pi$ of success and $P(Y = 0) = (1 - \pi)$ of failure. Its mean is $E(Y) = \pi$. For n independent observations, the number of successes has the binomial distribution specified by the index n and parameter π. The formula was shown in equation (1.1). Each binary observation is a binomial variate with $n = 1$.

This section introduces GLMs for binary responses. Although GLMs can have multiple explanatory variables, for simplicity we introduce them using a single x. The value of π can vary as the value of x changes, and we replace π by $\pi(x)$ when we want to describe its dependence on that value.

3.2.1 Linear Probability Model

In ordinary regression, $\mu = E(Y)$ is a linear function of x. For a binary response, an analogous model is

$$\pi(x) = \alpha + \beta x$$

This is called a *linear probability model*, because the probability of success changes linearly in x. The parameter β represents the change in the probability per unit change in x. This model is a GLM with binomial random component and identity link function.

This model is simple, but unfortunately it has a structural defect. Probabilities fall between 0 and 1, whereas linear functions take values over the entire real line. This model predicts $\pi(x) < 0$ and $\pi(x) > 1$ for sufficiently large or small x values. The model can fit adequately over a restricted range of x values. For most applications, however, especially with several predictors, we need a more complex model form.

The ML estimates for this model, like most GLMs, do not have closed form; that is, the likelihood function is sufficiently complex that no formula exists for its maximum. Software for GLMs finds the ML estimates using an algorithm for successively getting better and better approximations for the likelihood function near its maximum, as Section 3.5.1 describes. This fitting process fails when, at some stage, an estimated probability $\hat{\pi}(x) = \hat{\alpha} + \hat{\beta}x$ falls outside the 0–1 range at some observed x value. Software then usually displays an error message such as "lack of convergence."

If we ignored the binary nature of Y and used ordinary regression, the estimates of the parameters would be the *least squares* estimates. They are the ML estimates under the assumption of a normal response. These estimates exist, because for a normal response an estimated mean of Y can be any real number and is not restricted to the 0–1 range. Of course, an assumption of normality for a binary response is not sensible; when ML fitting with the binomial assumption fails, the least squares method is also likely to give estimated probabilities outside the 0–1 range for some x values.

3.2.2 Example: Snoring and Heart Disease

Table 3.1 is based on an epidemiological survey of 2484 subjects to investigate snoring as a possible risk factor for heart disease. The subjects were classified according to their snoring level, as reported by their spouses. The linear probability model states that the probability of heart disease $\pi(x)$ is a linear function of the snoring level x. We treat the rows of the table as independent binomial samples with probabilities $\pi(x)$. We use scores (0, 2, 4, 5) for $x =$ snoring level, treating the last two snoring categories as closer than the other adjacent pairs.

Software for GLMs reports the ML model fit, $\hat{\pi} = 0.0172 + 0.0198x$. For example, for nonsnorers ($x = 0$), the estimated probability of heart disease is $\hat{\pi} = 0.0172 + 0.0198(0) = 0.0172$. The estimated values of $E(Y)$ for a GLM are called *fitted values*. Table 3.1 shows the sample proportions and the fitted values for the linear probability model. Figure 3.1 graphs the sample proportions and fitted values. The table and graph suggest that the model fits these data well. (Section 5.2.2 discusses goodness-of-fit analyses for binary-response GLMs.)

The model interpretation is simple. The estimated probability of heart disease is about 0.02 (namely, 0.0172) for nonsnorers; it increases 2(0.0198) = 0.04 for occasional snorers, another 0.04 for those who snore nearly every night, and another 0.02 for those who always snore. This rather surprising effect is significant, as the standard error of $\hat{\beta} = 0.0198$ equals 0.0028.

Suppose we had chosen snoring-level scores with different relative spacings than the scores {0, 2, 4, 5}. Examples are {0, 2, 4, 4.5} or {0, 1, 2, 3}. Then the fitted values would change somewhat. They would not change if the relative spacings between

Table 3.1. Relationship Between Snoring and Heart Disease

	Heart Disease					
Snoring	Yes	No	Proportion Yes	Linear Fit	Logit Fit	Probit Fit
Never	24	1355	0.017	0.017	0.021	0.020
Occasional	35	603	0.055	0.057	0.044	0.046
Nearly every night	21	192	0.099	0.096	0.093	0.095
Every night	30	224	0.118	0.116	0.132	0.131

Note: Model fits refer to proportion of "yes" responses.

Source: P. G. Norton and E. V. Dunn, *Br. Med. J.*, **291**: 630–632, 1985, published by BMJ Publishing Group.

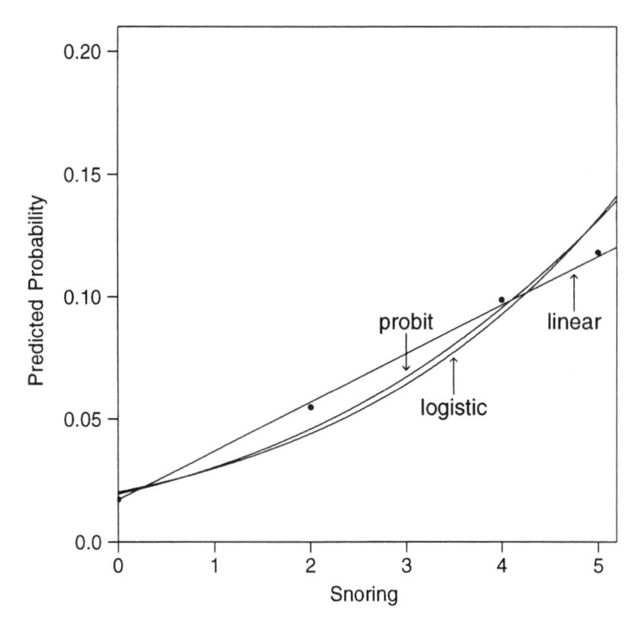

Figure 3.1. Fit of models for snoring and heart disease data.

scores were the same, such as $\{0, 4, 8, 10\}$ or $\{1, 3, 5, 6\}$. For these data, all these scores yield the conclusion that the probability of heart disease increases as snoring level increases.

Incidentally, if we entered the data as 2484 binary observations of 0 or 1 and fitted the model using ordinary least squares rather than ML, we would obtain $\hat{\pi} = 0.0169 + 0.0200x$. In practice, when the model fits well, least squares and ML estimates are similar.

3.2.3 Logistic Regression Model

Relationships between $\pi(x)$ and x are usually nonlinear rather than linear. A fixed change in x may have less impact when π is near 0 or 1 than when π is near the middle of its range. In the purchase of an automobile, for instance, consider the choice between buying new or used. Let $\pi(x)$ denote the probability of selecting a new car, when annual family income $= x$. An increase of \$10,000 in annual family income would likely have less effect when $x = \$1{,}000{,}000$ (for which π is near 1) than when $x = \$50{,}000$.

In practice, $\pi(x)$ often either increases continuously or decreases continuously as x increases. The S-shaped curves displayed in Figure 3.2 are often realistic shapes for the relationship. The most important mathematical function with this shape has formula

$$\pi(x) = \frac{\exp(\alpha + \beta x)}{1 + \exp(\alpha + \beta x)} = \frac{e^{\alpha+\beta x}}{1 + e^{\alpha+\beta x}}$$

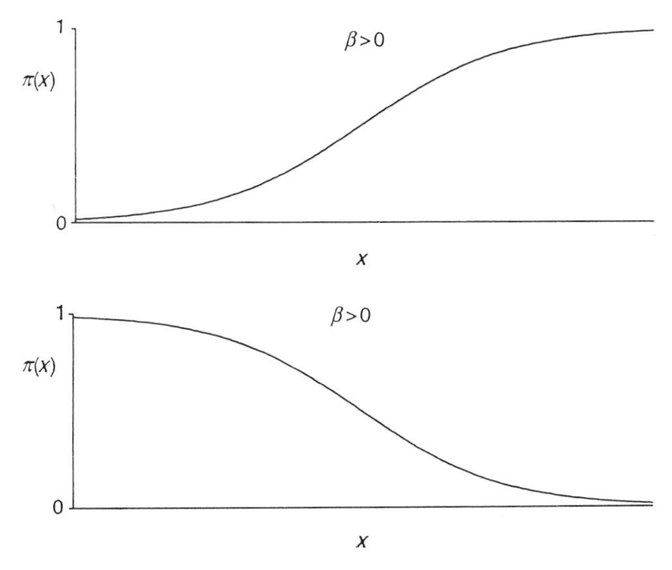

Figure 3.2. Logistic regression functions.

using the exponential function. This is called the *logistic regression* function. We will see in Chapter 4 that the corresponding logistic regression model form is

$$\log\left(\frac{\pi(x)}{1-\pi(x)}\right) = \alpha + \beta x \tag{3.2}$$

The logistic regression model (3.2) is a special case of a GLM. The random component for the (success, failure) outcomes has a binomial distribution. The link function is the *logit* function $\log[\pi/(1-\pi)]$ of π, symbolized by "logit(π)." Logistic regression models are often called *logit models*. Whereas π is restricted to the 0–1 range, the logit can be any real number. The real numbers are also the potential range for linear predictors (such as $\alpha + \beta x$) that form the systematic component of a GLM, so this model does not have the structural problem that the linear probability model has.

The parameter β in equation (3.2) determines the rate of increase or decrease of the curve. When $\beta > 0$, $\pi(x)$ increases as x increases, as in Figure 3.2(*a*). When $\beta < 0$, $\pi(x)$ decreases as x increases, as in Figure 3.2(*b*). The magnitude of β determines how fast the curve increases or decreases. As $|\beta|$ increases, the curve has a steeper rate of change. When $\beta = 0$, the curve flattens to a horizontal straight line.

For the logistic regression model for the snoring and heart disease data in Table 3.1, software reports the ML fit,

$$\text{logit}[\hat{\pi}(x)] = -3.87 + 0.40x$$

Since $\hat{\beta} = 0.40 > 0$, the estimated probability of heart disease increases as snoring level increases. Chapter 4 shows how to calculate estimated probabilities (the model *fitted values*) and presents ways of interpreting the model. Table 3.1 also reports these fitted values, and Figure 3.1 displays the fit. The fit is close to linear over this rather narrow range of estimated probabilities. Results are similar to those for the linear probability model.

3.2.4 Probit Regression Model

Another model that has the S-shaped curves of Figure 3.2 is called the *probit model*. The link function for the model, called the *probit link*, transforms probabilities to z-scores from the standard normal distribution. The probit model has expression

$$\text{probit}[\pi(x)] = \alpha + \beta x \tag{3.3}$$

The probit link function applied to $\pi(x)$ gives the standard normal z-score at which the left-tail probability equals $\pi(x)$. For instance, probit(0.05) $= -1.645$, because 5% of the standard normal distribution falls below -1.645. Likewise, probit(0.50) $= 0$, probit(0.95) $= 1.645$, and probit(0.975) $= 1.96$.

For the snoring and heart disease data with scores $\{0, 2, 4, 5\}$ for snoring level, software reports that the ML fit of the probit model is

$$\text{probit}[\hat{\pi}(x)] = -2.061 + 0.188x$$

At snoring level $x = 0$, the probit equals $-2.061 + 0.188(0) = -2.06$. The fitted probability $\hat{\pi}(0)$ is the left-tail probability for the standard normal distribution at $z = -2.06$, which equals 0.020. At snoring level $x = 5$, the probit equals $-2.061 + 0.188(5) = -1.12$, which corresponds to a fitted probability of 0.131.

The fitted values, shown in Table 3.1 and Figure 3.1, are similar to those for the linear probability and logistic regression models. In practice, probit and logistic regression models provide similar fits. If a logistic regression model fits well, then so does the probit model, and conversely.

3.2.5 Binary Regression and Cumulative Distribution Functions*

For a random variable X, the *cumulative distribution function* (*cdf*) $F(x)$ for X is defined as

$$F(x) = P(X \leq x), \quad -\infty < x < \infty$$

As x increases, $P(X \leq x)$ increases. Thus, as x increases over its range of values, $F(x)$ increases from 0 to 1. When X is a continuous random variable, the *cdf*, plotted as a function of x, has S-shaped appearance, like that in Figure 3.1(a). This suggests a class of models for binary responses whereby the dependence of $\pi(x)$ on x has the form

$$\pi(x) = F(x) \tag{3.4}$$

where F is a *cdf* for some continuous probability distribution.

When F is the *cdf* of a normal distribution, model type (3.4) is equivalent to the probit model (3.3). The probit link function transforms $\pi(x)$ so that the regression curve for $\pi(x)$ [or for $1 - \pi(x)$, when $\beta < 0$] has the appearance of the normal *cdf*. The parameters of the normal distribution relate to the parameters in the probit model by mean $\mu = -\alpha/\beta$ and standard deviation $\sigma = 1/|\beta|$. Each choice of α and of $\beta > 0$ corresponds to a different normal distribution.

For the snoring and heart disease data, probit$[\hat{\pi}(x)] = -2.061 + 0.188x$. This probit fit corresponds to a normal *cdf* having mean $-\hat{\alpha}/\hat{\beta} = 2.061/0.188 = 11.0$ and standard deviation $1/|\hat{\beta}| = 1/0.188 = 5.3$. The estimated probability of heart disease equals 1/2 at snoring level $x = 11.0$. That is, $x = 11.0$ has a fitted probit of $-2.061 + 0.188(11) = 0$, which is the z-score corresponding to a left-tail probability of 1/2. Since snoring level is restricted to the range 0–5 for these data, well below 11, the fitted probabilities over this range are quite small.

The logistic regression curve also has form (3.4). When $\beta > 0$ in model (3.2), the curve for $\pi(x)$ has the shape of the *cdf* $F(x)$ of a two-parameter *logistic distribution*. The logistic *cdf* corresponds to a probability distribution with a symmetric, bell shape. It looks similar to a normal distribution but with slightly thicker tails.

When both models fit well, parameter estimates in probit models have smaller magnitude than those in logistic regression models. This is because their link functions transform probabilities to scores from standard versions of the normal and logistic distribution, but those two distributions have different spread. The standard normal distribution has a mean of 0 and standard deviation of 1. The standard logistic distribution has a mean of 0 and standard deviation of 1.8. When both models fit well, parameter estimates in logistic regression models are approximately 1.8 times those in probit models.

The probit model and the *cdf* model form (3.4) were introduced in the mid1930s in toxicology studies. A typical experiment exposes animals (typically insects or mice) to various dosages of some potentially toxic substance. For each subject, the response is whether it dies. It is natural to assume a *tolerance distribution* for subjects' responses. For example, each insect may have a certain tolerance to an insecticide, such that it dies if the dosage level exceeds its tolerance and survives if the dosage level is less than its tolerance. Tolerances would vary among insects. If a *cdf* F describes the distribution of tolerances, then the model for the probability $\pi(x)$ of death at dosage level x has form (3.4). If the tolerances vary among insects according to a normal distribution, then $\pi(x)$ has the shape of a normal *cdf*. With sample data, model fitting determines *which* normal *cdf* best applies.

Logistic regression was not developed until the mid 1940s and not used much until the 1970s, but it is now more popular than the probit model. We will see in the next chapter that the logistic model parameters relate to odds ratios. Thus, one can fit the model to data from case–control studies, because one can estimate odds ratios for such data (Section 2.3.5).

3.3 GENERALIZED LINEAR MODELS FOR COUNT DATA

Many discrete response variables have *counts* as possible outcomes. Examples are Y = number of parties attended in the past month, for a sample of students, or Y = number of imperfections on each of a sample of silicon wafers used in manufacturing computer chips. Counts also occur in summarizing categorical variables with contingency tables. This section introduces GLMs for count data.

The simplest GLMs for count data assume a *Poisson distribution* for the random component. Like counts, Poisson variates can take any nonnegative integer value. We won't need to use the formula for the Poisson distribution here[1] but will merely state a few properties.

The Poisson distribution is unimodal and skewed to the right over the possible values $0, 1, 2, \ldots$. It has a single parameter $\mu > 0$, which is both its mean and its variance. That is,

$$E(Y) = \text{Var}(Y) = \mu, \quad \sigma(Y) = \sqrt{\mu}$$

Therefore, when the counts are larger, on the average, they also tend to be more variable. When the number of imperfections on a silicon wafer has the Poisson distribution with $\mu = 20$, we observe greater variability in y from wafer to wafer than when $\mu = 2$. As the mean increases the skew decreases and the distribution becomes more bell-shaped. Figure 3.3 shows Poisson distributions with means 2 and 6.

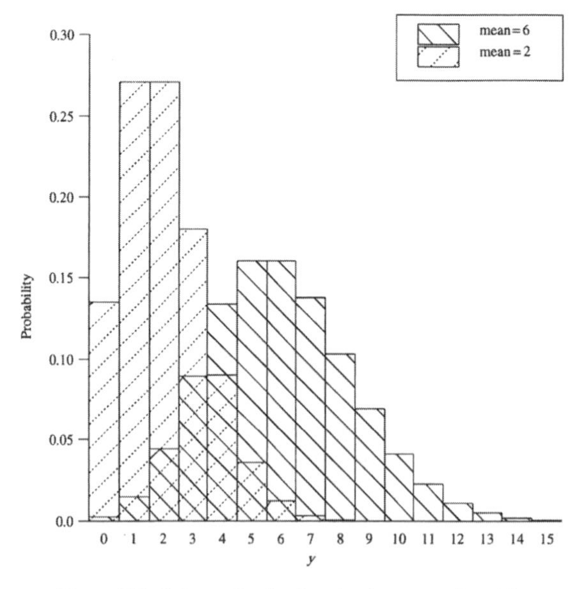

Figure 3.3. Poisson distributions having means 2 and 6.

[1]It is $P(y) = e^{-\mu}\mu^y/y!$, for $y = 0, 1, 2, \ldots$.

Chapter 7 presents Poisson GLMs for counts in contingency tables that cross classify categorical response variables. This section introduces Poisson GLMs for modeling counts for a single discrete response variable.

3.3.1 Poisson Regression

The Poisson distribution has a positive mean. GLMs for the Poisson mean can use the identity link, but it is more common to model the log of the mean. Like the linear predictor $\alpha + \beta x$, the log of the mean can take any real-number value. A *Poisson loglinear model* is a GLM that assumes a Poisson distribution for Y and uses the log link function.

For a single explanatory variable x, the Poisson loglinear model has form

$$\log \mu = \alpha + \beta x \tag{3.5}$$

The mean satisfies the exponential relationship

$$\mu = \exp(\alpha + \beta x) = e^{\alpha}(e^{\beta})^{x} \tag{3.6}$$

A one-unit increase in x has a multiplicative impact of e^{β} on μ: The mean of Y at $x + 1$ equals the mean of Y at x multiplied by e^{β}. If $\beta = 0$, then $e^{\beta} = e^{0} = 1$ and the multiplicative factor is 1. Then, the mean of Y does not change as x changes. If $\beta > 0$, then $e^{\beta} > 1$, and the mean of Y increases as x increases. If $\beta < 0$, the mean decreases as x increases.

3.3.2 Example: Female Horseshoe Crabs and their Satellites

Table 3.2 comes from a study of nesting horseshoe crabs (J. Brockmann, *Ethology*, **102**: 1–21, 1996). Each female horseshoe crab in the study had a male crab attached to her in her nest. The study investigated factors that affect whether the female crab had any other males, called *satellites*, residing nearby her. The response outcome for each female crab is her number of satellites. An explanatory variable thought possibly to affect this was the female crab's shell width, which is a summary of her size. In the sample, this shell width had a mean of 26.3 cm and a standard deviation of 2.1 cm.

Figure 3.4 plots the response counts against width (in centimeters), with numbered symbols indicating the number of observations at each point. To obtain a clearer picture of any trend, we grouped the female crabs into a set of width categories, ($\leq$23.25, 23.25–24.25, 24.25–25.25, 25.25–26.25, 26.25–27.25, 27.25–28.25, 28.25–29.25, $>$30.25), and calculated the sample mean number of satellites in each category. Figure 3.5 plots these sample means against the sample mean width for the female crabs in each category.

Table 3.2. Number of Crab Satellites by Female's Color, Spine Condition, Width, and Weight

C	S	W	Wt	Sa	C	S	W	Wt	Sa	C	S	W	Wt	Sa	C	S	W	Wt	Sa
2	3	28.3	3.05	8	3	3	22.5	1.55	0	1	1	26.0	2.30	9	3	3	24.8	2.10	0
3	3	26.0	2.60	4	2	3	23.8	2.10	0	3	2	24.7	1.90	0	2	1	23.7	1.95	0
3	3	25.6	2.15	0	3	3	24.3	2.15	0	2	3	25.8	2.65	0	2	3	28.2	3.05	11
4	2	21.0	1.85	0	2	1	26.0	2.30	14	1	1	27.1	2.95	8	2	3	25.2	2.00	1
2	3	29.0	3.00	1	4	3	24.7	2.20	0	2	3	27.4	2.70	5	2	2	23.2	1.95	4
1	2	25.0	2.30	3	2	1	22.5	1.60	1	3	3	26.7	2.60	2	4	3	25.8	2.00	3
4	3	26.2	1.30	0	2	3	28.7	3.15	3	2	1	26.8	2.70	5	4	3	27.5	2.60	0
2	3	24.9	2.10	0	1	1	29.3	3.20	4	1	3	25.8	2.60	0	2	2	25.7	2.00	0
2	1	25.7	2.00	8	2	1	26.7	2.70	5	4	3	23.7	1.85	0	2	3	26.8	2.65	0
2	3	27.5	3.15	6	4	3	23.4	1.90	0	2	3	27.9	2.80	6	3	3	27.5	3.10	3
1	1	26.1	2.80	5	1	1	27.7	2.50	6	2	1	30.0	3.30	5	3	1	28.5	3.25	9
3	3	28.9	2.80	4	2	3	28.2	2.60	6	2	3	25.0	2.10	4	2	3	28.5	3.00	3
2	1	30.3	3.60	3	4	3	24.7	2.10	5	2	3	27.7	2.90	5	1	1	27.4	2.70	6
2	3	22.9	1.60	4	2	1	25.7	2.00	5	2	3	28.3	3.00	15	2	3	27.2	2.70	3
3	3	26.2	2.30	3	2	1	27.8	2.75	0	4	3	25.5	2.25	0	3	3	27.1	2.55	0
3	3	24.5	2.05	5	3	1	27.0	2.45	3	2	3	26.0	2.15	5	2	3	28.0	2.80	1
2	3	30.0	3.05	8	2	3	29.0	3.20	10	2	3	26.2	2.40	0	2	1	26.5	1.30	0
2	3	26.2	2.40	3	3	3	25.6	2.80	7	3	3	23.0	1.65	1	3	3	23.0	1.80	0
2	3	25.4	2.25	6	3	3	24.2	1.90	0	2	2	22.9	1.60	0	3	2	26.0	2.20	3
2	3	25.4	2.25	4	3	3	25.7	1.20	0	2	3	25.1	2.10	5	3	2	24.5	2.25	0
4	3	27.5	2.90	0	3	3	23.1	1.65	0	3	1	25.9	2.55	4	2	3	25.8	2.30	0
4	3	27.0	2.25	3	2	3	28.5	3.05	0	4	1	25.5	2.75	0	4	3	23.5	1.90	0
2	2	24.0	1.70	0	2	1	29.7	3.85	5	2	1	26.8	2.55	0	4	3	26.7	2.45	0
2	1	28.7	3.20	0	3	3	23.1	1.55	0	2	1	29.0	2.80	1	3	3	25.5	2.25	0

3	3	26.5	1.97	1	3	3	24.5	2.20	1	3	3	28.5	3.00	1	2	3	28.2	2.87	1
2	3	24.5	1.60	1	2	3	27.5	2.55	1	2	2	24.7	2.55	4	2	1	25.2	2.00	1
3	3	27.3	2.90	1	2	3	26.3	2.40	1	2	3	29.0	3.10	1	2	3	25.3	1.90	2
2	3	26.5	2.30	4	2	3	27.8	3.25	3	2	3	27.0	2.50	6	3	3	25.7	2.10	0
2	3	25.0	2.10	2	2	3	31.9	3.33	2	4	3	23.7	1.80	0	4	3	29.3	3.23	12
3	3	22.0	1.40	0	2	3	25.0	2.40	5	3	3	27.0	2.50	6	3	3	23.8	1.80	6
1	1	30.2	3.28	2	3	3	26.2	2.22	0	2	3	24.2	1.65	2	2	3	27.4	2.90	3
2	2	25.4	2.30	0	3	3	28.4	3.20	3	4	3	22.5	1.47	4	2	3	26.2	2.02	2
2	1	24.9	2.30	6	1	2	24.5	1.95	6	2	3	25.1	1.80	0	2	1	28.0	2.90	4
4	3	25.8	2.25	10	2	3	27.9	3.05	7	2	3	24.9	2.20	0	2	1	28.4	3.10	5
3	3	27.2	2.40	5	2	2	25.0	2.25	6	2	3	27.5	2.63	6	2	1	33.5	5.20	7
2	3	30.5	3.32	3	3	3	29.0	2.92	3	2	1	24.3	2.00	0	2	3	25.8	2.40	0
4	3	25.0	2.10	8	2	1	31.7	3.73	4	2	3	29.5	3.02	4	3	3	24.0	1.90	10
2	3	30.0	3.00	9	2	3	27.6	2.85	4	2	3	26.2	2.30	0	2	1	23.1	2.00	0
2	1	22.9	1.60	0	4	3	24.5	1.90	0	2	3	24.7	1.95	4	2	3	28.3	3.20	0
2	3	23.9	1.85	2	3	3	23.8	1.80	0	3	2	29.8	3.50	4	2	3	26.5	2.35	4
2	3	26.0	2.28	3	2	3	28.2	3.05	8	4	3	25.7	2.15	0	2	3	26.5	2.75	7
2	3	25.8	2.20	0	3	3	24.1	1.80	0	3	3	26.2	2.17	2	3	3	26.1	2.75	3
3	3	29.0	3.28	4	1	1	28.0	2.62	0	4	3	27.0	2.63	0	2	2	24.5	2.00	0
1	1	26.5	2.35	0															

Note: C = Color (1 = light medium, 2 = medium, 3 = dark medium, 4 = dark), S = spine condition (1 = both good, 2 = one worn or broken, 3 = both worn or broken), W = carapace width (cm), Wt = weight (kg), Sa = number of satellites.

Source: Data provided by Dr. Jane Brockmann, Zoology Department, University of Florida; study described in *Ethology*, **102**: 1–21, 1996.

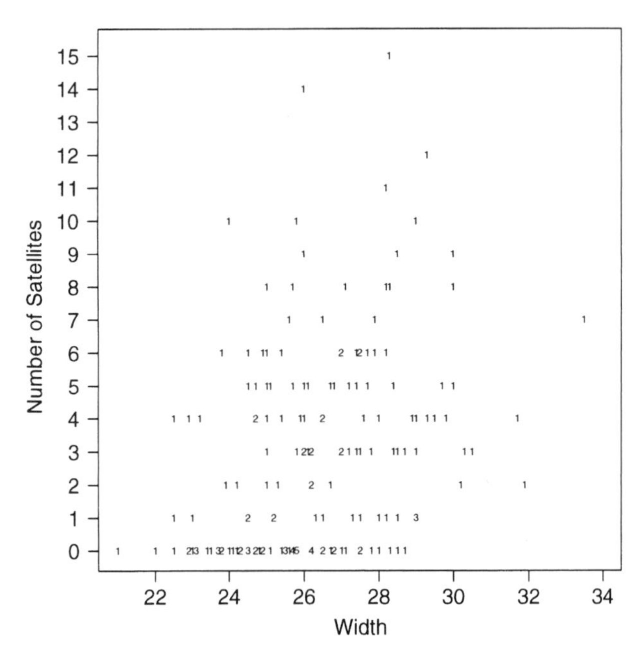

Figure 3.4. Number of satellites by width of female crab.

Most software has more sophisticated ways of smoothing the data, revealing the trend without grouping the width values. Smoothing methods based on *generalized additive models* do this by providing even more general structural form than GLMs. They find possibly complex functions of the explanatory variables that serve as the best predictors of a certain type. Figure 3.5 also shows a curve based on smoothing the data using this method. The sample means and the smoothed curve both show a strong increasing trend. (The means tend to fall above the curve, since the response counts in a category tend to be skewed to the right. The smoothed curve is less susceptible to outlying observations.) The trend seems approximately linear, and we next discuss models for which the mean or the log of the mean is linear in width.

Let μ denote the expected number of satellites for a female crab, and let x denote her width. From GLM software, the ML fit of the Poisson loglinear model (3.5) is

$$\log \hat{\mu} = \hat{\alpha} + \hat{\beta}x = -3.305 + 0.164x$$

The effect $\hat{\beta} = 0.164$ of width has $SE = 0.020$. Since $\hat{\beta} > 0$, width has a positive estimated effect on the number of satellites.

The model fit yields an estimated mean number of satellites $\hat{\mu}$, a *fitted value*, at any width. For instance, from equation (3.6), the fitted value at the mean width of $x = 26.3$ is

$$\hat{\mu} = \exp(\hat{\alpha} + \hat{\beta}x) = \exp[-3.305 + 0.164(26.3)] = e^{1.01} = 2.7$$

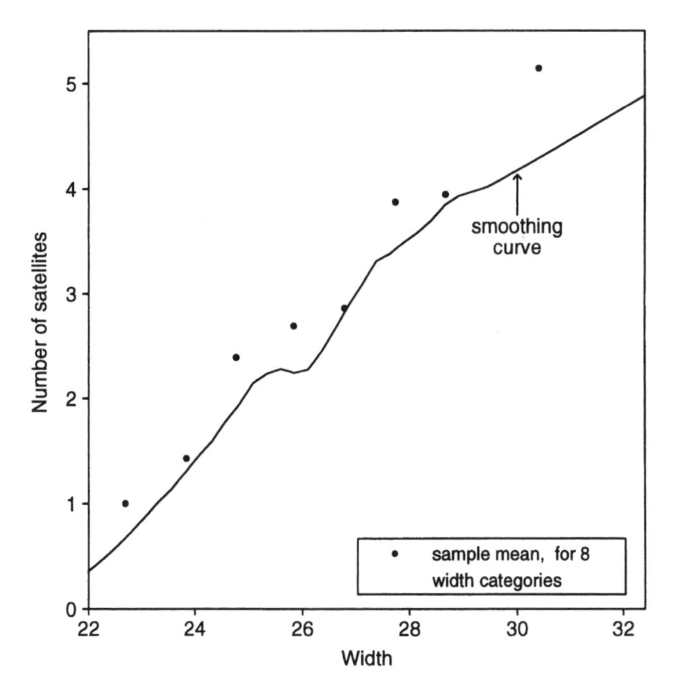

Figure 3.5. Smoothings of horseshoe crab counts.

For this model, $\exp(\hat{\beta}) = \exp(0.164) = 1.18$ represents the multiplicative effect on the fitted value for each 1-unit increase in x. For instance, the fitted value at $x = 27.3 = 26.3 + 1$ is $\exp[-3.305 + 0.164(27.3)] = 3.2$, which equals (1.18)(2.7). A 1 cm increase in width has an 18% increase in the estimated mean number of satellites.

Figure 3.5 suggests that the expected number of satellites may grow approximately linearly with width. The Poisson regression model with identity link function has ML fit

$$\hat{\mu} = -11.53 + 0.550x$$

where $\hat{\beta} = 0.550$ has $SE = 0.059$. The effect of x on μ in this model is additive, rather than multiplicative. A 1 cm increase in width has a predicted increase of $\hat{\beta} = 0.55$ in the expected number of satellites. For instance, the fitted value at the mean width of $x = 26.3$ is $\hat{\mu} = -11.53 + 0.550(26.3) = 2.93$; at $x = 27.3$, it is $2.93 + 0.55 = 3.48$. The fitted values are positive at all observed sample widths, and the model provides a simple description of the width effect: On average, a 2 cm increase in width corresponds to about an extra satellite.

Figure 3.6 plots the fitted number of satellites against width, for the models with log link and with identity link. Although they diverge somewhat for small and large widths, they provide similar predictions over the range of width values in which most observations occur.

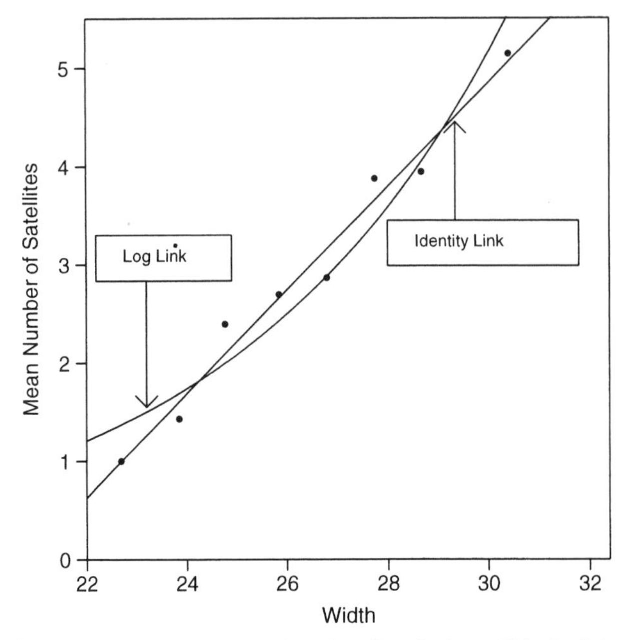

Figure 3.6. Estimated mean number of satellites for log and identity links.

3.3.3 Overdispersion: Greater Variability than Expected

Count data often vary more than we would expect if the response distribution truly were Poisson. For instance, for the grouped horseshoe crab data, Table 3.3 shows the sample mean and variance for the counts of the number of satellites for the female crabs in each width category. The variances are much larger than the means, whereas Poisson distributions have identical mean and variance. The phenomenon of the data having greater variability than expected for a GLM is called *overdispersion*.

A common cause of overdispersion is heterogeneity among subjects. For instance, suppose width, weight, color, and spine condition all affect a female crab's number

Table 3.3. Sample Mean and Variance of Number of Satellites

Width	No. Cases	No. Satellites	Sample Mean	Sample Variance
<23.25	14	14	1.00	2.77
23.25–24.25	14	20	1.43	8.88
24.25–25.25	28	67	2.39	6.54
25.25–26.25	39	105	2.69	11.38
26.25–27.25	22	63	2.86	6.88
27.25–28.25	24	93	3.87	8.81
28.25–29.25	18	71	3.94	16.88
>29.25	14	72	5.14	8.29

of satellites. Suppose the number of satellites has a Poisson distribution at each fixed combination of those four variables, but suppose the model uses width alone as a predictor. Crabs having a certain fixed width are a mixture of crabs of various weights, colors, and spine conditions. Thus, the population of crabs having that fixed width is a mixture of several Poisson populations, each having its own mean for the response. This heterogeneity in the crabs of a given width yields an overall response distribution at that width having greater variation than the Poisson predicts. If the variance equals the mean when *all* relevant variables are controlled, it exceeds the mean when only a *subset* of those variables is controlled.

Overdispersion is not an issue in ordinary regression models assuming normally distributed Y, because the normal has a separate parameter from the mean (i.e., the variance, σ^2) to describe variability. For Poisson distributions, however, the variance equals the mean. Overdispersion is common in applying Poisson GLMs to counts.

3.3.4 Negative Binomial Regression*

The *negative binomial* is another distribution that is concentrated on the nonnegative integers. Unlike the Poisson, it has an additional parameter such that the variance can exceed the mean. The negative binomial distribution has

$$E(Y) = \mu, \quad \text{Var}(Y) = \mu + D\mu^2$$

The index D, which is nonnegative, is called a *dispersion parameter*. The negative binomial distribution arises as a type of mixture of Poisson distributions.[2] Greater heterogeneity in the Poisson means results in a larger value of D. As $D \to 0$, $\text{Var}(Y) \to \mu$ and the negative binomial distribution converges to the Poisson distribution. The farther D falls above 0, the greater the overdispersion relative to Poisson variability.

Negative binomial GLMs for counts express μ in terms of explanatory variables. Most common is the log link, as in Poisson loglinear models, but sometimes the identity link is adequate. It is common to assume that the dispersion parameter D takes the same value at all predictor values, much as regression models for a normal response take the variance parameter to be constant. The model can be fitted with GLM software, such as PROC GENMOD in SAS (see Table A.4). The estimate of D summarizes the extent of overdispersion relative to the Poisson GLM.

Consider the horseshoe crab dataset in Section 3.3.2 above on Y = number of satellites, with x = shell width as predictor. The Poisson GLM with log link has

$$\log(\hat{\mu}) = -3.30 + 0.164x$$

[2]When the Poisson means follow a *gamma* distribution, unconditionally the distribution is the negative binomial.

with $SE = 0.020$ for $\hat{\beta}$. The negative binomial GLM has

$$\log(\hat{\mu}) = -4.05 + 0.192x$$

with $SE = 0.048$ for $\hat{\beta}$. Moreover, $\hat{D} = 1.1$, so at a predicted $\hat{\mu}$, the estimated variance is roughly $\hat{\mu} + \hat{\mu}^2$, compared with $\hat{\mu}$ for the Poisson GLM. Fitted values are similar, but the greater estimated variance in the negative binomial model and the resulting greater SE for $\hat{\beta}$ reflect the overdispersion uncaptured with the Poisson GLM. Inspection of Figure 3.4 shows that some zero counts occur even when the sample mean response is relatively large, reflecting this overdispersion.

For the Poisson model, the 95% Wald confidence interval for the effect of width (β) is $0.164 \pm 1.96(0.020)$, which is $(0.125, 0.203)$. For the negative binomial model, it is $0.192 \pm 1.96(0.048)$, which is $(0.099, 0.285)$. The confidence interval for the Poisson GLM is unrealistically narrow, because it does not allow for the overdispersion.

3.3.5 Count Regression for Rate Data*

When events occur over time, space, or some other index of size, models can focus on the *rate* at which the events occur. For example, in analyzing numbers of murders in 2006 for a sample of cities, we could form a rate for each city by dividing the number of murders by the city's population size. A model might describe how the rate depends on explanatory variables such as the city's unemployment rate, median income, and percentage of residents having completed high school.

When a response count Y has index (such as population size) equal to t, the sample rate is Y/t. The expected value of the rate is μ/t, where $\mu = E(Y)$. A loglinear model for the expected rate has form

$$\log(\mu/t) = \alpha + \beta x \tag{3.7}$$

This model has equivalent representation

$$\log \mu - \log t = \alpha + \beta x$$

The adjustment term, $-\log t$, to the log of the mean is called an *offset*. Standard GLM software can fit a model having an offset term.

For loglinear model (3.7), the expected number of outcomes satisfies

$$\mu = t \exp(\alpha + \beta x) \tag{3.8}$$

The mean μ is proportional to the index t, with proportionality constant depending on the value of the explanatory variable. For a fixed value of x, for example, doubling the population size t also doubles the expected number of murders μ.

3.3.6 Example: British Train Accidents over Time*

Table 3.4 lists the number of two types of train-related accidents in the UK between 1975 and 2003: (1) accidents involving trains alone (collisions, derailments, and overruns); and (2) collisions between trains and road vehicles. Here we consider only the second count. The table also shows the annual number of train-kilometers, which is a measure of railway activity indicating the millions of kilometers traveled by trains during the year.

Let μ denote the expected value of Y = annual number of collisions between trains and road vehicles, for t million kilometers of train travel. During the past decade, rail travel has become increasingly privatized in the UK, and some people have expressed fears that accidents have become more likely. To allow for a trend over time, we consider model (3.7) with x = number of years since 1975.

Assuming a Poisson distribution for Y, we get the ML fit

$$\log(\hat{\mu}) - \log(t) = -4.21 - 0.0329x$$

where the estimated slope has $SE = 0.011$. From equation (3.8), the estimated rate is $\exp(-4.21 - 0.0329x)$, which is $e^{-4.21}(e^{-0.0329})^x = (0.0148)(0.968)^x$. The estimated rate of train accidents decreases from 0.0148 in 1975 (take $x = 0$) to 0.0059 in 2003 (take $x = 28$). The estimated rate in 2003 of 0.0059 per million kilometers is roughly 6 per *billion* kilometers.

Table 3.4. Collisions Involving Trains in Great Britain

Year	Train-km	Train Collisions	Train-road Collisions	Year	Train-km	Train Collisions	Train-road Collisions
2003	518	0	3	1988	443	2	4
2002	516	1	3	1987	397	1	6
2001	508	0	4	1986	414	2	13
2000	503	1	3	1985	418	0	5
1999	505	1	2	1984	389	5	3
1998	487	0	4	1983	401	2	7
1997	463	1	1	1982	372	2	3
1996	437	2	2	1981	417	2	2
1995	423	1	2	1980	430	2	2
1994	415	2	4	1979	426	3	3
1993	425	0	4	1978	430	2	4
1992	430	1	4	1977	425	1	8
1991	439	2	6	1976	426	2	12
1990	431	1	2	1975	436	5	2
1989	436	4	4				

Source: British Department of Transport.

Similar results occur with a negative binomial model. The ML fit is

$$\log(\hat{\mu}) - \log(t) = -4.20 - 0.0337x$$

where the estimated slope has $SE = 0.013$. There is some sample overdispersion compared with the Poisson GLM, as the estimated dispersion parameter $\hat{D} = 0.099$ ($SE = 0.078$). Because of this, this model's evidence of a time effect is somewhat weaker (i.e., the SE of the time effect is larger). The estimated rate of train accidents is $e^{-4.20}(e^{-0.0337})^x = (0.0150)(0.967)^x$. This decreases from 0.0150 in 1975 to 0.0058 in 2003. For either sampling model, adding a quadratic time effect does not significantly improve the fit.

3.4 STATISTICAL INFERENCE AND MODEL CHECKING

For most GLMs, calculation of ML parameter estimates is computationally complex. Software uses an algorithm described in Section 3.5.1. The ML estimators are approximately normally distributed for large samples. Statistical inference based on the Wald statistic is simplest, but likelihood-ratio inference is more trustworthy.

3.4.1 Inference about Model Parameters

A Wald 95% confidence interval for a model parameter β equals $\hat{\beta} \pm 1.96(SE)$, where SE is the standard error of $\hat{\beta}$. To test H_0: $\beta = 0$, the Wald test statistic

$$z = \hat{\beta}/SE$$

has an approximate standard normal distribution when $\beta = 0$. Equivalently, z^2 has an approximate chi-squared distribution with $df = 1$.

For the likelihood-ratio approach, denote the maximized value of the likelihood function by ℓ_0 under H_0: $\beta = 0$ and by ℓ_1 when β need not equal 0. The *likelihood-ratio* test statistic equals

$$-2\log(\ell_0/\ell_1) = -2[\log(\ell_0) - \log(\ell_1)] = -2(L_0 - L_1)$$

where L_0 and L_1 denote the maximized log-likelihood functions. Under H_0: $\beta = 0$, this test statistic has a large-sample chi-squared distribution with $df = 1$. Software for GLMs can report the maximized log-likelihood values and the likelihood-ratio statistic.

The likelihood-ratio method can also determine a confidence interval for β. The 95% confidence interval consists of all β_0 values for which the P-value exceeds 0.05 in the likelihood-ratio test of H_0: $\beta = \beta_0$. For small n, this is preferable to the Wald interval. Some software (such as PROC GENMOD in SAS) can provide it for any GLM.

3.4.2 Example: Snoring and Heart Disease Revisited

Section 3.2.2 used a linear probability model to describe the probability of heart disease $\pi(x)$ as a linear function of snoring level x for data in Table 3.1. The ML model fit is $\hat{\pi} = 0.0172 + 0.0198x$, where the snoring effect $\hat{\beta} = 0.0198$ has $SE = 0.0028$.

The Wald test of H_0: $\beta = 0$ against H_a: $\beta \neq 0$ treats

$$z = \hat{\beta}/SE = 0.0198/0.0028 = 7.1$$

as standard normal, or $z^2 = 50.0$ as chi-squared with $df = 1$. This provides extremely strong evidence of a positive snoring effect on the probability of heart disease ($P < 0.0001$). We obtain similar strong evidence from a likelihood-ratio test comparing this model to the simpler one having $\beta = 0$. That chi-squared statistic equals $-2(L_0 - L_1) = 65.8$ with $df = 1$ ($P < 0.0001$). The likelihood-ratio 95% confidence interval for β is (0.0145, 0.0255).

3.4.3 The Deviance

Let L_M denote the maximized log-likelihood value for a model M of interest. Let L_S denote the maximized log-likelihood value for the most complex model possible. This model has a separate parameter for each observation, and it provides a perfect fit to the data. The model is said to be *saturated.*

For example, suppose M is the linear probability model, $\pi(x) = \alpha + \beta x$, applied to the 4×2 Table 3.1 on snoring and heart disease. The model has two parameters for describing how the probability of heart disease changes for the four levels of $x =$ snoring. The corresponding saturated model has a separate parameter for each of the four binomial observations: $\pi(x) = \pi_1$ for never snorers, $\pi(x) = \pi_2$ for occasional snorers, $\pi(x) = \pi_3$ for snoring nearly every night, $\pi(x) = \pi_4$ for snoring every night. The ML estimate for π_i is simply the sample proportion having heart disease at level i of snoring.

Because the saturated model has additional parameters, its maximized log likelihood L_S is at least as large as the maximized log likelihood L_M for a simpler model M. The *deviance* of a GLM is defined as

$$\text{Deviance} = -2[L_M - L_S]$$

The deviance is the likelihood-ratio statistic for comparing model M to the saturated model. It is a test statistic for the hypothesis that all parameters that are in the saturated model but not in model M equal zero. GLM software provides the deviance, so it is not necessary to calculate L_M or L_S.

For some GLMs, the deviance has approximately a chi-squared distribution. For example, in Section 5.2.2 we will see this happens for binary GLMs with a fixed number of explanatory levels in which each observation is a binomial variate having relatively large counts of successes and failures. For such cases, the deviance provides a goodness-of-fit test of the model, because it tests the hypothesis that all possible parameters not included in the model equal 0. The residual df equals the number of observations minus the number of model parameters. The P-value is the right-tail probability above the observed test statistic value, from the chi-squared distribution. Large test statistics and small P-values provide strong evidence of model lack of fit.

For the snoring and heart disease data, the linear probability model describes four binomial observations by two parameters. The deviance, which software reports to equal 0.1, has $df = 4 - 2 = 2$. For testing the null hypothesis that the model holds, the P-value is 0.97. The logistic regression model fitted in Section 3.2.3 has deviance = 2.8, with $df = 2$ (P-value = 0.25). Either model fits adequately.

3.4.4 Model Comparison Using the Deviance

Now consider two models, denoted by M_0 and M_1, such that M_0 is a special case of M_1. For normal-response models, the F-test comparison of the models decomposes a sum of squares representing the variability in the data. This *analysis of variance* for decomposing variability generalizes to an *analysis of deviance* for GLMs. Given that the more complex model holds, the likelihood-ratio statistic for testing that the simpler model holds is $-2[L_0 - L_1]$. Since

$$-2[L_0 - L_1] = -2[L_0 - L_S] - \{-2[L_1 - L_S]\} = \text{Deviance}_0 - \text{Deviance}_1$$

we can compare the models by comparing their deviances.

This test statistic is large when M_0 fits poorly compared with M_1. For large samples, the statistic has an approximate chi-squared distribution, with df equal to the difference between the residual df values for the separate models. This df value equals the number of additional parameters that are in M_1 but not in M_0. Large test statistics and small P-values suggest that model M_0 fits more poorly than M_1.

For the snoring and heart disease data, the deviance for the linear probability model is 0.1 with $df = 2$. The simpler model with no effect of snoring (i.e., taking $\beta = 0$) has deviance equal to 65.9 with $df = 3$. The difference between the deviances equals 65.8 with $df = 1$, but this is precisely the likelihood-ratio statistic for testing that $\beta = 0$ in the model. Hence, the model with $\beta = 0$ fits poorly.

3.4.5 Residuals Comparing Observations to the Model Fit

For any GLM, goodness-of-fit statistics only broadly summarize how well models fit data. We obtain further insight by comparing observed and fitted values individually.

For observation i, the difference $y_i - \hat{\mu}_i$ between an observed and fitted value has limited usefulness. For Poisson sampling, for instance, the standard deviation of a count is $\sqrt{\mu_i}$, so more variability tends to occur when μ_i is larger. The *Pearson residual* is a standardized difference

$$\text{Pearson residual} = \mathrm{e}_i = \frac{y_i - \hat{\mu}_i}{\sqrt{\widehat{\mathrm{Var}}(y_i)}} \tag{3.9}$$

For example, for Poisson GLMs the Pearson residual for count i equals

$$\mathrm{e}_i = \frac{y_i - \hat{\mu}_i}{\sqrt{\hat{\mu}_i}} \tag{3.10}$$

It divides by the estimated Poisson standard deviation. The reason for calling e_i a *Pearson residual* is that $\sum \mathrm{e}_i^2 = \sum_i (y_i - \hat{\mu}_i)^2/\hat{\mu}_i$. When the GLM is the model corresponding to independence for cells in a two-way contingency table, this is the Pearson chi-squared statistic X^2 for testing independence [equation (2.8)]. Therefore, X^2 decomposes into terms describing the lack of fit for separate observations. Components of the deviance, called *deviance residuals*, are alternative measures of lack of fit.

Pearson residuals fluctuate around zero, following approximately a normal distribution when μ_i is large. When the model holds, these residuals are less variable than standard normal, however, because the numerator must use the fitted value $\hat{\mu}_i$ rather than the true mean μ_i. Since the sample data determine the fitted value, $(y_i - \hat{\mu}_i)$ tends to be smaller than $y_i - \mu_i$.

The *standardized residual* takes $(y_i - \hat{\mu}_i)$ and divides it by its estimated standard error, that is

$$\text{Standardized residual} = \frac{y_i - \hat{\mu}_i}{SE}$$

It[3] does have an approximate standard normal distribution when μ_i is large. With standardized residuals it is easier to tell when a deviation $(y_i - \hat{\mu}_i)$ is "large." Standardized residuals larger than about 2 or 3 in absolute value are worthy of attention, although some values of this size occur by chance alone when the number of observations is large. Section 2.4.5 introduced standardized residuals that follow up tests of independence in two-way contingency tables. We will use standardized residuals with logistic regression in Chapter 5.

Other diagnostic tools from regression modeling are also helpful in assessing fits of GLMs. For instance, to assess the influence of an observation on the overall fit, one can refit the model with that observation deleted (Section 5.2.6).

[3] $SE = [\widehat{\mathrm{Var}}(y_i)(1 - h_i)]^{1/2}$, where h_i is the *leverage* of observation i (Section 5.2.6). The greater the value of h_i, the more potential that observation has for influencing the model fit.

3.5 FITTING GENERALIZED LINEAR MODELS

We finish this chapter by discussing model-fitting for GLMs. We first describe an algorithm that finds the ML estimates of model parameters. We then provide further details about how basic inference utilizes the likelihood function.

3.5.1 The Newton–Raphson Algorithm Fits GLMs

Software finds model parameter estimates using a numerical algorithm. The algorithm starts at an initial guess for the parameter values that maximize the likelihood function. Successive approximations produced by the algorithm tend to fall closer to the ML estimates. The *Fisher scoring* algorithm for doing this was first proposed by R. A. Fisher for ML fitting of probit models. For binomial logistic regression and Poisson loglinear models, Fisher scoring simplifies to a general-purpose method called the *Newton–Raphson* algorithm.

The Newton–Raphson algorithm approximates the log-likelihood function in a neighborhood of the initial guess by a polynomial function that has the shape of a concave (mound-shaped) parabola. It has the same slope and curvature at the initial guess as does the log-likelihood function. It is simple to determine the location of the maximum of this approximating polynomial. That location comprises the second guess for the ML estimates. The algorithm then approximates the log-likelihood function in a neighborhood of the second guess by another concave parabolic function, and the third guess is the location of its maximum. The process is called *iterative*, because the algorithm repeatedly uses the same type of step over and over until there is no further change (in practical terms) in the location of the maximum. The successive approximations converge rapidly to the ML estimates, often within a few cycles.

Each cycle in the Newton–Raphson method represents a type of weighted least squares fitting. This is a generalization of ordinary least squares that accounts for nonconstant variance of Y in GLMs. Observations that occur where the variability is smaller receive greater weight in determining the parameter estimates. The weights change somewhat from cycle to cycle, with revised approximations for the ML estimates and thus for variance estimates. ML estimation for GLMs is sometimes called *iteratively reweighted least squares*.

The Newton–Raphson method utilizes a matrix, called the *information matrix*, that provides SE values for the parameter estimates. That matrix is based on the curvature of the log likelihood function at the ML estimate. The greater the curvature, the greater the information about the parameter values. The standard errors are the square roots of the diagonal elements for the inverse of the information matrix. The greater the curvature of the log likelihood, the smaller the standard errors. This is reasonable, since large curvature implies that the log-likelihood drops quickly as β moves away from $\hat{\beta}$; hence, the data would have been much more likely to occur if β took value $\hat{\beta}$ than if it took some value not close to $\hat{\beta}$. Software for GLMs routinely calculates the information matrix and the associated standard errors.

3.5.2 Wald, Likelihood-Ratio, and Score Inference Use the Likelihood Function

Figure 3.7 shows a generic plot of a log-likelihood function L for a parameter β. This plot illustrates the Wald, likelihood-ratio, and score tests of H_0: $\beta = 0$. The log-likelihood function for some GLMs, including binomial logistic regression models and Poisson loglinear models, has concave (bowl) shape. The ML estimate $\hat{\beta}$ is the point at which the log-likelihood takes its highest value. The Wald test is based on the behavior of the log-likelihood function at the ML estimate $\hat{\beta}$, having chi-squared form $(\hat{\beta}/SE)^2$. The SE of $\hat{\beta}$ depends on the curvature of the log-likelihood function at the point where it is maximized, with greater curvature giving smaller SE values.

The score test is based on the behavior of the log-likelihood function at the null value for β of 0. It uses the size of the derivative (slope) of the log-likelihood function, evaluated at the null hypothesis value of the parameter. The derivative at $\beta = 0$ tends to be larger in absolute value when $\hat{\beta}$ is further from that null value. The score statistic also has an approximate chi-squared distribution with $df = 1$. We shall not present the general formula for score statistics, but many test statistics in this text are this type. An example is the Pearson statistic for testing independence. An advantage of the score statistic is that it exists even when the ML estimate $\hat{\beta}$ is infinite. In that case, one cannot compute the Wald statistic.

The likelihood-ratio test combines information about the log-likelihood function both at $\hat{\beta}$ and at the null value for β of 0. It compares the log-likelihood values L_1 at $\hat{\beta}$ and L_0 at $\beta = 0$ using the chi-squared statistic $-2(L_0 - L_1)$. In Figure 3.7, this statistic is twice the vertical distance between values of the log-likelihood function at $\hat{\beta}$ and at $\beta = 0$. In a sense, this statistic uses the most information of the three types of test statistic. It is usually more reliable than the Wald statistic, especially when n is small to moderate.

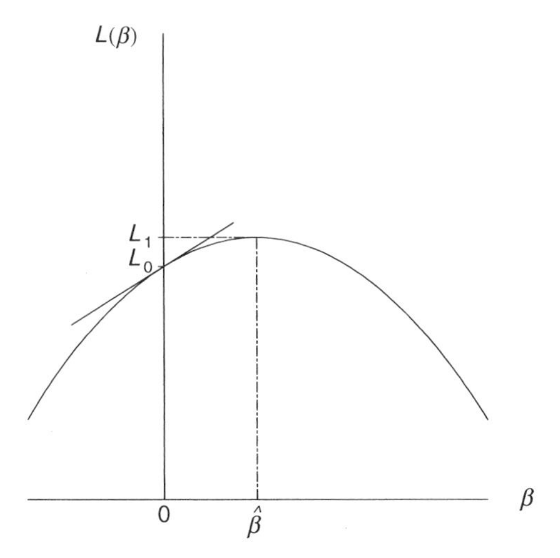

Figure 3.7. Information used in Wald, likelihood-ratio, and efficient score tests.

Table 3.5. Types of Generalized Linear Models for Statistical Analysis

Random Component	Link	Systematic Component	Model	Chapter
Normal	Identity	Continuous	Regression	
Normal	Identity	Categorical	Analysis of variance	
Normal	Identity	Mixed	Analysis of covariance	
Binomial	Logit	Mixed	Logistic regression	4–5, 8–10
Multinomial	Logits	Mixed	Multinomial response	6, 8–10
Poisson	Log	Mixed	Loglinear	7

3.5.3 Advantages of GLMs

The development of GLM theory in the mid-1970s unified important models for continuous and categorical response variables. Table 3.5 lists several popular GLMs for practical application.

A nice feature of GLMs is that the model-fitting algorithm, Fisher scoring, is the same for any GLM. This holds regardless of the choice of distribution for Y or link function. Therefore, GLM software can fit a very wide variety of useful models.

PROBLEMS

3.1 Describe the purpose of the link function of a GLM. Define the identity link, and explain why it is not often used with the binomial parameter.

3.2 In the 2000 US Presidential election, Palm Beach County in Florida was the focus of unusual voting patterns apparently caused by a confusing "butterfly ballot." Many voters claimed they voted mistakenly for the Reform party candidate, Pat Buchanan, when they intended to vote for Al Gore. Figure 3.8 shows the total number of votes for Buchanan plotted against the number of votes for the Reform party candidate in 1996 (Ross Perot), by county in Florida. (For details, see A. Agresti and B. Presnell, *Statist. Sci.*, **17**: 436–440, 2003.)

a. In county i, let π_i denote the proportion of the vote for Buchanan and let x_i denote the proportion of the vote for Perot in 1996. For the linear probability model fitted to all counties except Palm Beach County, $\hat{\pi}_i = -0.0003 + 0.0304x_i$. Give the value of P in the interpretation: The estimated proportion vote for Buchanan in 2000 was roughly $P\%$ of that for Perot in 1996.

b. For Palm Beach County, $\pi_i = 0.0079$ and $x_i = 0.0774$. Does this result appear to be an outlier? Investigate, by finding $\pi_i/\hat{\pi}_i$ and $\pi_i - \hat{\pi}_i$. (Analyses conducted by statisticians predicted that fewer than 900 votes were truly intended for Buchanan, compared with the 3407 he received. George W. Bush won the state by 537 votes and, with it, the Electoral College and

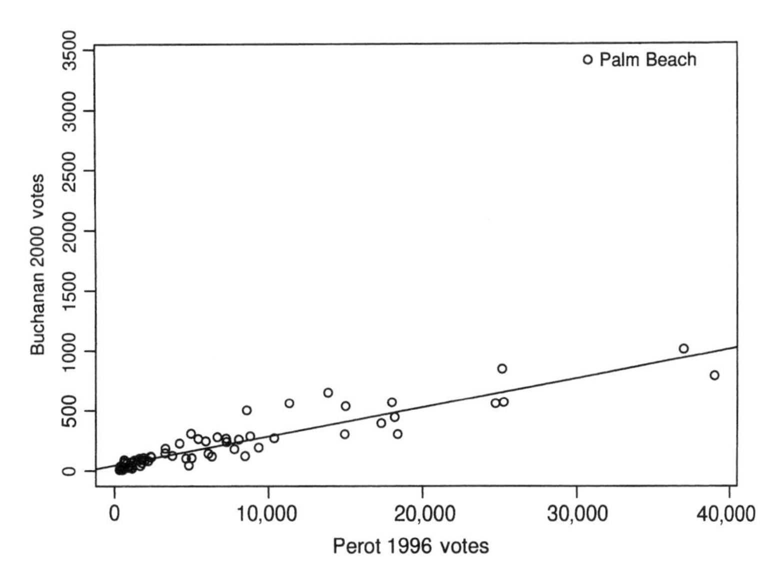

Figure 3.8. Total vote, by county in Florida, for Reform Party candidates Buchanan in 2000 and Perot in 1996.

the election. Other ballot design problems played a role in 110,000 disqualified "overvote" ballots, in which people mistakenly voted for more than one candidate, with Gore marked on 84,197 ballots and Bush on 37,731.)

3.3 Refer to Table 2.7 on x = mother's alcohol consumption and Y = whether a baby has sex organ malformation. WIth scores (0, 0.5, 1.5, 4.0, 7.0) for alcohol consumption, ML fitting of the linear probability model has the output:

Parameter	Estimate	Standard error	Likelihood ratio 95% confidence limits	
Intercept	0.00255	0.0003	0.0019	0.0032
Alcohol	0.00109	0.0007	−0.0001	0.0027

a. State the prediction equation, and interpret the intercept and slope.

b. Use the model fit to estimate the (i) probabilities of malformation for alcohol levels 0 and 7.0, (ii) relative risk comparing those levels.

3.4 Refer to the previous exercise and the solution to (**b**).

a. The sample proportion of malformations is much higher in the highest alcohol category than the others because, although it has only one malformation, its sample size is only 38. Is the result sensitive to this single malformation observation? Re-fit the model without it (using 0 malformations in

37 observations at that level), and re-evaluate estimated probabilities of malformation at alcohol levels 0 and 7 and the relative risk.

b. Is the result sensitive to the choice of scores? Re-fit the model using scores $(0, 1, 2, 3, 4)$, and re-evaluate estimated probabilities of malformation at the lowest and highest alcohol levels and the relative risk.

c. Fit a logistic regression or probit model. Report the prediction equation. Interpret the sign of the estimated effect.

3.5 For Table 3.1 on snoring and heart disease, re-fit the linear probability model or the logistic regression model using the scores (i) $(0, 2, 4, 6)$, (ii) $(0, 1, 2, 3)$, (iii) $(1, 2, 3, 4)$. Compare the model parameter estimates under the three choices. Compare the fitted values. What can you conclude about the effect of transformations of scores (called *linear*) that preserve relative sizes of spacings between scores?

3.6 In Section 3.2.2 on the snoring and heart disease data, refer to the linear probability model. Would the least squares fit differ from the ML fit for the 2484 binary observations? (Hint: The least squares fit is the same as the ML fit of the GLM assuming normal rather than binomial random component.)

3.7 Access the horseshoe crab data in Table 3.2 at www.stat.ufl.edu/~aa/intro-cda/appendix.html. Let $Y = 1$ if a crab has at least one satellite, and let $Y = 0$ otherwise. Using weight as the predictor, fit the linear probability model.

a. Use ordinary least squares. Interpret the parameter estimates. Find the predicted probability at the highest observed weight of 5.20 kg. Comment.

b. Attempt to fit the model using ML, treating Y as binomial. What does your software report? [The failure is due to a fitted probability falling outside the $(0, 1)$ range.]

c. Fit the logistic regression model. Show that the estimated logit at a weight of 5.20 kg equals 5.74. Show that $\hat{\pi} = 0.9968$ at that point by checking that $\log[\hat{\pi}/(1 - \hat{\pi})] = 5.74$ when $\hat{\pi} = 0.9968$.

3.8 Refer to the previous exercise for the horseshoe crab data.

a. Report the fit for the probit model, with weight predictor.

b. Find $\hat{\pi}$ at the highest observed weight, 5.20 kg.

c. Describe the weight effect by finding the difference between the $\hat{\pi}$ values at the upper and lower quartiles of weight, 2.85 and 2.00 kg.

d. Interpret the parameter estimates using characteristics of the normal *cdf* that describes the response curve.

3.9 Table 3.6 refers to a sample of subjects randomly selected for an Italian study on the relation between income and whether one possesses a travel credit card (such as American Express or Diners Club). At each level of annual income in millions of lira, the table indicates the number of subjects sampled and the number of them possessing at least one travel credit card. (Note: one million lira at the time of the study is currently worth aout 500 euros.) Software provides the following results of using logistic regression to relate the probability of having a travel credit card to income, treating these as independent binomial samples.

Parameter	Estimate	Standard error
Intercept	−3.5561	0.7169
Income	0.0532	0.0131

a. Report the prediction equation.

b. Interpret the sign of $\hat{\beta}$.

c. When $\hat{\pi} = 0.50$, show that the estimated logit value is 0. Based on this, for these data explain why the estimated probability of a travel credit card is 0.50 at income = 66.86 million lira.

Table 3.6. Data for Problem 3.9 on Italian Credit Cards

Inc.	No. Cases	Credit Cards	Inc.	No. Cases	Credit Cards	Inc.	No. Cases	Credit Cards	Inc.	No. Cases	Credit Cards
24	1	0	34	7	1	48	1	0	70	5	3
27	1	0	35	1	1	49	1	0	79	1	0
28	5	2	38	3	1	50	10	2	80	1	0
29	3	0	39	2	0	52	1	0	84	1	0
30	9	1	40	5	0	59	1	0	94	1	0
31	5	1	41	2	0	60	5	2	120	6	6
32	8	0	42	2	0	65	6	6	130	1	1
33	1	0	45	1	1	68	3	3			

Source: Based on data in *Categorical Data Analysis*, Quaderni del Corso Estivo di Statistica e Calcolo delle Probabilità, no. 4, Istituto di Metodi Quantitativi, Università Luigi Bocconi, by R. Piccarreta.

3.10 Refer to Problem 4.1 on cancer remission. Table 3.7 shows output for fitting a probit model. Interpret the parameter estimates (**a**) finding the remission value at which the estimated probability of remission equals 0.50, (**b**) finding the difference between the estimated probabilities of remission at the upper and lower quartiles of the labeling index, 14 and 28, and (**c**) using characteristics of the normal *cdf* response curve.

3.11 An experiment analyzes imperfection rates for two processes used to fabricate silicon wafers for computer chips. For treatment A applied to 10 wafers, the numbers of imperfections are 8, 7, 6, 6, 3, 4, 7, 2, 3, 4. Treatment B applied to 10 other wafers has 9, 9, 8, 14, 8, 13, 11, 5, 7, 6 imperfections. Treat the counts

Table 3.7. Table for Problem 3.10 on Cancer Remission

Parameter	Estimate	Standard Error	Likelihood Ratio 95% Confidence Limits		Chi-Square	Pr > ChiSq
Intercept	−2.3178	0.7795	−4.0114	−0.9084	8.84	0.0029
LI	0.0878	0.0328	0.0275	0.1575	7.19	0.0073

as independent Poisson variates having means μ_A and μ_B. Consider the model $\log \mu = \alpha + \beta x$, where $x = 1$ for treatment B and $x = 0$ for treatment A.

a. Show that $\beta = \log \mu_B - \log \mu_A = \log(\mu_B/\mu_A)$ and $e^\beta = \mu_b/\mu_A$.

b. Fit the model. Report the prediction equation and interpret $\hat{\beta}$.

c. Test H_0: $\mu_A = \mu_B$ by conducting the Wald or likelihood-ratio test of H_0: $\beta = 0$. Interpret.

d. Construct a 95% confidence interval for μ_B/μ_A. [Hint: Construct one for $\beta = \log(\mu_B/\mu_A)$ and then exponentiate.]

3.12 Refer to Problem 3.11. The wafers are also classified by thickness of silicon coating ($z = 0$, low; $z = 1$, high). The first five imperfection counts reported for each treatment refer to $z = 0$ and the last five refer to $z = 1$. Analyze these data, making inferences about the effects of treatment type and of thickness of coating.

3.13 Access the horseshoe crab data of Table 3.2 at www.stat.ufl.edu/~aa/intro-cda/appendix.html.

a. Using x = weight and Y = number of satellites, fit a Poisson loglinear model. Report the prediction equation.

b. Estimate the mean of Y for female crabs of average weight 2.44 kg.

c. Use $\hat{\beta}$ to describe the weight effect. Construct a 95% confidence interval for β and for the multiplicative effect of a 1 kg increase.

d. Conduct a Wald test of the hypothesis that the mean of Y is independent of weight. Interpret.

e. Conduct a likelihood-ratio test about the weight effect. Interpret.

3.14 Refer to the previous exercise. Allow overdispersion by fitting the negative binomial loglinear model.

a. Report the prediction equation and the estimate of the dispersion parameter and its SE. Is there evidence that this model gives a better fit than the Poisson model?

b. Construct a 95% confidence interval for β. Compare it with the one in (c) in the previous exercise. Interpret, and explain why the interval is wider with the negative binomial model.

3.15 A recent General Social Survey asked subjects, "Within the past 12 months, how many people have you known personally that were victims of homicide"? The sample mean for the 159 blacks was 0.522, with a variance of 1.150. The sample mean for the 1149 whites was 0.092, with a variance of 0.155.

a. Let y_{ij} denote the response for subject j of race i, and let $\mu_{ij} = E(Y_{ij})$. The Poisson model $\log(\mu_{ij}) = \alpha + \beta x_{ij}$ with $x_{1j} = 1$ (blacks) and $x_{2j} = 0$ (whites) has fit $\log(\hat{\mu}_{ij}) = -2.38 + 1.733x_{ij}$. Show that the estimated population means are 0.522 for blacks and 0.092 for whites, which are the sample means.

b. For the Poisson GLM, the standard error of $\hat{\beta}$ is 0.147. Show that the Wald 95% confidence interval for the ratio of means for blacks and whites is (4.2, 7.5). [Hint: Note that β is the log of the ratio of the means.]

c. The negative binomial loglinear model has the same estimates as in (a), but the standard error of $\hat{\beta}$ increases to 0.238 and the Wald 95% confidence interval for the ratio of means is (3.5, 9.0). Based on the sample means and variances, which confidence interval is more believeable? Why?

d. The negative binomial model has $\hat{D} = 4.94$ ($SE = 1.00$). Explain why this shows strong evidence that the negative binomial GLM is more appropriate than the Poisson GLM.

3.16 One question in a recent General Social Survey asked subjects how many times they had had sexual intercourse in the previous month.

a. The sample means were 5.9 for males and 4.3 for females; the sample variances were 54.8 and 34.4. Does an ordinary Poisson GLM seem appropriate? Explain.

b. The GLM with log link and a dummy variable for gender (1 = males, 0 = females) has gender estimate 0.308. The SE is 0.038 assuming a Poisson distribution and 0.127 assuming a negative binomial model. Why are the SE values so different?

c. The Wald 95% confidence interval for the ratio of means is (1.26, 1.47) for the Poisson model and (1.06, 1.75) for the negative binomial model. Which interval do you think is more appropriate? Why?

3.17 A study dealing with motor vehicle accident rates for elderly drivers (W. Ray et al., *Am. J. Epidemiol.*, **132**: 873–884, 1992) indicated that the entire cohort of elderly drivers had 495 injurious accidents in 38.7 thousand years of driving. Using a Poisson GLM, find a 95% confidence interval for the true rate. [Hint: Find a confidence interval first for the log rate by obtaining the estimate and standard error for the intercept term in a loglinear model that has no other predictor and uses log(38.7) as an offset.]

3.18 Table 3.8 lists total attendance (in thousands) and the total number of arrests in a season for soccer teams in the Second Division of the British football league.

a. Let Y denote the number of arrests for a team with total attendance t. Explain why the model $E(Y) = \mu t$ might be plausible. Show that it has alternative form $\log[E(Y)/t] = \alpha$, where $\alpha = \log(\mu)$, and express this model with an offset term.

b. Assuming Poisson sampling, fit the model. Report and interpret $\hat{\mu}$.

c. Plot arrests against attendance, and overlay the prediction equation. Use residuals to identify teams that had a much larger or smaller than expected number of arrests.

d. Now fit the model $\log[E(Y)/t] = \alpha$ by assuming a negative binomial distribution. Compare $\hat{\alpha}$ and its SE to what you got in (a). Based on this information and the estimate of the dispersion parameter and its SE, does the Poisson assumption seem appropriate?

Table 3.8. Data for Problem 3.18 on Soccer Game Arrests

Team	Attendance	Arrests	Team	Attendance	Arrests
Aston Villa	404	308	Shrewsbury	108	68
Bradford City	286	197	Swindon Town	210	67
Leeds United	443	184	Sheffield Utd	224	60
Bournemouth	169	149	Stoke City	211	57
West Brom	222	132	Barnsley	168	55
Hudderfield	150	126	Millwall	185	44
Middlesbro	321	110	Hull City	158	38
Birmingham	189	101	Manchester City	429	35
Ipswich Town	258	99	Plymouth	226	29
Leicester City	223	81	Reading	150	20
Blackburn	211	79	Oldham	148	19
Crystal Palace	215	78			

Source: The *Independent* (London), Dec. 21, 1988. Thanks to Dr. P. M. E. Altham for showing me these data.

3.19 Table 3.4 showed data on accidents involving trains.

a. Is it plausible that the collision counts are independent Poisson variates with constant rate over the 29 years? Respond by comparing a Poisson GLM for collision rates that contains only an intercept term to a Poisson GLM that contains also a time trend. The deviances of the two models are 35.1 and 23.5.

b. Section 3.3.6 fitted a negative binomial model. The estimated collision rate x years after 1975 was $e^{-4.20}(e^{-0.0337})^x = (0.015)(0.967)^x$. The

ML estimate $\hat{\beta} = -0.0337$ has $SE = 0.0130$. Conduct the Wald test of H_0: $\beta = 0$ against H_a: $\beta \neq 0$.

c. The likelihood-ratio 95% confidence interval for β is $(-0.060, -0.008)$. Find the interval for the multiplicative annual effect on the accident rate, and interpret.

3.20 Table 3.9, based on a study with British doctors conducted by R. Doll and A. Bradford Hill, was analyzed by N. R. Breslow in *A Celebration of Statistics*, Berlin: Springer, 1985.

a. For each age, compute the sample coronary death rates per 1000 person-years, for nonsmokers and smokers. To compare them, take their ratio and describe its dependence on age.

b. Specify a main-effects Poisson model for the log rates having four parameters for age and one for smoking. Explain why this model assumes a constant ratio of nonsmokers' to smokers' coronary death rates over levels of age. Based on (**a**), would you expect this model to be appropriate?

c. Based on (**a**), explain why it is sensible to add a quantitative interaction of age and smoking. Specify this model, and show that the log of the ratio of coronary death rates changes linearly with age.

d. Fit the model in (**b**). Assign scores to the levels of age for a product interaction term between age and smoking, and fit the model in (**c**). Compare the fits by comparing the deviances. Interpret.

Table 3.9. Data for Problem 3.20

	Person-Years		Coronary Deaths	
Age	Nonsmokers	Smokers	Nonsmokers	Smokers
35–44	18,793	52,407	2	32
45–54	10,673	43,248	12	104
55–64	5710	28,612	28	206
65–74	2585	12,663	28	186
75–84	1462	5317	31	102

Source: R. Doll and A. B. Hill, *Natl Cancer Inst. Monogr.*, **19**: 205–268, 1966.

3.21 For rate data, a GLM with identity link is

$$\mu/t = \alpha + \beta x$$

Explain why you could fit this model using t and tx as explanatory variables and with no intercept or offset terms.

3.22 True, or false?

a. An ordinary regression (or ANOVA) model that treats the response Y as normally distributed is a special case of a GLM, with normal random component and identity link function.

b. With a GLM, Y does not need to have a normal distribution and one can model a function of the mean of Y instead of just the mean itself, but in order to get ML estimates the variance of Y must be constant at all values of predictors.

c. The Pearson residual $e_i = (y_i - \hat{\mu}_i)/\sqrt{\hat{\mu}_i}$ for a GLM has a large-sample standard normal distribution.

CHAPTER 4

Logistic Regression

Let us now take a closer look at the statistical modeling of binary response variables, for which the response outcome for each subject is a "success" or "failure." Binary data are the most common form of categorical data, and the methods of this chapter are of fundamental importance. The most popular model for binary data is *logistic regression*. Section 3.2.3 introduced this model as a generalized linear model (GLM) with a binomial random component.

Section 4.1 interprets the logistic regression model. Section 4.2 presents statistical inference for the model parameters. Section 4.3 shows how to handle categorical predictors in the model, and Section 4.4 discusses the extension of the model for multiple explanatory variables. Section 4.5 presents ways of summarizing effects.

4.1 INTERPRETING THE LOGISTIC REGRESSION MODEL

To begin, suppose there is a single explanatory variable X, which is quantitative. For a binary response variable Y, recall that $\pi(x)$ denotes the "success" probability at value x. This probability is the parameter for the binomial distribution. The logistic regression model has linear form for the *logit* of this probability,

$$\text{logit}[\pi(x)] = \log\left(\frac{\pi(x)}{1-\pi(x)}\right) = \alpha + \beta x \tag{4.1}$$

The formula implies that $\pi(x)$ increases or decreases as an S-shaped function of x (recall Figure 3.2).

The logistic regression formula implies the following formula for the probability $\pi(x)$, using the exponential function (recall Section 2.3.3) $\exp(\alpha + \beta x) = e^{\alpha+\beta x}$,

$$\pi(x) = \frac{\exp(\alpha + \beta x)}{1 + \exp(\alpha + \beta x)} \tag{4.2}$$

This section shows ways of interpreting these model formulas.

4.1.1 Linear Approximation Interpretations

The logistic regression formula (4.1) indicates that the logit increases by β for every 1 cm increase in x. Most of us do not think naturally on a logit (logarithm of the odds) scale, so we need to consider alternative interpretations.

The parameter β in equations (4.1) and (4.2) determines the rate of increase or decrease of the S-shaped curve for $\pi(x)$. The sign of β indicates whether the curve ascends ($\beta > 0$) or descends ($\beta < 0$), and the rate of change increases as $|\beta|$ increases. When $\beta = 0$, the right-hand side of equation (4.2) simplifies to a constant. Then, $\pi(x)$ is identical at all x, so the curve becomes a horizontal straight line. The binary response Y is then independent of X.

Figure 4.1 shows the S-shaped appearance of the model for $\pi(x)$, as fitted for the example in the following subsection. Since it is curved rather than a straight line, the rate of change in $\pi(x)$ per 1-unit increase in x depends on the value of x. A straight line drawn tangent to the curve at a particular x value, such as shown in Figure 4.1,

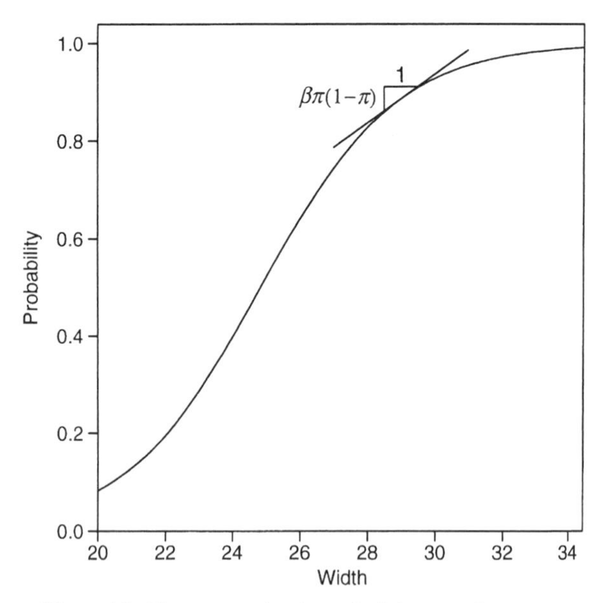

Figure 4.1. Linear approximation to logistic regression curve.

describes the rate of change at that point. For logistic regression parameter β, that line has slope equal to $\beta\pi(x)[1 - \pi(x)]$. For instance, the line tangent to the curve at x for which $\pi(x) = 0.50$ has slope $\beta(0.50)(0.50) = 0.25\beta$; by contrast, when $\pi(x) = 0.90$ or 0.10, it has slope 0.09β. The slope approaches 0 as the probability approaches 1.0 or 0.

The steepest slope occurs at x for which $\pi(x) = 0.50$. That x value relates to the logistic regression parameters by[1] $x = -\alpha/\beta$. This x value is sometimes called the *median effective level* and is denoted EL_{50}. It represents the level at which each outcome has a 50% chance.

4.1.2 Horseshoe Crabs: Viewing and Smoothing a Binary Outcome

To illustrate these interpretations, we re-analyze the horseshoe crab data introduced in Section 3.3.2 (Table 3.2). Here, we let Y indicate whether a female crab has any satellites (other males who could mate with her). That is, $Y = 1$ if a female crab has at least one satellite, and $Y = 0$ if she has no satellite. We first use the female crab's width (in cm) as the sole predictor. Section 5.1.2 uses additional predictors.

Figure 4.2 plots the data. The plot consists of a set of points at the level $y = 1$ and a second set of points at the level $y = 0$. The numbered symbols indicate the number

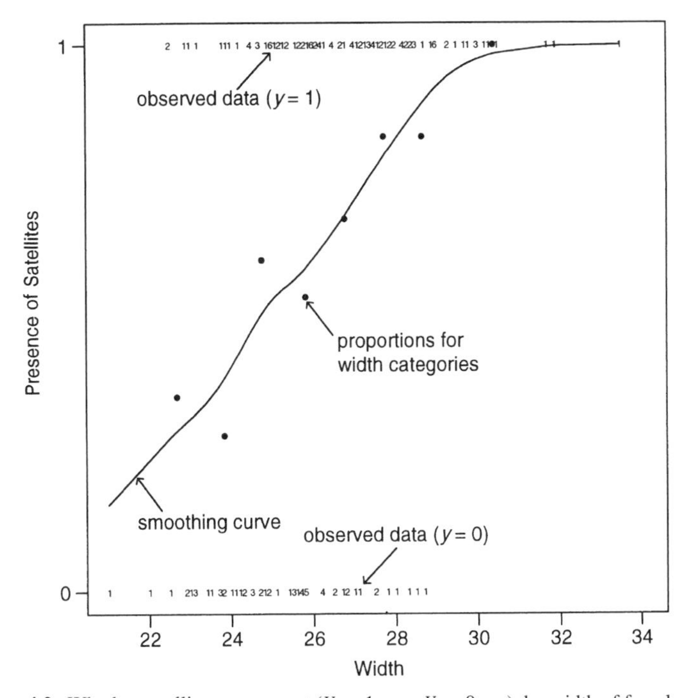

Figure 4.2. Whether satellites are present ($Y = 1$, yes; $Y = 0$, no), by width of female crab.

[1]One can check that $\pi(x) = 0.50$ at this point by substituting $-\alpha/\beta$ for x in equation (4.2), or by substituting $\pi(x) = 0.50$ in equation (4.1) and solving for x.

of observations at each point. It appears that $y = 1$ occurs relatively more often at higher x values. Since y takes only values 0 and 1, however, it is difficult to determine whether a logistic regression model is reasonable by plotting y against x.

Better information results from grouping the width values into categories and calculating a sample proportion of crabs having satellites for each category. This reveals whether the true proportions follow approximately the trend required by this model. Consider the grouping shown in Table 4.1. In each of the eight width categories, we computed the sample proportion of crabs having satellites and the mean width for the crabs in that category. Figure 4.2 contains eight dots representing the sample proportions of female crabs having satellites plotted against the mean widths for the eight categories.

Section 3.3.2 that introduced the horseshoe crab data mentioned that software can smooth the data without grouping observations. Figure 4.2 also shows a curve based on smoothing the data using *generalized additive models*, which allow the effect of x to be much more complex than linear. The eight plotted sample proportions and this smoothing curve both show a roughly increasing trend, so we proceed with fitting models that imply such trends.

4.1.3 Horseshoe Crabs: Interpreting the Logistic Regression Fit

For the ungrouped data in Table 3.2, let $\pi(x)$ denote the probability that a female horseshoe crab of width x has a satellite. The simplest model to interpret is the linear probability model, $\pi(x) = \alpha + \beta x$. During the ML fitting process, some predicted values for this GLM fall outside the legitimate 0–1 range for a binomial parameter, so ML fitting fails. Ordinary least squares fitting (such as GLM software reports when you assume a normal response and use the identity link function) yields $\hat{\pi}(x) = -1.766 + 0.092x$. The estimated probability of a satellite increases by 0.092

Table 4.1. Relation Between Width of Female Crab and Existence of Satellites, and Predicted Values for Logistic Regression Model

Width	Number of Cases	Number Having Satellites	Sample Proportion	Estimated Probability	Predicted Number of Crabs with Satellites
<23.25	14	5	0.36	0.26	3.6
23.25–24.25	14	4	0.29	0.38	5.3
24.25–25.25	28	17	0.61	0.49	13.8
25.25–26.25	39	21	0.54	0.62	24.2
26.25–27.25	22	15	0.68	0.72	15.9
27.25–28.25	24	20	0.83	0.81	19.4
28.25–29.25	18	15	0.83	0.87	15.6
>29.25	14	14	1.00	0.93	13.1

Note: The estimated probability is the predicted number (in the final column) divided by the number of cases.

for each 1 cm increase in width. This model provides a simple interpretation and realistic predictions over most of the width range, but it is inadequate for extreme values. For instance, at the maximum width in this sample of 33.5 cm, its estimated probability equals $-1.766 + 0.092(33.5) = 1.3$.

Table 4.2 shows some software output for logistic regression. The estimated probability of a satellite is the sample analog of formula (4.2),

$$\hat{\pi}(x) = \frac{\exp(-12.351 + 0.497x)}{1 + \exp(-12.351 + 0.497x)}$$

Since $\hat{\beta} > 0$, the estimated probability $\hat{\pi}$ is larger at larger width values. At the minimum width in this sample of 21.0 cm, the estimated probability is

$$\exp(-12.351 + 0.497(21.0))/[1 + \exp(-12.351 + 0.497(21.0))] = 0.129$$

At the maximum width of 33.5 cm, the estimated probability equals

$$\exp(-12.351 + 0.497(33.5))/[1 + \exp(-12.351 + 0.497(33.5))] = 0.987$$

The median effective level is the width at which $\hat{\pi}(x) = 0.50$. This is $x = EL_{50} = -\hat{\alpha}/\hat{\beta} = 12.351/0.497 = 24.8$. Figure 4.1 plots the estimated probabilities as a function of width.

At the sample mean width of 26.3 cm, $\hat{\pi}(x) = 0.674$. From Section 4.1.1, the incremental rate of change in the fitted probability at that point is $\hat{\beta}\hat{\pi}(x)[1 - \hat{\pi}(x)] = 0.497(0.674)(0.326) = 0.11$. For female crabs near the mean width, the estimated probability of a satellite increases at the rate of 0.11 per 1 cm increase in width. The estimated rate of change is greatest at the x value (24.8) at which $\hat{\pi}(x) = 0.50$; there, the estimated probability increases at the rate of $(0.497)(0.50)(0.50) = 0.12$ per 1 cm increase in width. Unlike the linear probability model, the logistic regression model permits the rate of change to vary as x varies.

To describe the fit further, for each category of width Table 4.1 reports the predicted number of female crabs having satellites (i.e., the fitted values). Each of these sums the $\hat{\pi}(x)$ values for all crabs in a category. For example, the estimated probabilities for the 14 crabs with widths below 23.25 cm sum to 3.6. The average estimated probability

Table 4.2. Computer Output for Logistic Regression Model with Horseshoe Crab Data

	Log Likelihood				−97.2263	
Parameter	Estimate	Standard Error	Likelihood Ratio 95% Conf. Limits		Wald Chi-Sq	Pr > ChiSq
Intercept	−12.3508	2.6287	−17.8097	−7.4573	22.07	<.0001
width	0.4972	0.1017	0.3084	0.7090	23.89	<.0001

for crabs in a given width category equals the fitted value divided by the number of crabs in that category. For the first width category, $3.6/14 = 0.26$ is the average estimated probability.

Table 4.1 reports the fitted values and the average estimated probabilities of a satellite, in grouped fashion. Figure 4.3 plots the sample proportions and the estimated probabilities against width. These comparisons suggest that the model fits decently. Section 5.2.2 presents objective criteria for making this comparison.

4.1.4 Odds Ratio Interpretation

An important interpretion of the logistic regression model uses the *odds* and the *odds ratio*. For model (4.1), the odds of response 1 (i.e., the odds of a success) are

$$\frac{\pi(x)}{1-\pi(x)} = \exp(\alpha + \beta x) = e^{\alpha}(e^{\beta})^{x} \tag{4.3}$$

This exponential relationship provides an interpretation for β: The odds multiply by e^{β} for every 1-unit increase in x. That is, the odds at level $x+1$ equal the odds at x multiplied by e^{β}. When $\beta = 0$, $e^{\beta} = 1$, and the odds do not change as x changes.

For the horseshoe crabs, $\text{logit}[\hat{\pi}(x)] = -12.35 + 0.497x$. So, the estimated odds of a satellite multiply by $\exp(\hat{\beta}) = \exp(0.497) = 1.64$ for each centimeter increase in

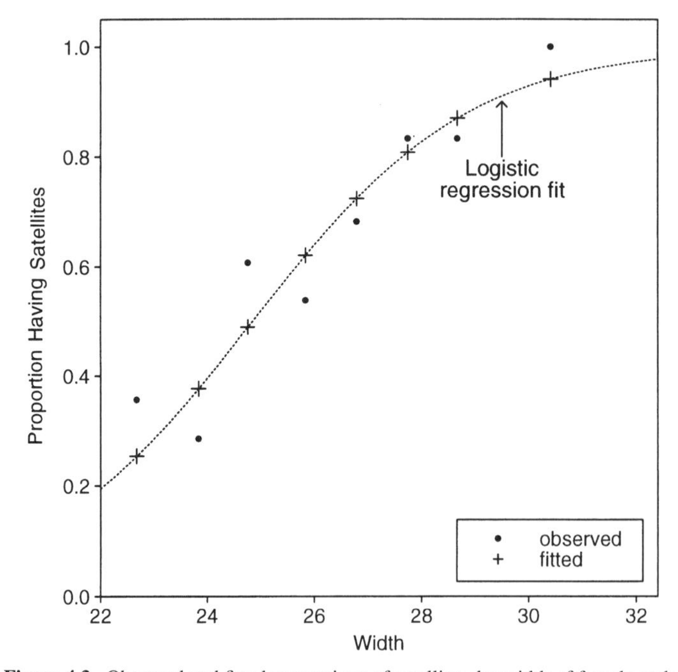

Figure 4.3. Observed and fitted proportions of satellites, by width of female crab.

width; that is, there is a 64% increase. To illustrate, the mean width value of $x = 26.3$ has $\hat{\pi}(x) = 0.674$, and odds $= 0.674/0.326 = 2.07$. At $x = 27.3 = 26.3 + 1.0$, you can check that $\hat{\pi}(x) = 0.773$ and odds $= 0.773/0.227 = 3.40$. However, this is a 64% increase; that is, $3.40 = 2.07(1.64)$.

4.1.5 Logistic Regression with Retrospective Studies

Another property of logistic regression relates to situations in which the explanatory variable X rather than the response variable Y is random. This occurs with retrospective sampling designs. Sometimes such designs are used because one of the response categories occurs rarely, and a prospective study might have too few cases to enable one to estimate effects of predictors well. For a given sample size, effect estimates have smaller SEs when the number of outcomes of the two types are similar than when they are very different.

Most commonly, retrospective designs are used with biomedical case-control studies (Section 2.3.5). For samples of subjects having $Y = 1$ (cases) and having $Y = 0$ (controls), the value of X is observed. Evidence exists of an association between X and Y if the distribution of X values differs between cases and controls. For case–control studies, it was noted in Section 2.3.5 that it is possible to estimate odds ratios but not other summary measures. Logistic regression parameters refer to odds and odds ratios. One can fit logistic regression models with data from case–control studies and estimate effects of explanatory variables. The intercept term α in the model is not meaningful, because it relates to the relative numbers of outcomes of $y = 1$ and $y = 0$. We do not estimate this, because the sample frequencies for $y = 1$ and $y = 0$ are fixed by the nature of the case–control study.

With case–control studies, it is not possible to estimate effects in binary models with link functions other than the logit. Unlike the odds ratio, the effect for the conditional distribution of X given Y does not then equal that for Y given X. This provides an important advantage of the logit link over links such as the probit. It is a major reason why logistic regression surpasses other models in popularity for biomedical studies.

Many case–control studies employ matching. Each case is matched with one or more control subjects. The controls are like the case on key characteristics such as age. The model and subsequent analysis should take the matching into account. Section 8.2.4 discusses logistic regression for matched case–control studies.

4.1.6 Normally Distributed X Implies Logistic Regression for Y

Regardless of the sampling mechanism, the logistic regression model may or may not describe a relationship well. In one special case, it does necessarily hold. Suppose the distribution of X for subjects having $Y = 1$ is normal $N(\mu_1, \sigma)$, and suppose the distribution of X for subjects having $Y = 0$ is normal $N(\mu_0, \sigma)$; that is, with different mean but the same standard deviation. Then, a Bayes theorem calculation converting from the distribution of X given $Y = y$ to the distribution of Y given $X = x$ shows that $P(Y = 1|x)$ satisfies the logistic regression curve. For that curve, the effect of x is $\beta = (\mu_1 - \mu_0)/\sigma^2$. In particular, β has the same sign as $\mu_1 - \mu_0$. For example, if those with $y = 1$ tend to have higher values of x, then $\beta > 0$.

For example, consider Y = heart disease (1 = yes, 0 = no) and X = cholesterol level. Suppose cholesterol levels have approximately a $N(\mu_0 = 160, \sigma = 50)$ distribution for those *without* heart disease and a $N(\mu_1 = 260, \sigma = 50)$ distribution for those *with* heart disease. Then, the probability of having heart disease satisfies the logistic regression function (4.2) with predictor x and $\beta = (260 - 160)/50^2 = 0.04$.

If the distributions of X are bell-shaped but with highly different spreads, then a logistic model containing also a quadratic term (i.e., both x and x^2) often fits well. In that case, the relationship is not monotone. Instead, $P(Y = 1)$ increases and then decreases, or the reverse (see Exercise 4.7).

4.2 INFERENCE FOR LOGISTIC REGRESSION

We have studied how logistic regression helps describe the effects of a predictor on a binary response variable. We next present statistical inference for the model parameters, to help judge the significance and size of the effects.

4.2.1 Binary Data can be Grouped or Ungrouped

Widely available software reports the ML estimates of parameters and their standard errors. Sometimes sets of observations have the same values of predictor variables, such as when explanatory variables are discrete. Then, ML model fitting can treat the observations as the binomial counts of successes out of certain sample sizes, at the various combinations of values of the predictors. We will refer to this case as *grouped binary data* and the case in which each observation is a single binary outcome as *ungrouped binary data*.

In Table 3.1 on snoring and heart disease in the previous chapter, 254 subjects reported snoring every night, of whom 30 had heart disease. If the data file has grouped binary data, a line in the data file reports these data as 30 cases of heart disease out of a sample size of 254. If the data file has ungrouped binary data, each line in the data file refers to a separate subject, so 30 lines contain a 1 for heart disease and 224 lines contain a 0 for heart disease. The ML estimates and *SE* values are the same for either type of data file.

When at least one explanatory variable is continuous, binary data are naturally ungrouped. An example is the data that Table 3.2 reports for the horseshoe crabs.

4.2.2 Confidence Intervals for Effects

A large-sample Wald confidence interval for the parameter β in the logistic regression model, $\text{logit}[\pi(x)] = \alpha + \beta x$, is

$$\hat{\beta} \pm z_{\alpha/2}(SE)$$

Exponentiating the endpoints yields an interval for e^{β}, the multiplicative effect on the odds of a 1-unit increase in x.

When n is small or fitted probabilities are mainly near 0 or 1, it is preferable to construct a confidence interval based on the likelihood-ratio test. This interval contains all the β_0 values for which the likelihood-ratio test of H_0: $\beta = \beta_0$ has P-value $>\alpha$. Some software can report this (such as PROC GENMOD in SAS with its LRCI option).

For the logistic regression analysis of the horseshoe crab data, the estimated effect of width on the probability of a satellite is $\hat{\beta} = 0.497$, with $SE = 0.102$. A 95% Wald confidence interval for β is $0.497 \pm 1.96(0.102)$, or $(0.298, 0.697)$. The likelihood-ratio-based confidence interval is $(0.308, 0.709)$. The likelihood-ratio interval for the effect on the odds per cm increase in width equals $(e^{0.308}, e^{0.709}) = (1.36, 2.03)$. We infer that a 1 cm increase in width has at least a 36 percent increase and at most a doubling in the odds that a female crab has a satellite.

From Section 4.1.1, a simpler interpretation uses a straight-line approximation to the logistic regression curve. The term $\beta\pi(x)[1 - \pi(x)]$ approximates the change in the probability per 1-unit increase in x. For instance, at $\pi(x) = 0.50$, the estimated rate of change is $0.25\hat{\beta} = 0.124$. A 95% confidence interval for 0.25β equals 0.25 times the endpoints of the interval for β. For the likelihood-ratio interval, this is $[0.25(0.308), 0.25(0.709)] = (0.077, 0.177)$. So, if the logistic regression model holds, then for values of x near the width value at which $\pi(x) = 0.50$, we infer that the rate of increase in the probability of a satellite per centimeter increase in width falls between about 0.08 and 0.18.

4.2.3 Significance Testing

For the logistic regression model, H_0: $\beta = 0$ states that the probability of success is independent of X. Wald test statistics (Section 1.4.1) are simple. For large samples,

$$z = \hat{\beta}/SE$$

has a standard normal distribution when $\beta = 0$. Refer z to the standard normal table to get a one-sided or two-sided P-value. Equivalently, for the two-sided H_a: $\beta \neq 0$, $z^2 = (\hat{\beta}/SE)^2$ has a large-sample chi-squared null distribution with $df = 1$.

Although the Wald test is adequate for large samples, the likelihood-ratio test is more powerful and more reliable for sample sizes often used in practice. The test statistic compares the maximum L_0 of the log-likelihood function when $\beta = 0$ to the maximum L_1 of the log-likelihood function for unrestricted β. The test statistic, $-2(L_0 - L_1)$, also has a large-sample chi-squared null distribution with $df = 1$.

For the horseshoe crab data, the Wald statistic $z = \hat{\beta}/SE = 0.497/0.102 = 4.9$. This shows strong evidence of a positive effect of width on the presence of satellites ($P < 0.0001$). The equivalent chi-squared statistic, $z^2 = 23.9$, has $df = 1$. Software reports that the maximized log likelihoods equal $L_0 = -112.88$ under H_0: $\beta = 0$ and $L_1 = -97.23$ for the full model. The likelihood-ratio statistic equals

$-2(L_0 - L_1) = 31.3$, with $df = 1$. This also provides extremely strong evidence of a width effect ($P < 0.0001$).

4.2.4 Confidence Intervals for Probabilities

Recall that the logistic regression estimate of $P(Y = 1)$ at a fixed setting x is

$$\hat{\pi}(x) = \exp(\hat{\alpha} + \hat{\beta}x)/[1 + \exp(\hat{\alpha} + \hat{\beta}x)] \tag{4.4}$$

Most software for logistic regression can report this estimate as well as a confidence interval for the true probability $\pi(x)$.

We illustrate by estimating the probability of a satellite for female crabs of width $x = 26.5$, which is near the mean width. The logistic regression fit yields

$$\hat{\pi} = \exp(-12.351 + 0.497(26.5))/[1 + \exp(-12.351 + 0.497(26.5))] = 0.695$$

From software, a 95% confidence interval for the true probability is (0.61, 0.77).

4.2.5 Why Use a Model to Estimate Probabilities?

Instead of finding $\hat{\pi}(x)$ using the model fit, as we just did at $x = 26.5$, we could simply use the sample proportion to estimate the probability. Six crabs in the sample had width 26.5, and four of them had satellites. The sample proportion estimate at $x = 26.5$ is $p = 4/6 = 0.67$, similar to the model-based estimate. From inverting small-sample tests using the binomial distribution, a 95% confidence interval based on these six observations equals (0.22, 0.96).

When the logistic regression model holds, the model-based estimator of $\pi(x)$ is much better than the sample proportion. It uses *all* the data rather than only the data at the fixed x value. The result is a more precise estimate. For instance, at $x = 26.5$, software reports an $SE = 0.04$ for the model-based estimate 0.695. By contrast, the SE for the sample proportion of 0.67 with only six observations is $\sqrt{[p(1-p)/n]} = \sqrt{[(0.67)(0.33)/6]} = 0.19$. The 95% confidence intervals are (0.61, 0.77) using the model vs (0.22, 0.96) using only the sample proportion at $x = 26.5$.

Reality is more complicated. In practice, any model will not *exactly* represent the true relationship between $\pi(x)$ and x. If the model approximates the true probabilities reasonably well, however, it performs well. The model-based estimator tends to be much closer than the sample proportion to the true value, unless the sample size on which that sample proportion is based is extremely large. The model smooths the sample data, somewhat dampening the observed variability.

4.2.6 Confidence Intervals for Probabilities: Details*

If your software does not report confidence intervals for probabilities, you can construct them by using the covariance matrix of the model parameter estimates.

The term $\hat{\alpha} + \hat{\beta}x$ in the exponents of the prediction equation (4.4) is the estimated linear predictor in the logit transform of $\pi(x)$. This estimated logit has large-sample *SE* given by the estimated square root of

$$\text{Var}(\hat{\alpha} + \hat{\beta}x) = \text{Var}(\hat{\alpha}) + x^2\text{Var}(\hat{\beta}) + 2x\,\text{Cov}(\hat{\alpha}, \hat{\beta})$$

A 95% confidence interval for the true logit is $(\hat{\alpha} + \hat{\beta}x) \pm 1.96(SE)$. Substituting the endpoints of this interval for $\alpha + \beta x$ in the two exponents in equation (4.4) gives a corresponding interval for the probability.

For example, at $x = 26.5$ for the horseshoe crab data, the estimated logit is $-12.351 + 0.497(26.5) = 0.825$. Software reports estimated covariance matrix for $(\hat{\alpha}, \hat{\beta})$ of

Estimated Covariance Matrix

Parameter	Intercept	width
Intercept	6.9102	−0.2668
width	−0.2668	0.0103

A covariance matrix has variances of estimates on the main diagonal and covariances off that diagonal. Here, $\widehat{\text{Var}}(\hat{\alpha}) = 6.9102$, $\widehat{\text{Var}}(\hat{\beta}) = 0.0103$, $\widehat{\text{Cov}}(\hat{\alpha}, \hat{\beta}) = -0.2668$. Therefore, the estimated variance of this estimated logit equals

$$\widehat{\text{Var}}(\hat{\alpha}) + x^2\,\widehat{\text{Var}}(\hat{\beta}) + 2x\,\widehat{\text{Cov}}(\hat{\alpha}, \hat{\beta}) = 6.9102 + (26.5)^2(0.0103) + 2(26.5)(-0.2668)$$

or 0.038. The 95% confidence interval for the true logit equals $0.825 \pm (1.96)\sqrt{0.038}$, or (0.44, 1.21). From equation (4.4), this translates to the confidence interval

$$\{\exp(0.44)/[1 + \exp(0.44)],\ \exp(1.21)/[1 + \exp(1.21)]\} = (0.61, 0.77)$$

for the probability of satellites at width 26.5 cm.

4.2.7 Standard Errors of Model Parameter Estimates*

We have used only a single explanatory variable so far, but the rest of the chapter allows additional predictors. The remarks of this subsection apply regardless of the number of predictors.

Software fits models and provides the ML parameter estimates. The standard errors of the estimates are the square roots of the variances from the main diagonal of the covariance matrix. For example, from the estimated covariance matrix reported above in Section 4.2.6, the estimated width effect of 0.497 in the logistic regression model has $SE = \sqrt{0.0103} = 0.102$.

The estimated covariance matrix for the ML parameter estimates is the inverse of the *information matrix* (see Section 3.5.1). This measures the curvature of the log likelihood function at the ML estimates. More highly curved log likelihood functions yield greater information about the parameter values. This results in smaller elements of the inverse of the information matrix and smaller standard errors. Software finds the information matrix as a by-product of fitting the model.

Let n_i denote the number of observations at setting i of the explanatory variables. (Note $n_i = 1$ when the binary data are ungrouped.) At setting i, let x_{ij} denote the value of explanatory variable j, and let $\hat{\pi}_i$ denote the estimated "success" probability based on the model fit. The element in row a and column b of the information matrix is

$$\sum_i x_{ia} x_{ib} n_i \hat{\pi}_i (1 - \hat{\pi}_i)$$

These elements increase, and thus the standard errors decrease, as the sample sizes $\{n_i\}$ increase. The standard errors also decrease by taking additional observations at other settings of the predictors (for ungrouped data).

For given $\{n_i\}$, the elements in the information matrix decrease, and the *SE* values increase, as the estimated probabilities $\{\hat{\pi}_i\}$ get closer to 0 or to 1. For example, it is harder to estimate effects of predictors well when nearly all the observations are "successes" compared to when there is a similar number of "successes" and "failures."

4.3 LOGISTIC REGRESSION WITH CATEGORICAL PREDICTORS

Logistic regression, like ordinary regression, can have multiple explanatory variables. Some or all of those predictors can be categorical, rather than quantitative. This section shows how to include categorical predictors, often called *factors*, and Section 4.4 presents the general form of multiple logistic regression models.

4.3.1 Indicator Variables Represent Categories of Predictors

Suppose a binary response Y has two binary predictors, X and Z. The data are then displayed in a $2 \times 2 \times 2$ contingency table, such as we'll see in the example in the next subsection.

Let x and z each take values 0 and 1 to represent the two categories of each explanatory variable. The model for $P(Y = 1)$,

$$\text{logit}[P(Y = 1)] = \alpha + \beta_1 x + \beta_2 z \tag{4.5}$$

has main effects for x and z. The variables x and z are called *indicator variables*. They indicate categories for the predictors. Indicator variables are also called *dummy variables*. For this coding, Table 4.3 shows the logit values at the four combinations of values of the two predictors.

Table 4.3. Logits Implied by Indicator Variables in Model, $\text{logit}[P(Y=1)] = \alpha + \beta_1 x + \beta_2 z$

x	z	Logit
0	0	α
1	0	$\alpha + \beta_1$
0	1	$\alpha + \beta_2$
1	1	$\alpha + \beta_1 + \beta_2$

This model assumes an absence of interaction. The effect of one factor is the same at each category of the other factor. At a fixed category z of Z, the effect on the logit of changing from $x = 0$ to $x = 1$ is

$$= [\alpha + \beta_1(1) + \beta_2 z] - [\alpha + \beta_1(0) + \beta_2 z] = \beta_1$$

This difference between two logits equals the difference of log odds. Equivalently, that difference equals the log of the odds ratio between X and Y, at that category of Z. Thus, $\exp(\beta_1)$ equals the conditional odds ratio between X and Y. Controlling for Z, the odds of "success" at $x = 1$ equal $\exp(\beta_1)$ times the odds of success at $x = 0$. This conditional odds ratio is the same at each category of Z. The lack of an interaction term implies a common value of the odds ratio for the partial tables at the two categories of Z. The model satisfies homogeneous association (Section 2.7.6).

Conditional independence exists between X and Y, controlling for Z, if $\beta_1 = 0$. In that case the common odds ratio equals 1. The simpler model,

$$\text{logit}[P(Y=1)] = \alpha + \beta_2 z \tag{4.6}$$

then applies to the three-way table.

4.3.2 Example: AZT Use and AIDS

We illustrate these models using Table 4.4, based on a study described in the *New York Times* (February 15, 1991) on the effects of AZT in slowing the development of AIDS symptoms. In the study, 338 veterans whose immune systems were beginning to falter after infection with the AIDS virus were randomly assigned either to receive AZT immediately or to wait until their T cells showed severe immune weakness. Table 4.4 is a $2 \times 2 \times 2$ cross classification of veteran's race, whether AZT was given immediately, and whether AIDS symptoms developed during the 3 year study. Let X = AZT treatment, Z = race, and Y = whether AIDS symptoms developed (1 = yes, 0 = no).

In model (4.5), let $x = 1$ for those who took AZT immediately and $x = 0$ otherwise, and let $z = 1$ for whites and $z = 0$ for blacks. Table 4.5 shows SAS output for the ML fit. The estimated effect of AZT is $\hat{\beta}_1 = -0.720$. The estimated conditional

Table 4.4. Development of AIDS Symptoms by AZT Use and Race

		Symptoms	
Race	AZT Use	Yes	No
White	Yes	14	93
	No	32	81
Black	Yes	11	52
	No	12	43

odds ratio between immediate AZT use and development of AIDS symptoms equals $\exp(-0.720) = 0.49$. For each race, the estimated odds of developing symptoms are half as high for those who took AZT immediately.

The hypothesis of conditional independence of AZT treatment and the development of AIDS symptoms, controlling for race, is H_0: $\beta_1 = 0$. The likelihood-ratio (LR) statistic $-2(L_0 - L_1)$ comparing models (4.6) and (4.5) equals 6.87, with $df = 1$, showing evidence of association ($P = 0.009$). The Wald statistic $(\hat{\beta}_1/SE)^2 = (-0.720/0.279)^2 = 6.65$ provides similar results ($P = 0.010$). The effect of race is not significant (Table 4.5 reports LR statistic = 0.04 and P-value = 0.85).

Table 4.5. Computer Output for Logit Model with AIDS Symptoms Data

Log Likelihood −167.5756

Analysis of Maximum Likelihood Estimates

Parameter	Estimate	Std Error	Wald Chi-Square	Pr > ChiSq
Intercept	−1.0736	0.2629	16.6705	<.0001
azt	−0.7195	0.2790	6.6507	0.0099
race	0.0555	0.2886	0.0370	0.8476

LR Statistics

Source	DF	Chi-Square	Pr > ChiSq
azt	1	6.87	0.0088
race	1	0.04	0.8473

Obs	race	azt	y	n	pi_hat	lower	upper
1	1	1	14	107	0.14962	0.09897	0.21987
2	1	0	32	113	0.26540	0.19668	0.34774
3	0	1	11	63	0.14270	0.08704	0.22519
4	0	0	12	55	0.25472	0.16953	0.36396

How do we know the model fits the data adequately? We will address model goodness of fit in the next chapter (Section 5.2.2).

4.3.3 ANOVA-Type Model Representation of Factors

A factor having two categories requires only a single indicator variable, taking value 1 or 0 to indicate whether an observation falls in the first or second category. A factor having I categories requires $I - 1$ indicator variables, as shown below and in Section 4.4.1.

An alternative representation of factors in logistic regression uses the way ANOVA models often express factors. The model formula

$$\text{logit}[P(Y = 1)] = \alpha + \beta_i^X + \beta_k^Z \tag{4.7}$$

represents the effects of X through parameters $\{\beta_i^X\}$ and the effects of Z through parameters $\{\beta_k^Z\}$. (The X and Z superscripts are merely labels and do not represent powers.) The term β_i^X denotes the effect on the logit of classification in category i of X. Conditional independence between X and Y, given Z, corresponds to $\beta_1^X = \beta_2^X = \cdots = \beta_I^X$.

Model form (4.7) applies for any numbers of categories for X and Z. Each factor has as many parameters as it has categories, but one is redundant. For instance, if X has I levels, it has $I - 1$ nonredundant parameters. To account for redundancies, most software sets the parameter for the last category equal to zero. The term β_i^X in this model then is a simple way of representing

$$\beta_1^X x_1 + \beta_2^X x_2 + \cdots + \beta_{I-1}^X x_{I-1}$$

where $\{x_1, \ldots, x_{I-1}\}$ are indicator variables for the first $I - 1$ categories of X. That is, $x_1 = 1$ when an observation is in category 1 and $x_1 = 0$ otherwise, and so forth. Category I does not need an indicator, because we know an observation is in that category when $x_1 = \cdots = x_{I-1} = 0$.

Consider model (4.7) when the predictor x is binary, as in Table 4.4. Although most software sets $\beta_2^X = 0$, some software sets $\beta_1^X = 0$ or $\beta_1^X + \beta_2^X = 0$. The latter corresponds to setting up the indicator variable so that $x = 1$ in category 1 and $x = -1$ in category 2. For any coding scheme, the difference $\beta_1^X - \beta_2^X$ is the same and represents the conditional log odds ratio between X and Y, given Z. For example, the estimated common odds ratio between immediate AZT use and development of symptoms, for each race, is $\exp(\hat{\beta}_1^X - \hat{\beta}_2^X) = \exp(-0.720) = 0.49$.

By itself, the parameter estimate for a single category of a factor is irrelevant. Different ways of handling parameter redundancies result in different values for that estimate. An estimate makes sense only by comparison with one for another category. Exponentiating a difference between estimates for two categories determines the odds ratio relating to the effect of classification in one category rather than the other.

4.3.4 The Cochran–Mantel–Haenszel Test for 2 × 2 × K Contingency Tables*

In many examples with two categorical predictors, X identifies two groups to compare and Z is a control variable. For example, in a clinical trial X might refer to two treatments and Z might refer to several centers that recruited patients for the study. Problem 4.20 shows such an example. The data then can be presented in several 2×2 tables.

With K categories for Z, model (4.7) refers to a $2 \times 2 \times K$ contingency table. That model can then be expressed as

$$\text{logit}[P(Y = 1)] = \alpha + \beta x + \beta_k^Z \tag{4.8}$$

where x is an indicator variable for the two categories of X. Then, $\exp(\beta)$ is the common XY odds ratio for each of the K partial tables for categories of Z. This is the *homogeneous association* structure for multiple 2×2 tables, introduced in Section 2.7.6.

In this model, conditional independence between X and Y, controlling for Z, corresponds to $\beta = 0$. When $\beta = 0$, the XY odds ratio equals 1 for each partial table. Given that model (4.8) holds, one can test conditional independence by the Wald test or the likelihood-ratio test of H_0: $\beta = 0$.

The *Cochran–Mantel–Haenszel test* is an alternative test of XY conditional independence in $2 \times 2 \times K$ contingency tables. This test conditions on the row totals and the column totals in each partial table. Then, as in Fisher's exact test, the count in the first row and first column in a partial table determines all the other counts in that table. Under the usual sampling schemes (e.g., binomial for each row in each partial table), the conditioning results in a hypergeometric distribution (Section 2.6.1) for the count n_{11k} in the cell in row 1 and column 1 of partial table k. The test statistic utilizes this cell in each partial table.

In partial table k, the row totals are $\{n_{1+k}, n_{2+k}\}$, and the column totals are $\{n_{+1k}, n_{+2k}\}$. Given these totals, under H_0,

$$\mu_{11k} = E(n_{11k}) = n_{1+k}n_{+1k}/n_{++k}$$

$$\text{Var}(n_{11k}) = n_{1+k}n_{2+k}n_{+1k}n_{+2k}/n_{++k}^2(n_{++k} - 1)$$

The *Cochran–Mantel–Haenszel* (*CMH*) test statistic summarizes the information from the K partial tables using

$$CMH = \frac{\left[\sum_k (n_{11k} - \mu_{11k})\right]^2}{\sum_k \text{Var}(n_{11k})} \tag{4.9}$$

This statistic has a large-sample chi-squared null distribution with $df = 1$. The approximation improves as the total sample size increases, regardless of whether the number of strata K is small or large.

When the true odds ratio exceeds 1.0 in partial table k, we expect $(n_{11k} - \mu_{11k}) > 0$. The test statistic combines these differences across all K tables, and we then expect the sum of such differences to be a relatively large positive number. When the odds ratio is less than 1.0 in each table, the sum of such differences tends to be a relatively large negative number. The *CMH* statistic takes larger values when $(n_{11k} - \mu_{11k})$ is consistently positive or consistently negative for all tables, rather than positive for some and negative for others. The test works best when the XY association is similar in each partial table.

This test was proposed in 1959, well before logistic regression was popular. The formula for the *CMH* test statistic seems to have nothing to do with modeling. In fact, though, the *CMH* test is the score test (Section 1.4.1) of *XY* conditional independence for model (4.8). Recall that model assumes a common odds ratio for the partial tables (i.e., homogeneous association). Similarity of results for the likelihood-ratio, Wald, and *CMH* (score) tests usually happens when the sample size is large.

For Table 4.4 from the AZT and AIDS study, consider H_0: conditional independence between immediate AZT use and AIDS symptom development. Section 4.3.2 noted that the likelihood-ratio test statistic is $-2(L_0 - L_1) = 6.9$ and the Wald test statistic is $(\hat{\beta}_1/SE)^2 = 6.6$, each with $df = 1$. The *CMH* statistic (4.9) equals 6.8, also with $df = 1$, giving similar results ($P = 0.01$).

4.3.5 Testing the Homogeneity of Odds Ratios*

Model (4.8) and its special case (4.5) when Z is also binary have the homogeneous association property of a common *XY* odds ratio at each level of Z. Sometimes it is of interest to test the hypothesis of homogeneous association (although it is not necessary to do so to justify using the *CMH* test). A test of homogeneity of the odds ratios is, equivalently, a test of the goodness of fit of model (4.8). Section 5.2.2 will show how to do this.

Some software reports a test, called the *Breslow–Day test*, that is a chi-squared test specifically designed to test homogeneity of odds ratios. It has the form of a Pearson chi-squared statistic, comparing the observed cell counts to estimated expected frequencies that have a common odds ratio. This test is an alternative to the goodness-of-fit tests of Section 5.2.2.

4.4 MULTIPLE LOGISTIC REGRESSION

Next we will consider the general logistic regression model with multiple explanatory variables. Denote the k predictors for a binary response Y by $x_1, x_2, \ldots, x_k$. The model for the log odds is

$$\text{logit}[P(Y = 1)] = \alpha + \beta_1 x_1 + \beta_2 x_2 + \cdots + \beta_k x_k \tag{4.10}$$

The parameter β_i refers to the effect of x_i on the log odds that $Y = 1$, controlling the other xs. For example, $\exp(\beta_i)$ is the multiplicative effect on the odds of a 1-unit increase in x_i, at fixed levels of the other xs.

4.4.1 Example: Horseshoe Crabs with Color and Width Predictors

We continue the analysis of the horseshoe crab data (Sections 3.3.2 and 4.1.3) by using both the female crab's shell width and color as predictors. Color has five categories: light, medium light, medium, medium dark, dark. Color is a surrogate for age, older crabs tending to have darker shells. The sample contained no light crabs, so we use only the other four categories.

To treat color as a nominal-scale predictor, we use three indicator variables for the four categories. The model is

$$\text{logit}[P(Y=1)] = \alpha + \beta_1 c_1 + \beta_2 c_2 + \beta_3 c_3 + \beta_4 x, \tag{4.11}$$

where x denotes width and

$c_1 = 1$ for color = medium light, 0 otherwise
$c_2 = 1$ for color = medium, 0 otherwise
$c_3 = 1$ for color = medium dark, 0 otherwise

The crab color is dark (category 4) when $c_1 = c_2 = c_3 = 0$. Table 4.6 shows the ML parameter estimates. For instance, for dark crabs, $c_1 = c_2 = c_3 = 0$, and the prediction equation is $\text{logit}[\hat{P}(Y=1)] = -12.715 + 0.468x$. By contrast, for medium-light crabs, $c_1 = 1$, and

$$\text{logit}[\hat{P}(Y=1)] = (-12.715 + 1.330) + 0.468x = -11.385 + 0.468x$$

The model assumes a lack of interaction between color and width. Width has the same effect (coefficient 0.468) for all colors. This implies that the shapes of the four curves relating width to $P(Y = 1)$ (for the four colors) are identical. For each color,

Table 4.6. Computer Output for Model for Horseshoe Crabs with Width and Color Predictors

Parameter	Estimate	Std. Error	Like. Ratio Confidence	95% Limits	Chi Square	Pr > ChiSq
intercept	−12.7151	2.7618	−18.4564	−7.5788	21.20	<.0001
c1	1.3299	0.8525	−0.2738	3.1354	2.43	0.1188
c2	1.4023	0.5484	0.3527	2.5260	6.54	0.0106
c3	1.1061	0.5921	−0.0279	2.3138	3.49	0.0617
width	0.4680	0.1055	0.2713	0.6870	19.66	<.0001

LR Statistics

Source	DF	Chi-Square	Pr > ChiSq
width	1	24.60	<.0001
color	3	7.00	0.0720

a 1 cm increase in width has a multiplicative effect of $\exp(0.468) = 1.60$ on the odds that $Y = 1$. Figure 4.4 displays the fitted model. Any one curve is any other curve shifted to the right or to the left.

The parallelism of curves in the horizontal dimension implies that two curves never cross. At all width values, for example, color 4 (dark) has a lower estimated probability of a satellite than the other colors. To illustrate, a dark crab of average width (26.3 cm) has estimated probability

$$\exp[-12.715 + 0.468(26.3)]/\{1 + \exp[-12.715 + 0.468(26.3)]\} = 0.399.$$

By contrast, a medium-light crab of average width has estimated probability

$$\exp[-11.385 + 0.468(26.3)]/\{1 + \exp[-11.385 + 0.468(26.3)]\} = 0.715.$$

The exponentiated difference between two color parameter estimates is an odds ratio comparing those colors. For example, the difference in color parameter estimates between medium-light crabs and dark crabs equals 1.330. So, at any given width, the estimated odds that a medium-light crab has a satellite are $\exp(1.330) = 3.8$ times the estimated odds for a dark crab. Using the probabilities just calculated at width 26.3, the odds equal $0.399/0.601 = 0.66$ for a dark crab and $0.715/0.285 = 2.51$ for a medium-light crab, for which $2.51/0.66 = 3.8$.

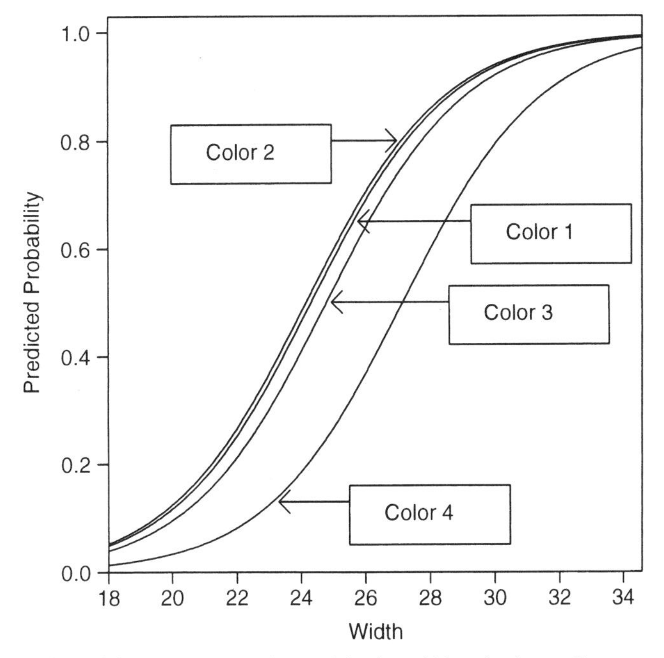

Figure 4.4. Logistic regression model using width and color predictors.

4.4.2 Model Comparison to Check Whether a Term is Needed

Are certain terms needed in a model? To test this, we can compare the maximized log-likelihood values for that model and the simpler model without those terms.

To test whether color contributes to model (4.11), we test H_0: $\beta_1 = \beta_2 = \beta_3 = 0$. This hypothesis states that, controlling for width, the probability of a satellite is independent of color. The likelihood-ratio test compares the maximized log-likelihood L_1 for the full model (4.11) to the maximized log-likelihood L_0 for the simpler model in which those parameters equal 0. Table 4.6 shows that the test statistic is $-2(L_0 - L_1) = 7.0$. Under H_0, this test statistic has an approximate chi-squared distribution with $df = 3$, the difference between the numbers of parameters in the two models. The P-value of 0.07 provides slight evidence of a color effect. Since the analysis in the previous subsection noted that estimated probabilities are quite different for dark-colored crabs, it seems safest to leave the color predictor in the model.

4.4.3 Quantitative Treatment of Ordinal Predictor

Color has a natural ordering of categories, from lightest to darkest. Model (4.11) ignores this ordering, treating color as nominal scale. A simpler model treats color in a quantitative manner. It supposes a linear effect, on the logit scale, for a set of scores assigned to its categories.

To illustrate, we use scores $c = \{1, 2, 3, 4\}$ for the color categories and fit the model

$$\text{logit}[P(Y = 1)] = \alpha + \beta_1 c + \beta_2 x \tag{4.12}$$

The prediction equation is

$$\text{logit}[\hat{P}(Y = 1)] = -10.071 - 0.509c + 0.458x$$

The color and width estimates have *SE* values of 0.224 and 0.104, showing strong evidence of an effect for each. At a given width, for every one-category increase in color darkness, the estimated odds of a satellite multiply by $\exp(-0.509) = 0.60$. For example, the estimated odds of a satellite for dark colored crabs are 60% of those for medium-dark crabs.

A likelihood-ratio test compares the fit of this model to the more complex model (4.11) that has a separate parameter for each color. The test statistic equals $-2(L_0 - L_1) = 1.7$, based on $df = 2$. This statistic tests that the simpler model (4.12) holds, given that model (4.11) is adequate. It tests that the color parameters in equation (4.11), when plotted against the color scores, follow a linear trend. The simplification seems permissible ($P = 0.44$).

The estimates of the color parameters in the model (4.11) that treats color as nominal scale are (1.33, 1.40, 1.11, 0). The 0 value for the dark category reflects the lack of an indicator variable for that category. Though these values do not depart significantly from a linear trend, the first three are similar compared to the last one.

This suggests that another potential color scoring for model (4.12) is $\{1, 1, 1, 0\}$; that is, $c = 0$ for dark-colored crabs, and $c = 1$ otherwise. The likelihood-ratio statistic comparing model (4.12) with these binary scores to model (4.11) with color treated as nominal scale equals 0.5, based on $df = 2$. So, this simpler model is also adequate ($P = 0.78$). This model has a color estimate of 1.300 ($SE = 0.525$). At a given width, the estimated odds that a lighter-colored crab has a satellite are $\exp(1.300) = 3.7$ times the estimated odds for a dark crab.

In summary, the nominal-scale model, the quantitative model with color scores $\{1, 2, 3, 4\}$, and the model with binary color scores $\{1, 1, 1, 0\}$ all suggest that dark crabs are least likely to have satellites. When the sample size is not very large, it is not unusual that several models fit adequately.

It is advantageous to treat ordinal predictors in a quantitative manner, when such models fit well. The model is simpler and easier to interpret, and tests of the effect of the ordinal predictor are generally more powerful when it has a single parameter rather than several parameters.

4.4.4 Allowing Interaction

The models we have considered so far assume a lack of interaction between width and color. Let us check now whether this is sensible. We can allow interaction by adding cross products of terms for width and color. Each color then has a different-shaped curve relating width to the probability of a satellite, so a comparison of two colors varies according to the value of width.

For example, consider the model just discussed that has a dummy variable $c = 0$ for dark-colored crabs and $c = 1$ otherwise. The model with an interaction term has the prediction equation

$$\text{logit}[\hat{P}(Y = 1)] = -5.854 - 6.958c + 0.200x + 0.322(c \times x)$$

Let us see what this implies about the prediction equations for each color. For dark crabs, $c = 0$ and

$$\text{logit}[\hat{P}(Y = 1)] = -5.854 + 0.200x$$

For lighter crabs, $c = 1$ and

$$\text{logit}[\hat{P}(Y = 1)] = -12.812 + 0.522x$$

The curve for lighter crabs has a faster rate of increase. The curves cross at x such that $-5.854 + 0.200x = -12.812 + 0.522x$, that is, at $x = 21.6$ cm. The sample widths range between 21.0 and 33.5 cm, so the lighter-colored crabs have a higher estimated probability of a satellite over essentially the entire range.

We can compare this to the simpler model without interaction to analyze whether the fit is significantly better. The likelihood-ratio statistic comparing the models equals

1.2, based on $df = 1$. The evidence of interaction is not strong ($P = 0.28$). Although the sample slopes for the width effect are quite different for the two colors, the sample had only 24 crabs of dark color. So, effects involving it have relatively large standard errors.

Fitting the interaction model is equivalent to fitting the logistic regression model with width as the predictor separately for the crabs of each color. The reduced model has the advantage of simpler interpretations.

4.5 SUMMARIZING EFFECTS IN LOGISTIC REGRESSION

We have interpreted effects in logistic regression using multiplicative effects on the odds, which correspond to odds ratios. However, many find it difficult to understand odds ratios.

4.5.1 Probability-Based Interpretations

For a relatively small change in a quantitative predictor, Section 4.1.1 used a straight line to approximate the change in the probability. This simpler interpretation applies also with multiple predictors.

Consider a setting of predictors at which $\hat{P}(Y = 1) = \hat{\pi}$. Then, controlling for the other predictors, a 1-unit increase in x_j corresponds approximately to a $\hat{\beta}_j \hat{\pi}(1 - \hat{\pi})$ change in $\hat{\pi}$. For example, for the horseshoe crab data with predictors x = width and an indicator c that is 0 for dark crabs and 1 otherwise, $\text{logit}(\hat{\pi}) = -12.98 + 1.300c + 0.478x$. When $\hat{\pi} = 0.50$, the approximate effect on $\hat{\pi}$ of a 1 cm increase in x is $(0.478)(0.50)(0.50) = 0.12$. This is considerable, since a 1 cm change in width is less than half its standard deviation (which is 2.1 cm).

This straight-line approximation deteriorates as the change in the predictor values increases. More precise interpretations use the probability formula directly. One way to describe the effect of a predictor x_j sets the other predictors at their sample means and finds $\hat{\pi}$ at the smallest and largest x_j values. The effect is summarized by reporting those $\hat{\pi}$ values or their difference. However, such summaries are sensitive to outliers on x_j. To obtain a more robust summary, it is more sensible to use the quartiles of the x_j values.

For the prediction equation $\text{logit}(\hat{\pi}) = -12.98 + 1.300c + 0.478x$, the sample means are 26.3 cm for x = width and 0.873 for c = color. The lower and upper quartiles of x are $LQ = 24.9$ cm and $UQ = 27.7$ cm. At $x = 24.9$ and $c = \bar{c}$, $\hat{\pi} = 0.51$. At $x = 27.7$ and $c = \bar{c}$, $\hat{\pi} = 0.80$. The change in $\hat{\pi}$ from 0.51 to 0.80 over the middle 50% of the range of width values reflects a strong width effect. Since c takes only values 0 and 1, one could instead report this effect separately for each value of c rather than just at its mean.

To summarize the effect of an indicator explanatory variable, it makes sense to report the estimated probabilities at its two values rather than at quartiles, which could be identical. For example, consider the color effect in the prediction equation

Table 4.7. Summary of Effects in Model with Crab Width and Whether Color is Dark as Predictors of Presence of Satellites

Variable	Estimate	*SE*	Comparison	Change in Probability
No interaction model				
Intercept	−12.980	2.727		
Color (0 = dark, 1 = other)	1.300	0.526	$(1, 0)$ at $\bar{x}$	$0.31 = 0.71 - 0.40$
Width (x)	0.478	0.104	(UQ, LQ) at $\bar{c}$	$0.29 = 0.80 - 0.51$
Interaction model				
Intercept	−5.854	6.694		
Color (0 = dark, 1 = other)	−6.958	7.318		
Width (x)	0.200	0.262	(UQ, LQ) at $c = 0$	$0.13 = 0.43 - 0.30$
Width*color	0.322	0.286	(UQ, LQ) at $c = 1$	$0.29 = 0.84 - 0.55$

$\text{logit}(\hat{\pi}) = -12.98 + 1.300c + 0.478x$. At $\bar{x} = 26.3$, $\hat{\pi} = 0.40$ when $c = 0$ and $\hat{\pi} = 0.71$ when $c = 1$. This color effect, differentiating dark crabs from others, is also substantial.

Table 4.7 summarizes effects using estimated probabilities. It also shows results for the extension of the model permitting interaction. The estimated width effect is then greater for the lighter colored crabs. However, the interaction is not significant.

4.5.2 Standardized Interpretations

With multiple predictors, it is tempting to compare magnitudes of $\{\hat{\beta}_j\}$ to compare effects of predictors. For binary predictors, this gives a comparison of conditional log odds ratios, given the other predictors in the model. For quantitative predictors, this is relevant if the predictors have the same units, so a 1-unit change means the same thing for each. Otherwise, it is not meaningful.

An alternative comparison of effects of quantitative predictors having different units uses *standardized* coefficients. The model is fitted to standardized predictors, replacing each x_j by $(x_j - \bar{x}_j)/s_{x_j}$. A 1-unit change in the standardized predictor is a standard deviation change in the original predictor. Then, each regression coefficient represents the effect of a standard deviation change in a predictor, controlling for the other variables. The standardized estimate for predictor x_j is the unstandardized estimate $\hat{\beta}_j$ multiplied by s_{x_j}. See Problem 4.27.

PROBLEMS

4.1 A study used logistic regression to determine characteristics associated with Y = whether a cancer patient achieved remission (1 = yes). The most important explanatory variable was a labeling index (*LI*) that measures proliferative activity of cells after a patient receives an injection of tritiated thymidine. It represents the percentage of cells that are "labeled." Table 4.8 shows the grouped data. Software reports Table 4.9 for a logistic regression model using LI to predict $\pi = P(Y = 1)$.

a. Show how software obtained $\hat{\pi} = 0.068$ when $LI = 8$.

b. Show that $\hat{\pi} = 0.50$ when $LI = 26.0$.

c. Show that the rate of change in $\hat{\pi}$ is 0.009 when $LI = 8$ and is 0.036 when $LI = 26$.

d. The lower quartile and upper quartile for LI are 14 and 28. Show that $\hat{\pi}$ increases by 0.42, from 0.15 to 0.57, between those values.

e. When LI increases by 1, show the estimated odds of remission multiply by 1.16.

Table 4.8. Data for Exercise 4.1 on Cancer Remission

LI	Number of Cases	Number of Remissions	LI	Number of Cases	Number of Remissions	LI	Number of Cases	Number of Remissions
8	2	0	18	1	1	28	1	1
10	2	0	20	3	2	32	1	0
12	3	0	22	2	1	34	1	1
14	3	0	24	1	0	38	3	2
16	3	0	26	1	1			

Source: Reprinted with permission from E. T. Lee, *Computer Prog. Biomed.*, **4**: 80–92, 1974.

Table 4.9. Computer Output for Problem 4.1

```
                         Standard    Likelihood Ratio
Parameter    Estimate      Error     95% Conf. Limits      Chi-Square
Intercept    -3.7771      1.3786    -6.9946   -1.4097         7.51
li            0.1449      0.0593     0.0425    0.2846         5.96

                                 LR Statistic
             Source     DF       Chi-Square      Pr > ChiSq
             li          1          8.30           0.0040

Obs      li      remiss       n      pi_hat       lower        upper
1         8        0          2     0.06797     0.01121     0.31925
2        10        0          2     0.08879     0.01809     0.34010
....
```

4.2 Refer to the previous exercise. Using information from Table 4.9:

a. Conduct a Wald test for the LI effect. Interpret.

b. Construct a Wald confidence interval for the odds ratio corresponding to a 1-unit increase in LI. Interpret.

c. Conduct a likelihood-ratio test for the LI effect. Interpret.

d. Construct the likelihood-ratio confidence interval for the odds ratio. Interpret.

4.3 In the first nine decades of the twentieth century in baseball's National League, the percentage of times the starting pitcher pitched a complete game were: 72.7 (1900–1909), 63.4, 50.0, 44.3, 41.6, 32.8, 27.2, 22.5, 13.3 (1980–1989) (*Source*: George Will, *Newsweek*, April 10, 1989).

a. Treating the number of games as the same in each decade, the linear probability model has ML fit $\hat{\pi} = 0.7578 - 0.0694x$, where $x =$ decade ($x = 1, 2, \ldots, 9$). Interpret -0.0694.

b. Substituting $x = 12$, predict the percentage of complete games for 2010–2019. Is this prediction plausible? Why?

c. The logistic regression ML fit is $\hat{\pi} = \exp(1.148 - 0.315x)/[1 + \exp(1.148 - 0.315x)]$. Obtain $\hat{\pi}$ for $x = 12$. Is this more plausible than the prediction in (**b**)?

4.4 Consider the snoring and heart disease data of Table 3.1 in Section 3.2.2. With scores $\{0, 2, 4, 5\}$ for snoring levels, the logistic regression ML fit is $\text{logit}(\hat{\pi}) = -3.866 + 0.397x$.

a. Interpret the sign of the estimated effect of x.

b. Estimate the probabilities of heart disease at snoring levels 0 and 5.

c. Describe the estimated effect of snoring on the odds of heart disease.

4.5 For the 23 space shuttle flights before the Challenger mission disaster in 1986, Table 4.10 shows the temperature (°F) at the time of the flight and whether at least one primary O-ring suffered thermal distress.

a. Use logistic regression to model the effect of temperature on the probability of thermal distress. Interpret the effect.

b. Estimate the probability of thermal distress at 31°F, the temperature at the time of the Challenger flight.

c. At what temperature does the estimated probability equal 0.50? At that temperature, give a linear approximation for the change in the estimated probability per degree increase in temperature.

d. Interpret the effect of temperature on the odds of thermal distress.

e. Test the hypothesis that temperature has no effect, using (i) the Wald test, (ii) the likelihood-ratio test.

4.6 Refer to Exercise 3.9. Use the logistic regression output reported there to (**a**) interpret the effect of income on the odds of possessing a travel credit card, and conduct a (**b**) significance test and (**c**) confidence interval about that effect.

Table 4.10. Data for Problem 4.5 on Space Shuttle

Ft	Temperature	TD	Ft	Temperature	TD
1	66	0	13	67	0
2	70	1	14	53	1
3	69	0	15	67	0
4	68	0	16	75	0
5	67	0	17	70	0
6	72	0	18	81	0
7	73	0	19	76	0
8	70	0	20	79	0
9	57	1	21	75	1
10	63	1	22	76	0
11	70	1	23	58	1
12	78	0			

Note: Ft = flight no., TD = thermal distress (1 = yes, 0 = no).

Source: Data based on Table 1 in S. R. Dalal, E. B. Fowlkes and B. Hoadley, *J. Am. Statist. Assoc.*, **84**: 945–957, 1989. Reprinted with the permission of the American Statistical Association.

4.7 Hastie and Tibshirani (1990, p. 282) described a study to determine risk factors for kyphosis, which is severe forward flexion of the spine following corrective spinal surgery. The age in months at the time of the operation for the 18 subjects for whom kyphosis was present were 12, 15, 42, 52, 59, 73, 82, 91, 96, 105, 114, 120, 121, 128, 130, 139, 139, 157 and for the 22 subjects for whom kyphosis was absent were 1, 1, 2, 8, 11, 18, 22, 31, 37, 61, 72, 81, 97, 112, 118, 127, 131, 140, 151, 159, 177, 206.

a. Fit a logistic regression model using age as a predictor of whether kyphosis is present. Test whether age has a significant effect.

b. Plot the data. Note the difference in dispersion of age at the two levels of kyphosis.

c. Fit the model $\text{logit}[\pi(x)] = \alpha + \beta_1 x + \beta_2 x^2$. Test the significance of the squared age term, plot the fit, and interpret. (The final paragraph of Section 4.1.6 is relevant to these results.)

4.8 For the horseshoe crab data (Table 3.2, available at www.stat.ufl.edu/~aa/intro-cda/appendix.html), fit the logistic regression model for π = probability of a satellite, using weight as the predictor.

a. Report the ML prediction equation.

b. Find $\hat{\pi}$ at the weight values 1.20, 2.44, and 5.20 kg, which are the sample minimum, mean, and maximum.

c. Find the weight at which $\hat{\pi} = 0.50$.

d. At the weight value found in (**c**), give a linear approximation for the estimated effect of (i) a 1 kg increase in weight. This represents a relatively large increase, so convert this to the effect of (ii) a 0.10 kg increase, and (iii) a standard deviation increase in weight (0.58 kg).

e. Construct a 95% confidence interval to describe the effect of weight on the odds of a satellite. Interpret.

f. Conduct the Wald or likelihood-ratio test of the hypothesis that weight has no effect. Report the P-value, and interpret.

4.9 For the horseshoe crab data, fit a logistic regression model for the probability of a satellite, using color alone as the predictor.

a. Treat color as nominal scale (qualitative). Report the prediction equation, and explain how to interpret the coefficient of the first indicator variable.

b. For the model in (**a**), conduct a likelihood-ratio test of the hypothesis that color has no effect. Interpret.

c. Treating color in a quantitative manner, obtain a prediction equation. Interpret the coefficient of color.

d. For the model in (**c**), test the hypothesis that color has no effect. Interpret.

e. When we treat color as quantitative instead of qualitative, state an advantage relating to power and a potential disadvantage relating to model lack of fit.

4.10 An international poll quoted in an Associated Press story (December 14, 2004) reported low approval ratings for President George W. Bush among traditional allies of the United States, such as 32% in Canada, 30% in Britain, 19% in Spain, and 17% in Germany. Let Y indicate approval of Bush's performance ($1 = \text{yes}, 0 = \text{no}$), $\pi = P(Y = 1)$, $c_1 = 1$ for Canada and 0 otherwise, $c_2 = 1$ for Britain and 0 otherwise, and $c_3 = 1$ for Spain and 0 otherwise.

a. Explain why these results suggest that for the identity link function, $\hat{\pi} = 0.17 + 0.15c_1 + 0.13c_2 + 0.02c_3$.

b. Show that the prediction equation for the logit link function is $\text{logit}(\hat{\pi}) = -1.59 + 0.83c_1 + 0.74c_2 + 0.14c_3$.

4.11 Moritz and Satariano (*J. Clin. Epidemiol.*, **46**: 443–454, 1993) used logistic regression to predict whether the stage of breast cancer at diagnosis was advanced or local for a sample of 444 middle-aged and elderly women. A table referring to a particular set of demographic factors reported the estimated odds ratio for the effect of living arrangement (three categories) as 2.02 for spouse vs alone and 1.71 for others vs alone; it reported the effect of income (three categories) as 0.72 for \$10,000–24,999 vs <\$10,000 and 0.41 for \$25,000+ vs <\$10,000. Estimate the odds ratios for the third pair of categories for each factor.

4.12 Section 2.7 mentioned a study in Florida that stated that the death penalty was given in 53 out of 467 cases in which a white killed a white, in 0 out

Table 4.11. Computer Output for Problem 4.12 on Death Penalty

Parameter	Estimate	Standard Error	Likelihood Ratio 95% Conf. Limits		Chi-Square
Intercept	−3.5961	0.5069	−4.7754	−2.7349	50.33
def	−0.8678	0.3671	−1.5633	−0.1140	5.59
vic	2.4044	0.6006	1.3068	3.7175	16.03

Source	DF	LR Statistics Chi-Square	Pr > ChiSq
def	1	5.01	0.0251
vic	1	20.35	<.0001

of 16 cases in which a white killed a black, in 11 out of 48 cases in which a black killed a white, and in 4 out of 143 cases in which a black killed a black. Table 4.11 shows results of fitting a logit model for death penalty as the response (1 = yes), with defendant's race (1 = white) and victims' race (1 = white) as indicator predictors.

a. Based on the parameter estimates, which group is most likely to have the "yes" response? Estimate the probability in that case.

b. Interpret the parameter estimate for victim's race.

c. Using information shown, construct and interpret a 95% likelihood-ratio confidence interval for the conditional odds ratio between the death penalty verdict and victim's race.

d. Test the effect of victim's race, controlling for defendant's race, using a Wald test or likelihood-ratio test. Interpret.

4.13 Refer to (**d**) in the previous exercise. The Cochran–Mantel–Haenszel test statistic for this hypothesis equals 7.00.

a. Report the null sampling distribution of the statistic and the P-value.

b. Under H_0, find the expected count for the cell in which white defendants who had black victims received the death penalty. Based on comparing this to the observed count, interpret the result of the test.

4.14 Refer to the results that Table 4.5 shows for model (4.5) fitted to the data from the AZT and AIDS study in Table 4.4.

a. For black veterans without immediate AZT use, use the prediction equation to estimate the probability of AIDS symptoms.

b. Construct a 95% confidence interval for the conditional odds ratio between AZT use and the development of symptoms.

c. Describe and test for the effect of race in this model.

4.15 Table 4.12 refers to ratings of agricultural extension agents in North Carolina. In each of five districts, agents were classified by their race and by whether they qualified for a merit pay increase.

a. Conduct the Cochran–Mantel–Haenszel test of the hypothesis that the merit pay decision is independent of race, conditional on the district. Interpret.

b. Show how you could alternatively test the hypothesis in (**a**) using a test about a parameter in a logistic regression model.

c. What information can you get from a model-based analysis that you do not get from the *CMH* test?

Table 4.12. Data for Problem 4.15 on Merit Pay and Race

	Blacks, Merit Pay		Whites, Merit Pay	
District	Yes	No	Yes	No
NC	24	9	47	12
NE	10	3	45	8
NW	5	4	57	9
SE	16	7	54	10
SW	7	4	59	12

Source: J. Gastwirth, *Statistical Reasoning in Law and Public Policy*, Vol. 1, 1988, p. 268.

4.16 Table 4.13 shows the result of cross classifying a sample of people from the MBTI Step II National Sample (collected and compiled by CPP, Inc.) on whether they report drinking alcohol frequently (1 = yes,

Table 4.13. Data for Problem 4.16 on Drinking Frequently and Four Scales of Myers–Briggs Personality Test

Extroversion/Introversion		E				I			
Sensing/iNtuitive		S		N		S		N	
		Alcohol Frequently							
Thinking/Feeling	Judging/Perceiving	Yes	No	Yes	No	Yes	No	Yes	No
T	J	10	67	3	20	17	123	1	12
	P	8	34	2	16	3	49	5	30
F	J	5	101	4	27	6	132	1	30
	P	7	72	15	65	4	102	6	73

Source: Reproduced with special permission of CPP Inc., Mountain View, CA 94043.

0 = no) and on the four binary scales of the Myers–Briggs personality test: Extroversion/Introversion (E/I), Sensing/iNtuitive (S/N), Thinking/Feeling (T/F) and Judging/Perceiving (J/P). The 16 predictor combinations correspond to the 16 personality types: ESTJ, ESTP, ESFJ, ESFP, ENTJ, ENTP, ENFJ, ENFP, ISTJ, ISTP, ISFJ, ISFP, INTJ, INTP, INFJ, INFP.

a. Fit a model using the four scales as predictors of π = the probability of drinking alcohol frequently. Report the prediction equation, specifying how you set up the indicator variables.

b. Find $\hat{\pi}$ for someone of personality type ESTJ.

c. Based on the model parameter estimates, explain why the personality type with the highest $\hat{\pi}$ is ENTP.

4.17 Refer to the previous exercise. Table 4.14 shows the fit of the model with only E/I and T/F as predictors.

a. Find $\hat{\pi}$ for someone of personality type introverted and feeling.

b. Report and interpret the estimated conditional odds ratio between E/I and the response.

c. Use the limits reported to construct a 95% likelihood-ratio confidence interval for the conditional odds ratio between E/I and the response. Interpret.

d. The estimates shown use E for the first category of the E/I scale. Suppose you instead use I for the first category. Then, report the estimated conditional odds ratio and the 95% likelihood-ratio confidence interval. Interpret.

e. Show steps of a test of whether E/I has an effect on the response, controlling for T/F. Indicate whether your test is a Wald or a likelihood-ratio test.

Table 4.14. Output for Problem 4.17 on Fitting Model to Table 4.13

Analysis Of Parameter Estimates

Parameter		DF	Estimate	Standard Error	Likelihood Ratio 95% Conf. Limits		Wald Chi-Square
Intercept		1	−2.8291	0.1955	−3.2291	−2.4614	209.37
EI	e	1	0.5805	0.2160	0.1589	1.0080	7.22
TF	t	1	0.5971	0.2152	0.1745	1.0205	7.69

LR Statistics

Source	DF	Chi-Square	Pr > ChiSq
EI	1	7.28	0.0070
TF	1	7.64	0.0057

4.18 A study used the 1998 Behavioral Risk Factors Social Survey to consider factors associated with American women's use of oral contraceptives. Table 4.15

Table 4.15. Table for Problem 4.18 on Oral Contraceptive Use

Variable	Coding = 1 if:	Estimate	*SE*
Age	35 or younger	−1.320	0.087
Race	White	0.622	0.098
Education	$\geq$1 year college	0.501	0.077
Marital status	Married	−0.460	0.073

Source: Debbie Wilson, College of Pharmacy, University of Florida.

summarizes effects for a logistic regression model for the probability of using oral contraceptives. Each predictor uses an indicator variable, and the table lists the category having value 1.

a. Interpret effects.

b. Construct and interpret a confidence interval for the conditional odds ratio between contraceptive use and education.

4.19 A sample of subjects were asked their opinion about current laws legalizing abortion (support, oppose). For the explanatory variables gender (female, male), religious affiliation (Protestant, Catholic, Jewish), and political party affiliation (Democrat, Republican, Independent), the model for the probability π of supporting legalized abortion,

$$\text{logit}(\pi) = \alpha + \beta_h^{\text{G}} + \beta_i^{\text{R}} + \beta_j^{\text{P}}$$

has reported parameter estimates (setting the parameter for the last category of a variable equal to 0.0) $\hat{\alpha} = -0.11$, $\hat{\beta}_1^{\text{G}} = 0.16$, $\hat{\beta}_2^{\text{G}} = 0.0$, $\hat{\beta}_1^{\text{R}} = -0.57$, $\hat{\beta}_2^{\text{R}} = -0.66$, $\hat{\beta}_3^{\text{R}} = 0.0$, $\hat{\beta}_1^{\text{P}} = 0.84$, $\hat{\beta}_2^{\text{P}} = -1.67$, $\hat{\beta}_3^{\text{P}} = 0.0$.

a. Interpret how the odds of supporting legalized abortion depend on gender.

b. Find the estimated probability of supporting legalized abortion for (i) male Catholic Republicans and (ii) female Jewish Democrats.

c. If we defined parameters such that the *first* category of a variable has value 0, then what would $\hat{\beta}_2^{\text{G}}$ equal? Show then how to obtain the odds ratio that describes the conditional effect of gender.

d. If we defined parameters such that they sum to 0 across the categories of a variable, then what would $\hat{\beta}_1^{\text{G}}$ and $\hat{\beta}_2^{\text{G}}$ equal? Show then how to obtain the odds ratio that describes the conditional effect of gender.

4.20 Table 4.16 shows results of an eight-center clinical trial to compare a drug to placebo for curing an infection. At each center, subjects were randomly assigned to groups.

a. Analyze these data, describing and making inference about the group effect, using logistic regression.

Table 4.16. Clinical Trial Data for Problem 4.20

Center	Treatment	Response: Success	Response: Failure	Sample Odds Ratio
1	Drug	11	25	1.19
	Control	10	27	
2	Drug	16	4	1.82
	Control	22	10	
3	Drug	14	5	4.80
	Control	7	12	
4	Drug	2	14	2.29
	Control	1	16	
5	Drug	6	11	∞
	Control	0	12	
6	Drug	1	10	∞
	Control	0	10	
7	Drug	1	4	2.0
	Control	1	8	
8	Drug	4	2	0.33
	Control	6	1	

Source: P. J. Beitler and J. R. Landis, *Biometrics*, **41**: 991–1000, 1985.

b. Conduct the Cochran–Mantel–Haenszel test. Specify the hypotheses, report the P-value, and interpret.

4.21 In a study designed to evaluate whether an educational program makes sexually active adolescents more likely to obtain condoms, adolescents were randomly assigned to two experimental groups. The educational program, involving a lecture and videotape about transmission of the HIV virus, was provided to one group but not the other. In logistic regression models, factors observed to influence a teenager to obtain condoms were gender, socioeconomic status, lifetime number of partners, and the experimental group. Table 4.17 summarizes study results.

a. Interpret the odds ratio and the related confidence interval for the effect of group.

b. Find the parameter estimates for the fitted model, using $(1, 0)$ indicator variables for the first three predictors. Based on the corresponding confidence interval for the log odds ratio, determine the standard error for the group effect.

c. Explain why either the estimate of 1.38 for the odds ratio for gender or the corresponding confidence interval is incorrect. Show that, if the reported interval is correct, then 1.38 is actually the *log* odds ratio, and the estimated odds ratio equals 3.98.

Table 4.17. Table for Problem 4.21 on Condom Use

Variables	Odds Ratio	95% Confidence Interval
Group (education vs none)	4.04	(1.17, 13.9)
Gender (males vs females)	1.38	(1.23, 12.88)
SES (high vs low)	5.82	(1.87, 18.28)
Lifetime no. of partners	3.22	(1.08, 11.31)

Source: V. I. Rickert et al., *Clin. Pediat.*, **31**: 205–210, 1992.

4.22 Refer to model (4.11) with width and color effects for the horseshoe crab data. Using the data at www.stat.ufl.edu/~aa/intro-cda/appendix.html:

a. Fit the model, treating color as nominal-scale but with weight instead of width as x. Interpret the parameter estimates.

b. Controlling for weight, conduct a likelihood-ratio test of the hypothesis that having a satellite is independent of color. Interpret.

c. Using models that treat color in a quantitative manner with scores $\{1, 2, 3, 4\}$, repeat the analyses in (**a**) and (**b**).

4.23 Table 4.18 shows estimated effects for a fitted logistic regression model with squamous cell esophageal cancer ($1 = \text{yes}$, $0 = \text{no}$) as the response variable Y. Smoking status (S) equals 1 for at least one pack per day and 0 otherwise, alcohol consumption (A) equals the average number of alcoholic drinks consumed per day, and race (R) equals 1 for blacks and 0 for whites.

a. To describe the race-by-smoking interaction, construct the prediction equation when $R = 1$ and again when $R = 0$. Find the fitted *YS* conditional odds ratio for each case. Similarly, construct the prediction equation when $S = 1$ and again when $S = 0$. Find the fitted *YR* conditional odds ratio for each case. Note that, for each association, the coefficient of the cross-product term is the difference between the log odds ratios at the two fixed levels for the other variable.

Table 4.18. Table for Problem 4.23 on Effects on Esophageal Cancer

Variable	Effect	P-value
Intercept	−7.00	<0.01
Alcohol use	0.10	0.03
Smoking	1.20	<0.01
Race	0.30	0.02
Race × smoking	0.20	0.04

b. In Table 4.18, explain what the coefficients of R and S represent, for the coding as given above. What hypotheses do the P-values refer to for these variables?

c. Suppose the model also contained an $A \times R$ interaction term, with coefficient 0.04. In the prediction equation, show that this represents the difference between the effect of A for blacks and for whites.

4.24 Table 4.19 shows results of a study about Y = whether a patient having surgery with general anesthesia experienced a sore throat on waking (1 = yes) as a function of D = duration of the surgery (in minutes) and T = type of device used to secure the airway (0 = laryngeal mask airway, 1 = tracheal tube).

a. Fit a main effects model using these predictors. Interpret parameter estimates.

b. Conduct inference about the D effect in (**a**).

c. Fit a model permitting interaction. Report the prediction equation for the effect of D when (i) $T = 1$, (ii) $T = 0$. Interpret.

d. Conduct inference about whether you need the interaction term in (**c**).

Table 4.19. Data for Problem 4.24 on Sore Throat after Surgery

Patient	D	T	Y	Patient	D	T	Y	Patient	D	T	Y
1	45	0	0	13	50	1	0	25	20	1	0
2	15	0	0	14	75	1	1	26	45	0	1
3	40	0	1	15	30	0	0	27	15	1	0
4	83	1	1	16	25	0	1	28	25	0	1
5	90	1	1	17	20	1	0	29	15	1	0
6	25	1	1	18	60	1	1	30	30	0	1
7	35	0	1	19	70	1	1	31	40	0	1
8	65	0	1	20	30	0	1	32	15	1	0
9	95	0	1	21	60	0	1	33	135	1	1
10	35	0	1	22	61	0	0	34	20	1	0
11	75	0	1	23	65	0	1	35	40	1	0
12	45	1	1	24	15	1	0				

Source: Data from D. Collett, in *Encyclopedia of Biostatistics*, Wiley, New York, 1998, pp. 350–358. Predictors are D = duration of surgery, T = type of device.

4.25 For model (4.11) for the horseshoe crabs with color and width predictors, add three terms to permit interaction between color and width.

a. Report the prediction equations relating width to the probability of a satellite, for each color. Plot or sketch them, and interpret.

b. Test whether the interaction model gives a better fit than the simpler model lacking the interaction terms. Interpret.

4.26 Model (4.11) for the probability π of a satellite for horseshoe crabs with color and width predictors has fit

$$\text{logit}(\hat{\pi}) = -12.715 + 1.330c_1 + 1.402c_2 + 1.106c_3 + 0.468x$$

Consider this fit for crabs of width $x = 20$ cm.

a. Estimate π for medium-dark crabs ($c_3 = 1$) and for dark crabs ($c_1 = c_2 = c_3 = 0$). Then, estimate the ratio of probabilities.

b. Estimate the odds of a satellite for medium-dark crabs and the odds for dark crabs. Show that the odds ratio equals $\exp(1.106) = 3.02$. When each probability is close to zero, the odds ratio is similar to the ratio of probabilities, providing another interpretation for logistic regression parameters. For widths at which $\hat{\pi}$ is small, $\hat{\pi}$ for medium-dark crabs is about three times that for dark crabs.

4.27 The prediction equation for the horseshoe crab data using width and quantitative color (scores 1, 2, 3, 4) is $\text{logit}(\hat{\pi}) = -10.071 - 0.509c + 0.458x$. Color has mean = 2.44 and standard deviation = 0.80, and width has mean = 26.30 and standard deviation = 2.11.

a. For standardized versions of the predictors, explain why the estimated coefficients equal $(0.80)(-.509) = -0.41$ and $(2.11)(.458) = 0.97$. Interpret these by comparing the partial effects on the odds of a one standard deviation increase in each predictor.

b. Section 4.5.1 interpreted the width effect by finding the change in $\hat{\pi}$ over the middle 50% of width values, between 24.9 cm and 27.7 cm. Do this separately for each value of c, and interpret the width effect for each color.

4.28 For recent General Social Survey data, a prediction equation relating Y = whether attended college (1 = yes) to x = family income (thousands of dollars, using scores for grouped categories), m = whether mother attended college (1 = yes, 0 = no), f = whether father attended college (1 = yes, 0 = no), was $\text{logit}[\hat{P}(Y = 1)] = -1.90 + 0.02x + 0.82m + 1.33f$. To summarize the cumulative effect of the predictors, report the range of $\hat{\pi}$ values between their lowest levels ($x = 0.5$, $m = 0$, $f = 0$) and their highest levels ($x = 130$, $m = 1$, $f = 1$).

4.29 Table 4.20 appeared in a national study of 15- and 16-year-old adolescents. The event of interest is ever having sexual intercourse. Analyze these data and summarize in a one-page report, including description and inference about the effects of both gender and race.

4.30 The US National Collegiate Athletic Association (NCAA) conducted a study of graduation rates for student athletes who were freshmen during the

Table 4.20. Data for Problem 4.29 on Teenagers and Sex

		Intercourse	
Race	Gender	Yes	No
White	Male	43	134
	Female	26	149
Black	Male	29	23
	Female	22	36

Source: S. P. Morgan and J. D. Teachman, *J. Marriage Fam.*, **50**: 929–936, 1988. Reprinted with permission of The National Council on Family Relations.

1984–1985 academic year. Table 4.21 shows the data. Analyze and interpret in a one-page report, including description and inference.

Table 4.21. Data for Problem 4.30 on Graduation of NCAA Athletes

Athlete Group	Sample Size	Graduates
White females	796	498
White males	1625	878
Black females	143	54
Black males	660	197

Source: J. J. McArdle and F. Hamagami, *J. Am. Statist. Assoc.*, **89**: 1107–1123, 1994. Reprinted with permission of the American Statistical Association.

4.31 Refer to Table 7.3, treating marijuana use as the response variable. Analyze these data. Prepare a one-page report summarizing your descriptive and inferential results.

4.32 See http://bmj.com/cgi/content/full/317/7153/235 for a meta analysis of studies about whether administering albumin to critically ill patients increases or decreases mortality. Analyze the data for the three studies with burn patients using logistic regression methods. Summarize your analyses in a one-page report.

4.33 Fowlkes et al. (*J. Am. Statist. Assoc.*, **83**: 611–622, 1988) reported results of a survey of employees of a large national corporation to determine how satisfaction depends on race, gender, age, and regional location. The data are at www.stat.ufl.edu/~aa/cda/cda.html. Fowlkes et al. reported "The least-satisfied employees are less than 35 years of age, female, other (race), and work in the Northeast; ... The most satisfied group is greater than 44 years of age, male, other, and working in the Pacific or Mid-Atlantic regions; the odds of such employees being satisfied are about 3.5 to 1." Analyze the data, and show how you would make this interpretation.

4.34 For the model, $\text{logit}[\pi(x)] = \alpha + \beta x$, show that e^{α} equals the odds of success when $x = 0$. Construct the odds of success when $x = 1$, $x = 2$, and $x = 3$. Use this to provide an interpretation of β. Generalize these results to the multiple logistic regression model (4.10).

4.35 The slope of the line drawn tangent to the probit regression curve at a particular x value equals $(0.40)\beta \exp[-(\alpha + \beta x)^2/2]$.

a. Show this is highest when $x = -\alpha/\beta$, where it equals 0.40β. At this point, $\pi(x) = 0.50$.

b. The fit of the probit model to the horseshoe crab data using x = width is $\text{probit}[\hat{\pi}(x)] = -7.502 + 0.302x$. At which x-value does the estimated probability of a satellite equal 0.50?

c. Find the rate of change in $\hat{\pi}(x)$ per 1 cm increase in width at the x-value found in (**b**). Compare the results with those obtained with logistic regression in Section 4.1.3, for which $\hat{\pi}(x) = 1/2$ at $x = 24.8$, where the rate of change is 0.12. (Probit and logistic models give very similar fits to data.)

4.36 When $\beta > 0$, the logistic regression curve (4.1) has the shape of the *cdf* of a logistic distribution with mean $\mu = -\alpha/\beta$ and standard deviation $\sigma = 1.814/\beta$. Section 4.1.3 showed that the horseshoe crab data with x = width has fit, $\text{logit}[\hat{\pi}(x)] = -12.351 + 0.497x$.

a. Show that the curve for $\hat{\pi}(x)$ has the shape of a logistic *cdf* with mean 24.8 and standard deviation 3.6.

b. Since about 95% of a bell-shaped distribution occurs within two standard deviations of the mean, argue that the probability of a satellite increases from near 0 to near 1 as width increases from about 17 to 32 cm.

4.37 For data from Florida on Y = whether someone convicted of multiple murders receives the death penalty (1 = yes, 0 = no), the prediction equation is $\text{logit}(\hat{\pi}) = -2.06 + .87d - 2.40v$, where d and v are defendant's race and victims' race (1 = black, 0 = white). The following are true–false questions based on the prediction equation.

a. The estimated probability of the death penalty is lowest when the defendant is white and victims are black.

b. Controlling for victims' race, the estimated odds of the death penalty for white defendants equal 0.87 times the estimated odds for black defendants. If we instead let $d = 1$ for white defendants and 0 for black defendants, the estimated coefficient of d would be $1/0.87 = 1.15$ instead of 0.87.

c. The lack of an interaction term means that the estimated odds ratio between the death penalty outcome and defendant's race is the same for each category of victims' race.

d. The intercept term -2.06 is the estimated probability of the death penalty when the defendant and victims were white (i.e., $d = v = 0$).

e. If there were 500 cases with white victims and defendants, then the model fitted count (i.e., estimated expected frequency) for the number who receive the death penalty equals $500e^{-2.06}/(1 + e^{-2.06})$.

CHAPTER 5

Building and Applying Logistic Regression Models

Having learned the basics of logistic regression, we now study issues relating to building a model with multiple predictors and checking its fit. Section 5.1 discusses strategies for model selection. After choosing a preliminary model, model checking explores possible lack of fit. Section 5.2 presents goodness-of-fit tests and diagnostics, such as residuals, for doing this. In practice, large-sample methods of inference are not always appropriate. Section 5.3 discusses how parameter estimates can be infinite with small or unbalanced samples, and Section 5.4 presents small-sample inference methods. Section 5.5 addresses power and sample size determination for logistic regression.

5.1 STRATEGIES IN MODEL SELECTION

For a given data set with a binary response, how do we select a logistic regression model? The same issues arise as with ordinary regression. The selection process becomes more challenging as the number of explanatory variables increases, because of the rapid increase in possible effects and interactions. There are two competing goals: The model should be complex enough to fit the data well, but simpler models are easier to interpret.

Most studies are designed to answer certain questions, which motivates including certain terms in the model. To answer those questions, confirmatory analyses use a restricted set of models. A study's theory about an effect may be tested by comparing models with and without that effect. In the absence of underlying theory, some studies are *exploratory* rather than *confirmatory*. Then, a search among many models may provide clues about which predictors are associated with the response and suggest questions for future research.

An Introduction to Categorical Data Analysis, Second Edition. By Alan Agresti

5.1.1 How Many Predictors Can You Use?

Data are unbalanced on Y if $y = 1$ occurs relatively few times or if $y = 0$ occurs relatively few times. This limits the number of predictors for which effects can be estimated precisely. One guideline[1] suggests there should ideally be at least 10 outcomes of each type for every predictor. For example, if $y = 1$ only 30 times out of $n = 1000$ observations, the model should have no more than about three predictors even though the overall sample size is large.

This guideline is approximate. When not satisfied, software still fits the model. In practice, often the number of variables is large, sometimes even of similar magnitude as the number of observations. However, when the guideline is violated, ML estimates may be quite biased and estimates of standard errors may be poor. From results to be discussed in Section 5.3.1, as the number of model predictors increases, it becomes more likely that some ML model parameter estimates are infinite.

Cautions that apply to building ordinary regression models hold for any GLM. For example, models with several predictors often suffer from *multicollinearity* – correlations among predictors making it seem that no one variable is important when all the others are in the model. A variable may seem to have little effect because it overlaps considerably with other predictors in the model, itself being predicted well by the other predictors. Deleting such a redundant predictor can be helpful, for instance to reduce standard errors of other estimated effects.

5.1.2 Example: Horseshoe Crabs Revisited

The horseshoe crab data set of Table 3.2 analyzed in Sections 3.3.2, 4.1.3, and 4.4.1 has four predictors: color (four categories), spine condition (three categories), weight, and width of the carapace shell. We now fit logistic regression models using all these to predict whether the female crab has satellites (males that could mate with her). Again we let $y = 1$ if there is at least one satellite, and $y = 0$ otherwise.

Consider a model with all the main effects. Let $\{c_1, c_2, c_3\}$ be indicator variables for the first three (of four) colors and let $\{s_1, s_2\}$ be indicator variables for the first two (of three) spine conditions. The model

$$\text{logit}[P(Y = 1)] = \alpha + \beta_1\text{weight} + \beta_2\text{width} + \beta_3 c_1 + \beta_4 c_2 + \beta_5 c_3 + \beta_6 s_1 + \beta_7 s_2$$

treats color and spine condition as nominal-scale factors. Table 5.1 shows the results.

A likelihood-ratio test that Y is jointly independent of these predictors simultaneously tests H_0: $\beta_1 = \cdots = \beta_7 = 0$. The test statistic is $-2(L_0 - L_1) = 40.6$ with $df = 7$ ($P < 0.0001$). This shows extremely strong evidence that at least one predictor has an effect.

Although this overall test is highly significant, the Table 5.1 results are discouraging. The estimates for weight and width are only slightly larger than their

[1] See P. Peduzzi et al., *J. Clin. Epidemiol.*, **49**: 1373–1379, 1996.

Table 5.1. Parameter Estimates for Main Effects Model with Horseshoe Crab Data

Parameter	Estimate	*SE*
Intercept	−9.273	3.838
Color(1)	1.609	0.936
Color(2)	1.506	0.567
Color(3)	1.120	0.593
Spine(1)	−0.400	0.503
Spine(2)	−0.496	0.629
Weight	0.826	0.704
Width	0.263	0.195

SE values. The estimates for the factors compare each category to the final one as a baseline. For color, the largest difference is less than two standard errors. For spine condition, the largest difference is less than a standard error.

The small P-value for the overall test yet the lack of significance for individual effects is a warning sign of multicollinearity. Section 4.2.3 showed strong evidence of a width effect. Controlling for weight, color, and spine condition, little evidence remains of a width effect. However, weight and width have a strong correlation (0.887). For practical purposes they are equally good predictors, but it is nearly redundant to use them both. Our further analysis uses width (W) with color (C) and spine condition (S) as predictors.

For simplicity below, we symbolize models by their highest-order terms, regarding C and S as factors. For instance, $(C + S + W)$ denotes the model with main effects, whereas $(C + S * W)$ denotes the model with those main effects plus an $S \times W$ interaction. It is not sensible to use a model with interaction but not the main effects that make up that interaction. A reason for including lower-order terms is that, otherwise, the statistical significance and practical interpretation of a higher-order term depends on how the variables are coded. This is undesirable. By including all the lower-order effects that make up an interaction, the same results occur no matter how variables are coded.

5.1.3 Stepwise Variable Selection Algorithms

As in ordinary regression, algorithms can select or delete predictors from a model in a stepwise manner. In exploratory studies, such model selection methods can be informative if used cautiously. Forward selection adds terms sequentially until further additions do not improve the fit. Backward elimination begins with a complex model and sequentially removes terms. At a given stage, it eliminates the term in the model that has the largest P-value in the test that its parameters equal zero. We test only the highest-order terms for each variable. It is inappropriate, for instance, to remove a main effect term if the model contains higher-order interactions involving that term. The process stops when any further deletion leads to a significantly poorer fit.

With either approach, for categorical predictors with more than two categories, the process should consider the entire variable at any stage rather than just individual indicator variables. Otherwise, the result depends on how you choose the baseline category for the indicator variables. Add or drop the entire variable rather than just one of its indicators.

Variable selection methods need not yield a meaningful model. Use them with caution! When you evaluate many terms, one or two that are not truly important may look impressive merely due to chance.

In any case, statistical significance should not be the sole criterion for whether to include a term in a model. It is sensible to include a variable that is important for the purposes of the study and report its estimated effect even if it is not statistically significant. Keeping it in the model may help reduce bias in estimating effects of other predictors and may make it possible to compare results with other studies where the effect is significant (perhaps because of a larger sample size). Likewise, with a very large sample size sometimes a term might be statistically significant but not practically significant. You might then exclude it from the model because the simpler model is easier to interpret – for example, when the term is a complex interaction.

5.1.4 Example: Backward Elimination for Horseshoe Crabs

When one model is a special case of another, we can test the null hypothesis that the simpler model is adequate against the alternative hypothesis that the more complex model fits better. According to the alternative, at least one of the extra parameters in the more complex model is nonzero. Recall that the *deviance* of a GLM is the likelihood-ratio statistic for comparing the model to the saturated model, which has a separate parameter for each observation (Section 3.4.3). As Section 3.4.4 showed, the likelihood-ratio test statistic $-2(L_0 - L_1)$ for comparing the models is the difference between the deviances for the models. This test statistic has an approximate chi-squared null distribution.

Table 5.2 summarizes results of fitting and comparing several logistic regression models. To select a model, we use a modified *backward elimination* procedure. We start with a complex model, check whether the interaction terms are needed, and then successively take out terms.

We begin with model (1) in Table 5.2, symbolized by $C * S + C * W + S * W$. It contains all the two-factor interactions and main effects. We test all the interactions simultaneously by comparing it to model (2) containing only the main effects. The likelihood-ratio statistic equals the difference in deviances, which is $186.6 - 173.7 = 12.9$, with $df = 166 - 155 = 11$. This does not suggest that the interactions terms are needed ($P = 0.30$). If they were, we could check individual interactions to see whether they could be eliminated (see Problem 5.3).

The next stage considers dropping a term from the main effects model. Table 5.2 shows little consequence from removing spine condition S (model 3c). Both remaining variables (C and W) then have nonnegligible effects. For instance, removing C increases the deviance (comparing models 4b and 3c) by 7.0 on $df = 3$ ($P = 0.07$). The analysis in Section 4.4.3 revealed a noticeable difference between dark crabs

Table 5.2. Results of Fitting Several Logistic Regression Models to Horseshoe Crab Data

Model	Predictors	Deviance	df	AIC	Models Compared	Deviance Difference
1	$C * S + C * W + S * W$	173.7	155	209.7	–	
2	$C + S + W$	186.6	166	200.6	(2)–(1)	12.9 ($df = 11$)
3a	$C + S$	208.8	167	220.8	(3a)–(2)	22.2 ($df = 1$)
3b	$S + W$	194.4	169	202.4	(3b)–(2)	7.8 ($df = 3$)
3c	$C + W$	187.5	168	197.5	(3c)–(2)	0.9 ($df = 2$)
4a	C	212.1	169	220.1	(4a)–(3c)	24.6 ($df = 1$)
4b	W	194.5	171	198.5	(4b)–(3c)	7.0 ($df = 3$)
5	$C = \text{dark} + W$	188.0	170	194.0	(5)–(3c)	0.5 ($df = 2$)
6	None	225.8	172	227.8	(6)–(5)	37.8 ($df = 2$)

Note: C = color, S = spine condition, W = width.

(category 4) and the others. The simpler model that has a single dummy variable for color, equaling 0 for dark crabs and 1 otherwise, fits essentially as well [the deviance difference between models (5) and (3c) equals 0.5, with $df = 2$]. Further simplification results in large increases in the deviance and is unjustified.

5.1.5 AIC, Model Selection, and the "Correct" Model

In selecting a model, you should not think that you have found the "correct" one. Any model is a simplification of reality. For example, you should not expect width to have an *exactly* linear effect on the logit probability of satellites. However, a simple model that fits adequately has the advantages of model parsimony. If a model has relatively little bias, describing reality well, it provides good estimates of outcome probabilities and of odds ratios that describe effects of the predictors.

Other criteria besides significance tests can help select a good model. The best known is the *Akaike information criterion* (AIC). It judges a model by how close its fitted values tend to be to the true expected values, as summarized by a certain expected distance between the two. The optimal model is the one that tends to have its fitted values closest to the true outcome probabilities. This is the model that minimizes

$$\text{AIC} = -2(\text{log likelihood} - \text{number of parameters in model})$$

We illustrate this criterion using the models that Table 5.2 lists. For the model $C + W$, having main effects of color and width, software (PROC LOGISTIC in SAS) reports a -2 log likelihood value of 187.5. The model has five parameters – an intercept and a width effect and three coefficients of dummy variables for color. Thus, AIC $= 187.5 + 2(5) = 197.5$.

Of models in Table 5.2 using some or all of the three basic predictors, AIC is smallest (AIC = 197.5) for $C + W$. The simpler model replacing C by an indicator variable for whether a crab is dark fares better yet (AIC $= 194.0$). Either model seems reasonable. Although the simpler model has lower AIC, that model was suggested by inspecting the parameter estimates for model $C + W$.

The AIC penalizes a model for having many parameters. Even though a simple model is farther than a more complex model from the true relationship, for a sample the simple model may provide better estimates of the true expected values. For example, because the model $\text{logit}[\pi(x)] = \alpha + \beta_1 x + \beta_2 x^2 + \cdots + \beta_{10} x^{10}$ contains the model $\text{logit}[\pi(x)] = \alpha + \beta_1 x$ as a special case, it is closer to the true relationship. If the true relationship is approximately linear, however, with sample data we would get better estimates of $\pi(x)$ by fitting the simpler model.

5.1.6 Summarizing Predictive Power: Classification Tables*

Sometimes it is useful to summarize the predictive power of a binary regression model. One way to do this is with a *classification table*. This cross classifies the binary outcome y with a prediction of whether $y = 0$ or 1. The prediction is $\hat{y} = 1$ when $\hat{\pi}_i > \pi_0$ and $\hat{y} = 0$ when $\hat{\pi}_i \leq \pi_0$, for some cutoff π_0. One possibility is to take $\pi_0 = 0.50$. However, if a low (high) proportion of observations have $y = 1$, the model fit may never (always) have $\hat{\pi}_i > 0.50$, in which case one never (always) predicts $\hat{y} = 1$. Another possibility takes π_0 as the sample proportion of 1 outcomes, which is $\hat{\pi}_i$ for the model containing only an intercept term.

We illustrate for the model using width and color as predictors of whether a horseshoe crab has a satellite. Of the 173 crabs, 111 had a satellite, for a sample proportion of 0.642. Table 5.3 shows classification tables using $\pi_0 = 0.50$ and $\pi_0 = 0.642$.

Table 5.3. Classification Tables for Horseshoe Crab Data

	Prediction, $\pi_0 = 0.64$		Prediction, $\pi_0 = 0.50$		
Actual	$\hat{y} = 1$	$\hat{y} = 0$	$\hat{y} = 1$	$\hat{y} = 0$	Total
$y = 1$	74	37	94	17	111
$y = 0$	20	42	37	25	62

Two useful summaries of predictive power are

$$\text{sensitivity} = P(\hat{y} = 1 \mid y = 1), \quad \text{specificity} = P(\hat{y} = 0 \mid y = 0)$$

Section 2.1.3 introduced these measures for predictions with diagnostic medical tests. When $\pi_0 = 0.642$, from Table 5.3 the estimated sensitivity $= 74/111 = 0.667$ and specificity $= 42/62 = 0.677$.

Another summary of predictor power from the classification table is the overall proportion of correct classifications. This estimates

$$\begin{aligned} P(\text{correct classification}) &= P(y = 1 \text{ and } \hat{y} = 1) + P(y = 0 \text{ and } \hat{y} = 0) \\ &= P(\hat{y} = 1 \mid y = 1)P(y = 1) + P(\hat{y} = 0 \mid y = 0)P(y = 0) \end{aligned}$$

which is a weighted average of sensitivity and specificity. For Table 5.3 with $\pi_0 = 0.64$, the proportion of correct classifications is $(74 + 42)/173 = 0.671$.

A classification table has limitations: It collapses continuous predictive values $\hat{\pi}$ into binary ones. The choice of π_0 is arbitrary. Results are sensitive to the relative numbers of times that $y = 1$ and $y = 0$.

5.1.7 Summarizing Predictive Power: ROC Curves*

A *receiver operating characteristic* (ROC) curve is a plot of sensitivity as a function of (1 – specificity) for the possible cutoffs π_0. An ROC curve is more informative than a classification table, because it summarizes predictive power for all possible π_0. When π_0 gets near 0, almost all predictions are $\hat{y} = 1$; then, sensitivity is near 1, specificity is near 0, and the point for (1 – specificity, sensitivity) has coordinates near (1, 1). When π_0 gets near 1, almost all predictions are $\hat{y} = 0$; then, sensitivity is near 0, specificity is near 1, and the point for (1 – specificity, sensitivity) has coordinates near (0, 0). The ROC curve usually has a concave shape connecting the points (0, 0) and (1, 1).

For a given specificity, better predictive power correspond to higher sensitivity. So, the better the predictive power, the higher the ROC curve. Figure 5.1 shows how SAS (PROC LOGISTIC) reports the ROC curve for the model for the horseshoe

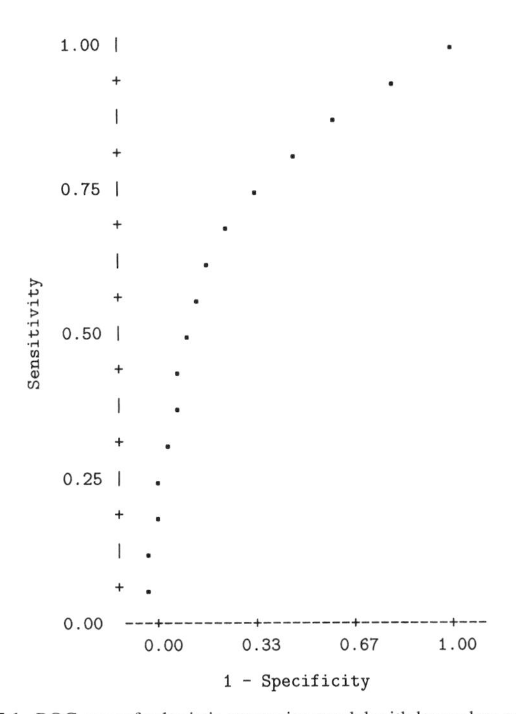

Figure 5.1. ROC curve for logistic regression model with horseshoe crab data.

crabs using width and color as predictors. When $\pi_0 = 0.642$, specificity $= 0.68$, sensitivity $= 0.67$, and the point plotted has coordinates (0.32, 0.67).

The area under the ROC curve is identical to the value of a measure of predictive power called the *concordance index*. Consider all pairs of observations (i, j) such that $y_i = 1$ and $y_j = 0$. The concordance index c estimates the probability that the predictions and the outcomes are *concordant*, which means that the observation with the larger y also has the larger $\hat{\pi}$. A value $c = 0.50$ means predictions were no better than random guessing. This corresponds to a model having only an intercept term. Its ROC curve is a straight line connecting the points (0, 0) and (1, 1). For the horseshoe crab data, $c = 0.639$ with color alone as a predictor, 0.742 with width alone, 0.771 with width and color, and 0.772 with width and an indicator for whether a crab has dark color.

5.1.8 Summarizing Predictive Power: A Correlation*

For a GLM, a way to summarize prediction power is by the correlation R between the observed responses $\{y_i\}$ and the model's fitted values $\{\hat{\mu}_i\}$. For least squares regression, R represents the *multiple correlation* between the response variable and the predictors. Then, R^2 describes the proportion of the variation in Y that is explained by the predictors. An advantage of R compared with R^2 is that it uses the original scale and it has value approximately proportional to the effect size (for instance, with a single predictor, the correlation is the slope multiplied by the ratio of standard deviations of the two variables).

For a binary regression model, R is the correlation between the n binary $\{y_i\}$ observations (1 or 0 for each) and the estimated probabilities $\{\hat{\pi}_i\}$. The highly discrete nature of Y can suppress the range of possible R values. Also, like any correlation measure, its value depends on the range of values observed for the explanatory variables. Nevertheless, R is useful for comparing fits of different models for the same data.

According to the correlation between observed responses and estimated probabilities for the horseshoe crab data, using color alone does not do nearly as well as using width alone ($R = 0.285$ vs $R = 0.402$). Using both predictors together increases R to 0.452. The simpler model that uses color merely to indicate whether a crab is dark does essentially as well, with $R = 0.447$.

5.2 MODEL CHECKING

For any particular logistic regression model, there is no guarantee that the model fits the data well. We next consider ways of checking the model fit.

5.2.1 Likelihood-Ratio Model Comparison Tests

One way to detect lack of fit uses a likelihood-ratio test to compare the model with more complex ones. A more complex model might contain a nonlinear effect, such as a quadratic term to allow the effect of a predictor to change directions as its value increases. Models with multiple predictors would consider interaction terms. If more complex models do not fit better, this provides some assurance that a chosen model is adequate.

We illustrate for the model in Section 4.1.3 that used x = width alone to predict the probability that a female crab has a satellite,

$$\text{logit}[\pi(x)] = \alpha + \beta x$$

One check compares this model with the more complex model that contains a quadratic term,

$$\text{logit}[\hat{\pi}(x)] = \alpha + \beta_1 x + \beta_2 x^2$$

For that model, $\hat{\beta}_2 = 0.040$ has $SE = 0.046$. There is not much evidence to support adding that term. The likelihood-ratio statistic for testing H_0: $\beta_2 = 0$ equals 0.83 ($df = 1$, P-value $= 0.36$).

The model in Section 4.4.1 used width and color predictors, with three dummy variables for color. Section 4.4.2 noted that an improved fit did not result from adding three cross-product terms for the interaction between width and color in their effects.

5.2.2 Goodness of Fit and the Deviance

A more general way to detect lack of fit searches for *any* way the model fails. A goodness-of-fit test compares the model fit with the data. This approach regards the data as representing the fit of the most complex model possible – the saturated model, which has a separate parameter for each observation.

Denote the working model by M. In testing the fit of M, we test whether *all* parameters that are in the saturated model but not in M equal zero. In GLM terminology, the likelihood-ratio statistic for this test is the deviance of the model (Section 3.4.3). In certain cases, this test statistic has a large-sample chi-squared null distribution.

When the predictors are solely categorical, the data are summarized by counts in a contingency table. For the n_i subjects at setting i of the predictors, multiplying the estimated probabilities of the two outcomes by n_i yields estimated expected frequencies for $y = 0$ and $y = 1$. These are the *fitted values* for that setting. The deviance statistic then has the G^2 form introduced in equation (2.7), namely

$$G^2(M) = 2 \sum \text{observed} \left[\log(\text{observed}/\text{fitted})\right]$$

for all the cells in that table. The corresponding Pearson statistic is

$$X^2(M) = \sum (\text{observed} - \text{fitted})^2/\text{fitted}$$

For a fixed number of settings, when the fitted counts are all at least about 5, $X^2(M)$ and $G^2(M)$ have approximate chi-squared null distributions. The degrees of freedom, called the *residual df* for the model, subtract the number of parameters in the model from the number of parameters in the saturated model. The number of parameters in the saturated model equals the number of settings of the predictors, which is the number of binomial observations for the data in the grouped form of the contingency table. Large $X^2(M)$ or $G^2(M)$ values provide evidence of lack of fit. The P-value is the right-tail probability.

We illustrate by checking the model Section 4.3.2 used for the data on AIDS symptoms ($y = 1$, yes), AZT use, and race, shown again in Table 5.4. Let $x = 1$ for those who took AZT immediately and $x = 0$ otherwise, and let $z = 1$ for whites and $z = 0$ for blacks. The ML fit is

$$\text{logit}(\hat{\pi}) = -1.074 - 0.720x + 0.056z$$

The model assumes homogeneous association (Section 2.7.6), with odds ratio between each predictor and the response the same at each category of the other variable. Is this assumption plausible?

Table 5.4. Development of AIDS Symptoms by AZT Use and Race

		Symptoms		
Race	AZT Use	Yes	No	Total
White	Yes	14	93	107
	No	32	81	113
Black	Yes	11	52	63
	No	12	43	55

For this model fit, white veterans with immediate AZT use had estimated probability 0.150 of developing AIDS symptoms during the study. Since 107 white veterans took AZT, the fitted number developing symptoms is $107(0.150) = 16.0$, and the fitted number not developing symptoms is $107(0.850) = 91.0$. Similarly, one can obtain fitted values for all eight cells in Table 5.4. Substituting these and the cell counts into the goodness-of-fit statistics, we obtain $G^2(M) = 1.38$ and $X^2(M) = 1.39$. The model applies to four binomial observations, one at each of the four combinations of AZT use and race. The model has three parameters, so the residual $df = 4 - 3 = 1$. The small G^2 and X^2 values suggest that the model fits decently ($P = 0.24$).

5.2.3 Checking Fit: Grouped Data, Ungrouped Data, and Continuous Predictors

The beginning of Section 4.2 noted that, with categorical predictors, the data file can have the form of *ungrouped data* or *grouped data*. The ungrouped data are the raw 0 and 1 observations. The grouped data are the totals of successes and failures at each combination of the predictor values. Although the ML estimates of parameters are the same for either form of data, the X^2 and G^2 statistics are not. These goodness-of-fit tests only make sense for the grouped data. The large-sample theory for X^2 and G^2 applies for contingency tables when the fitted counts mostly exceed about 5.

When calculated for logistic regression models fitted with continuous or nearly continuous predictors, the X^2 and G^2 statistics do *not* have approximate chi-squared distributions. How can we check the adequacy of a model for such data? One way creates categories for each predictor (e.g., four categories according to where a value falls relative to the quartiles) and then applies X^2 or G^2 to observed and fitted counts for the grouped data. As the number of explanatory variables increases, however, simultaneous grouping of values for each variable produces a contingency table with a very large number of cells. Most cells then have fitted values that are too small for the chi-squared approximation to be good.

An alternative way of grouping the data forms observed and fitted values based on a partitioning of the estimated probabilities. With 10 groups of equal size, the first pair of observed counts and corresponding fitted counts refers to the $n/10$ observations having the highest estimated probabilities, the next pair refers to the $n/10$ observations having the second decile of estimated probabilities, and so forth. Each group has an observed count of subjects with each outcome and a fitted value for each outcome. The fitted value for an outcome is the sum of the estimated probabilities for that outcome for all observations in that group.

The *Hosmer–Lemeshow test* uses a Pearson test statistic to compare the observed and fitted counts for this partition. The test statistic does not have exactly a limiting chi-squared distribution. However, Hosmer and Lemeshow (2000, pp. 147–156) noted that, when the number of distinct patterns of covariate values (for the original data) is close to the sample size, the null distribution is approximated by chi-squared with $df =$ number of groups -2.

For the fit to the horseshoe crab data of the logistic regression model with width (which is continuous) as the sole predictor, SAS (PROC LOGISTIC) reports that the Hosmer–Lemeshow statistic with 10 groups equals 6.6, with $df = 10 - 2 = 8$. It indicates an adequate fit (P-value $= 0.58$). For the model having width and color as predictors, the Hosmer–Lemeshow statistic equals 4.4 ($df = 8$), again indicating an adequate fit.

5.2.4 Residuals for Logit Models

From a scientific perspective, the approach of comparing a working model to more complex models is more useful than a global goodness-of-fit test. A large goodness-of-fit statistic merely indicates *some* lack of fit, but provides no insight about its nature. Comparing a model to a more complex model, on the other hand, indicates whether lack of fit exists of a particular type. For either approach, when the fit is poor, diagnostic measures describe the influence of individual observations on the model fit and highlight reasons for the inadequacy.

With categorical predictors, we can use residuals to compare observed and fitted counts. This should be done with the grouped form of the data. Let y_i denote the number of "successes" for n_i trials at setting i of the explanatory variables. Let $\hat{\pi}_i$ denote the estimated probability of success for the model fit. Then, the estimated binomial mean $n_i\hat{\pi}_i$ is the fitted number of successes.

For a GLM with binomial random component, the *Pearson residual* (3.9) comparing y_i to its fit is

$$\text{Pearson residual} = e_i = \frac{y_i - n_i\hat{\pi}_i}{\sqrt{[n_i\hat{\pi}_i(1-\hat{\pi}_i)]}}$$

Each Pearson residual divides the difference between an observed count and its fitted value by the estimated binomial standard deviation of the observed count. When n_i is large, e_i has an approximate normal distribution. When the model holds, $\{e_i\}$ has an approximate expected value of zero but a smaller variance than a standard normal variate.

The *standardized residual* divides $(y_i - n_i\hat{\pi}_i)$ by its *SE*,

$$\text{standardized residual} = \frac{y_i - n_i\hat{\pi}_i}{SE} = \frac{y_i - n_i\hat{\pi}_i}{\sqrt{[n_i\hat{\pi}_i(1-\hat{\pi}_i)(1-h_i)]}}$$

The term h_i in this formula is the observation's *leverage*, its element from the diagonal of the so-called *hat matrix*. (Roughly speaking, the hat matrix is a matrix that, when applied to the sample logits, yields the predicted logit values for the model.) The greater an observation's leverage, the greater its potential influence on the model fit.

The standardized residual equals $e_i/\sqrt{(1-h_i)}$, so it is larger in absolute value than the Pearson residual e_i. It *is* approximately standard normal when the model holds. We prefer it. An absolute value larger than roughly 2 or 3 provides evidence of lack of fit. This serves the same purpose as the standardized residual (2.9) defined in Section 2.4.5 for detecting patterns of dependence in two-way contingency tables. It is a special case of the standardized residual presented in Section 3.4.5 for describing lack of fit in GLMs.

When fitted values are very small, we have noted that X^2 and G^2 do not have approximate null chi-squared distributions. Similarly, residuals have limited meaning in that case. For ungrouped binary data and often when explanatory variables are continuous, each $n_i = 1$. Then, y_i can equal only 0 or 1, and a residual can assume only two values and is usually uninformative. Plots of residuals also then have limited use, consisting merely of two parallel lines of dots. The deviance itself is then completely uninformative about model fit. When data can be grouped into sets of observations having common predictor values, it is better to compute residuals for the grouped data than for individual subjects.

5.2.5 Example: Graduate Admissions at University of Florida

Table 5.5 refers to graduate school applications to the 23 departments in the College of Liberal Arts and Sciences at the University of Florida, during the 1997–98 academic year. It cross-classifies whether the applicant was admitted (Y), the applicant's gender (G), and the applicant's department (D). For the n_{ik} applications by gender i in department k, let y_{ik} denote the number admitted and let π_{ik} denote the probability of admission. We treat $\{Y_{ik}\}$ as independent binomial variates for $\{n_{ik}\}$ trials with success probabilities $\{\pi_{ik}\}$.

Other things being equal, one would hope the admissions decision is independent of gender. The model with no gender effect, given department, is

$$\text{logit}(\pi_{ik}) = \alpha + \beta_k^{\text{D}}$$

However, the model may be inadequate, perhaps because a gender effect exists in some departments or because the binomial assumption of an identical probability of admission for all applicants of a given gender to a department is unrealistic. Its goodness-of-fit statistics are $G^2 = 44.7$ and $X^2 = 40.9$, both with $df = 23$. This model fits rather poorly (P-values $= 0.004$ and 0.012).

Table 5.5 also reports standardized residuals for the number of females who were admitted, for this model. For instance, the Astronomy department admitted six females, which was 2.87 standard deviations higher than predicted by the model. Each department has $df = 1$ (the df for independence in a 2×2 table) and only a single nonredundant standardized residual. The standardized residuals are identical

Table 5.5. Table Relating Whether Admitted to Graduate School at Florida to Gender and Department, Showing Standardized Residuals for Model with no Gender Effect

	Females		Males		Std. Res		Females		Males		Std. Res
Dept	Yes	No	Yes	No	(Fem,Yes)	Dept	Yes	No	Yes	No	(Fem,Yes)
anth	32	81	21	41	−0.76	ling	21	10	7	8	1.37
astr	6	0	3	8	2.87	math	25	18	31	37	1.29
chem	12	43	34	110	−0.27	phil	3	0	9	6	1.34
clas	3	1	4	0	−1.07	phys	10	11	25	53	1.32
comm	52	149	5	10	−0.63	poli	25	34	39	49	−0.23
comp	8	7	6	12	1.16	psyc	2	123	4	41	−2.27
engl	35	100	30	112	0.94	reli	3	3	0	2	1.26
geog	9	1	11	11	2.17	roma	29	13	6	3	0.14
geol	6	3	15	6	−0.26	soci	16	33	7	17	0.30
germ	17	0	4	1	1.89	stat	23	9	36	14	−0.01
hist	9	9	21	19	−0.18	zool	4	62	10	54	−1.76
lati	26	7	25	16	1.65						

Note: Thanks to Dr. James Booth for showing me these data.

in absolute value for males and females but of different sign. Astronomy admitted three males, and their standardized residual was −2.87; the number admitted was 2.87 standard deviations lower than predicted.[2]

Departments with large standardized residuals are responsible for the lack of fit. Significantly more females were admitted than the model predicts in the Astronomy and Geography departments, and fewer were admitted in the Psychology department. Without these three departments, the model fits adequately ($G^2 = 24.4$, $X^2 = 22.8$, $df = 20$).

For the complete data, next we consider the model that also has a gender effect. It does not provide an improved fit ($G^2 = 42.4$, $X^2 = 39.0$, $df = 22$), because the departments just described have associations in different directions and of greater magnitude than other departments. This model has an ML estimate of 1.19 for the GY conditional odds ratio: The estimated odds of admission were 19% higher for females than males, given department. By contrast, the marginal table collapsed over department has a GY sample odds ratio of 0.94, the overall odds of admission being 6% lower for females. This illustrates Simpson's paradox (Section 2.7.3), because the conditional association has a different direction than the marginal association.

5.2.6 Influence Diagnostics for Logistic Regression

As in ordinary regression, some observations may have too much influence in determining the parameter estimates. The fit could be quite different if they were deleted. Whenever a residual indicates that a model fits an observation poorly, it can be informative to delete the observation and re-fit the model to the remaining ones. However, a single observation can have a more exorbitant influence in ordinary regression than in logistic regression, since ordinary regression has no bound on the distance of y_i from its expected value.

Several diagnostics describe various aspects of influence. Many of them relate to the effect on certain characteristics of removing the observation from the data set. In logistic regression, the observation could be a single binary response or a binomial response for a set of subjects all having the same predictor values (i.e., grouped data). These diagnostics are algebraically related to an observation's leverage. Influence diagnostics for each observation include:

1. For each model parameter, the change in the parameter estimate when the observation is deleted. This change, divided by its standard error, is called *Dfbeta*.
2. A measure of the change in a joint confidence interval for the parameters produced by deleting the observation. This confidence interval displacement diagnostic is denoted by c.

[2]This is an advantage of standardized residuals. Only one bit of information ($df = 1$) exists about how the data depart from independence, yet the Pearson residuals for males and for females normally do not have the same absolute value.

3. The change in X^2 or G^2 goodness-of-fit statistics when the observation is deleted.

For each diagnostic, the larger the value, the greater the influence. Some software for logistic regression (such as PROC LOGISTIC in SAS) produces them.

5.2.7 Example: Heart Disease and Blood Pressure

Table 5.6 is from an early analysis of data from the Framingham study, a longitudinal study of male subjects in Framingham, Massachusetts. In this analysis, men aged 40–59 were classified on x = blood pressure and y = whether developed heart disease during a 6 year follow-up period. Let π_i be the probability of heart disease for blood pressure category i. The table shows the fit for the linear logit model,

$$\text{logit}(\pi_i) = \alpha + \beta x_i$$

with scores $\{x_i\}$ for blood pressure level. We used scores (111.5, 121.5, 131.5, 141.5, 151.5, 161.5, 176.5, 191.5). The nonextreme scores are midpoints for the intervals of blood pressure.

Table 5.6 also reports standardized residuals and approximations reported by SAS (PROC LOGISTIC) for the *Dfbeta* measure for the coefficient of blood pressure, the confidence interval diagnostic c, the change in X^2 (This is the square of the standardized residual), and the change in G^2 (LR difference). All their values show that deleting the second observation has the greatest effect. One relatively large diagnostic is not surprising, however. With many observations, a small percentage may be large purely by chance.

For these data, the overall fit statistics ($G^2 = 5.9$, $X^2 = 6.3$ with $df = 6$) do not indicate lack of fit. In analyzing diagnostics, we should be cautious about attributing

Table 5.6. Diagnostic Measures for Logistic Regression Models Fitted to Heart Disease Data

Blood Pressure	Sample Size	Observed Disease	Fitted Disease	Standardized Residual	*Dfbeta*	c	Pearson Difference	LR Difference
111.5	156	3	5.2	−1.11	0.49	0.34	1.22	1.39
121.5	252	17	10.6	2.37	−1.14	2.26	5.64	5.04
131.5	284	12	15.1	−0.95	0.33	0.31	0.89	0.94
141.5	271	16	18.1	−0.57	0.08	0.09	0.33	0.34
151.5	139	12	11.6	0.13	0.01	0.00	0.02	0.02
161.5	85	8	8.9	−0.33	−0.07	0.02	0.11	0.11
176.5	99	16	14.2	0.65	0.40	0.26	0.42	0.42
191.5	43	8	8.4	−0.18	−0.12	0.02	0.03	0.03

Source: J. Cornfield, *Fed. Proc.*, **21**(suppl. 11): 58–61, 1962.

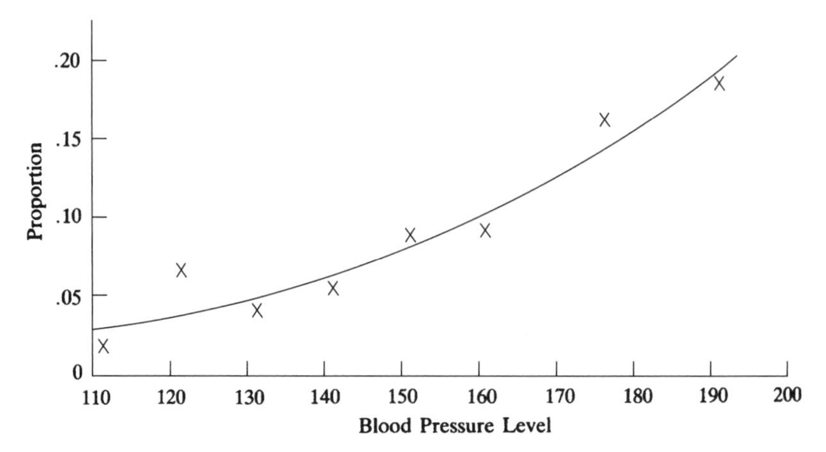

Figure 5.2. Observed proportion (x) and estimated probability of heart disease (curve) for linear logit model.

patterns to what might be chance variation from a model. Also, these deletion diagnostics all relate to removing an entire binomial sample at a blood pressure level instead of removing a single subject's binary observation. Such subject-level deletions have little effect for this model.

Another useful graphical display for showing lack of fit compares observed and fitted proportions by plotting them against each other, or by plotting both of them against explanatory variables. For the linear logit model, Figure 5.2 plots both the observed proportions and the estimated probabilities of heart disease against blood pressure. The fit seems decent.

5.3 EFFECTS OF SPARSE DATA

The log likelihood function for logistic regression models has a concave (bowl) shape. Because of this, the algorithm for finding the ML estimates (Section 3.5.1) usually converges quickly to the correct values. However, certain data patterns present difficulties, with the ML estimates being infinite or not existing. For quantitative or categorical predictors, this relates to observing only successes or only failures over certain ranges of predictor values.

5.3.1 Infinite Effect Estimate: Quantitative Predictor

Consider first the case of a single quantitative predictor. The ML estimate for its effect is infinite when the predictor values having $y = 0$ are completely below or completely above those having $y = 1$, as Figure 5.3 illustrates. In it, $y = 0$ at $x = 10$, 20, 30, 40, and $y = 1$ at $x = 60$, 70, 80, 90. An ideal (perfect) fit has $\hat{\pi} = 0$ for $x \leq 40$ and $\hat{\pi} = 1$ for $x \geq 60$. One can get a sequence of logistic curves that gets closer and closer to this ideal by letting $\hat{\beta}$ increase without limit, with $\hat{\alpha} = -50\hat{\beta}$. (This $\hat{\alpha}$ value yields

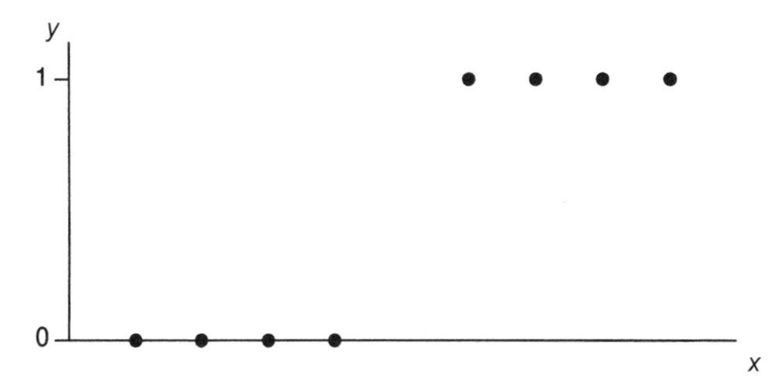

Figure 5.3. Perfect discrimination resulting in infinite logistic regression parameter estimate.

$\hat{\pi} = 0.50$ at $x = 50$.) In fact, the likelihood function keeps increasing as $\hat{\beta}$ increases, and its ML estimate is ∞.

In practice, most software fails to recognize when $\hat{\beta} = \infty$. After a few cycles of the iterative fitting process, the log likelihood looks flat at the working estimate, and convergence criteria are satisfied. Because the log likelihood is so flat and because standard errors of parameter estimates become greater when the curvature is less, software then reports huge standard errors. A danger is that you might not realize when reported estimated effects and results of statistical inferences are invalid. For the data in Figure 5.3, for instance, SAS (PROC GENMOD) reports $\text{logit}(\hat{\pi}) = -192.2 + 3.84x$ with standard errors of 0.80×10^7 and 1.56×10^7. Some software (such as PROC LOGISTIC in SAS) does provide warnings when infinite estimates occur.

With several predictors, consider the multidimensional space for displaying the data. Suppose you could pass a plane through the space of predictor values such that on one side of that plane $y = 0$ for all observations, whereas on the other side $y = 1$ always. There is then *perfect discrimination*: You can predict the sample outcomes perfectly by knowing the predictor values (except possibly at boundary points between the two regions). Again, at least one estimate will be infinite. When the spaces overlap where $y = 1$ and where $y = 0$, the ML estimates are finite.

5.3.2 Infinite Effect Estimate: Categorical Predictors

Infinite estimates also can occur with categorical predictors. A simple case of this is a single binary predictor, so the data are counts in a 2×2 contingency table. The logistic regression model has an indicator variable for the binary predictor. Then, the ML estimate of the effect is the sample log odds ratio. When one of the cell counts is 0, that estimate is plus or minus infinity.

With two or more categorical predictors, the data are counts in a multiway contingency table. When the table has a large number of cells, most cell counts are usually small and many may equal 0. Contingency tables having many cells with small counts are said to be *sparse*. Sparseness is common in contingency tables with many variables or with classifications having several categories.

A cell with a count of 0 is said to be *empty*. Although empty, in the population the cell's true probability is almost always positive. That is, it is theoretically possible to have observations in the cell, and a positive count would occur if the sample size were sufficiently large. To emphasize this, such an empty cell is often called a *sampling zero*.

Depending on the model, sampling zeroes can cause ML estimates of model parameters to be infinite. When all cell counts are positive, all parameter estimates are necessarily finite. When any marginal counts corresponding to terms in a model equal zero, infinite estimates occur for that term. For instance, consider a three-way table with binary predictors X_1 and X_2 for a binary response Y. When a marginal total equals zero in the 2×2 table relating Y to X_1, then the ML estimate of the effect of X_1 in the logistic regression model is infinite.

ML estimates are finite when all the marginal totals corresponding to terms in the model are positive. For example suppose a logit model has main effects for X_1, X_2, and X_3, so the data are counts in a four-way table. The effect of X_1 will be finite if the X_1Y two-way marginal table has positive counts. If there is also an X_1X_2 interaction term, that effect will be finite if the X_1X_2Y three-way marginal table has positive counts.

Empty cells and sparse tables can also cause bias in estimators of odds ratios. As noted at the end of Section 2.3.3, one remedy is first to add $1/2$ to cell counts. However, doing this before fitting a model smooths the data too much, causing havoc with sampling distributions of test statistics. Also, when a ML parameter estimate is infinite, this is not fatal to data analysis. For example, when the ML estimate of an odds ratio is $+\infty$, a likelihood-ratio confidence interval has a finite lower bound. Thus, one can determine how small the true effect may plausibly be.

When your software's fitting processes fail to converge because of infinite estimates, adding a very small constant (such as 10^{-8}) is adequate for ensuring convergence. One can then estimate parameters for which the true estimates are finite and are not affected by the empty cells, as the following example shows. For each possibly influential observation, delete it or move it to another cell to check how much the results vary with small perturbations to the data. Often, some associations are not affected by the empty cells and give stable results for the various analyses, whereas others that are affected are highly unstable. Use caution in making conclusions about an association if small changes in the data are influential. Sometimes it makes sense to fit the model by excluding part of the data containing empty cells, or by combining that part with other parts of the data, or by using fewer predictors so the data are less sparse.

The Bayesian approach to statistical inference typically provides a finite estimate in cases for which an ML estimate is infinite. However, that estimate may depend strongly on the choice of prior distribution. See O'Hagan and Forster (2004) for details.

5.3.3 Example: Clinical Trial with Sparse Data

Table 5.7 shows results of a randomized clinical trial conducted at five centers. The purpose was to compare an active drug to placebo for treating fungal infections

Table 5.7. Clinical Trial Relating Treatment (X) to Response (Y) for Five Centers (Z), with XY and YZ Marginal Tables

		Response (Y)		YZ Marginal	
Center (Z)	Treatment (X)	Success	Failure	Success	Failure
1	Active drug	0	5	0	14
	Placebo	0	9		
2	Active drug	1	12	1	22
	Placebo	0	10		
3	Active drug	0	7	0	12
	Placebo	0	5		
4	Active drug	6	3	8	9
	Placebo	2	6		
5	Active drug	5	9	7	21
	Placebo	2	12		
XY Marginal	Active drug	12	36		
	Placebo	4	42		

Source: Diane Connell, Sandoz Pharmaceuticals Corp.

(1 = success, 0 = failure). For these data, let Y = Response, X = Treatment (Active drug or Placebo), and Z = Center. Centers 1 and 3 had no successes. Thus, the 5×2 marginal table relating center to response, collapsed over treatment, contains zero counts. This marginal table is shown in the last two columns of Table 5.7.

For these data, consider the model

$$\text{logit}[P(Y = 1) = \alpha + \beta x + \beta_k^Z$$

Because centers 1 and 3 had no successes, the ML estimates of the terms β_1^Z and β_3^Z pertaining to their effects equal $-\infty$. The fitted logits for those centers equal $-\infty$, for which the fitted probability of success is 0.

In practice, software notices that the likelihood function continually increases as β_1^Z and β_3^Z decrease toward $-\infty$, but the fitting algorithm may "converge" at large negative values. For example, SAS (PROC GENMOD) reports $\hat{\beta}_1^Z$ and $\hat{\beta}_3^Z$ to both be about -26 with standard errors of about 200,000. Since the software uses default coding that sets $\beta_5^Z = 0$, these estimates refer to contrasts of each center with the last one. If we remove α from the model, then one of $\{\beta_k^Z\}$ is no longer redundant. Each center parameter then refers to that center alone rather than a contrast with another center. Most software permits fitting a model parameterized in this way by using a "no intercept" option. When SAS (PROC GENMOD) does this, $\hat{\beta}_1^Z$ and $\hat{\beta}_3^Z$ are both about -28 with standard errors of about 200,000.

The counts in the 2×2 marginal table relating treatment to response, shown in the bottom panel of Table 5.7, are all positive. The empty cells in Table 5.7 affect the center estimates, but not the treatment estimates, for this model. The estimated log odds ratio equals 1.55 for the treatment effect ($SE = 0.70$). The deviance (G^2) goodness-of-fit statistic equals 0.50 ($df = 4$, $P = 0.97$).

The treatment log odds ratio estimate of 1.55 also results from deleting centers 1 and 3 from the analysis. In fact, when a center has outcomes of only one type, it provides no information about the odds ratio between treatment and response. Such tables also make no contribution to the Cochran–Mantel–Haenszel test (Section 4.3.4) or to a small-sample, exact test of conditional independence between treatment and response (Section 5.4.2).

An alternative strategy in multi-center analyses combines centers of a similar type. Then, if each resulting partial table has responses with both outcomes, the inferences use all data. For estimating odds ratios, however, this usually has little impact. For Table 5.7, perhaps centers 1 and 3 are similar to center 2, since the success rate is very low for that center. Combining these three centers and re-fitting the model to this table and the tables for the other two centers yields an estimated treatment effect of 1.56 ($SE = 0.70$). Centers with no successes or with no failures can be useful for estimating some parameters, such as the difference of proportions, but they do not help us estimate odds ratios for logistic regression models or give us information about whether a treatment effect exists in the population.

5.3.4 Effect of Small Samples on X^2 and G^2 Tests

When a model for a binary response has only categorical predictors, the true sampling distributions of goodness-of-fit statistics are approximately chi-squared, for large sample size n. The adequacy of the chi-squared approximation depends both on n and on the number of cells. It tends to improve as the average number of observations per cell increases.

The quality of the approximation has been studied carefully for the Pearson X^2 test of independence for two-way tables (Section 2.4.3). Most guidelines refer to the fitted values. When $df > 1$, a minimum fitted value of about 1 is permissible as long as no more than about 20% of the cells have fitted values below 5. However, the chi-squared approximation can be poor for sparse tables containing both very small and very large fitted values. Unfortunately, a single rule cannot cover all cases. When in doubt, it is safer to use a small-sample, exact test (Section 2.6.1).

The X^2 statistic tends to be valid with smaller samples and sparser tables than G^2. The distribution of G^2 is usually poorly approximated by chi-squared when n/(number of cells) is less than 5. Depending on the sparseness, P-values based on referring G^2 to a chi-squared distribution can be too large or too small. When most fitted values are smaller than 0.50, treating G^2 as chi-squared gives a highly conservative test; that is, when H_0 is true, reported P-values tend to be much larger than true ones. When most fitted values are between about 0.50 and 5, G^2 tends to be too liberal; the reported P-value tends to be too small.

For fixed values of n and the number of cells, the chi-squared approximation is better for tests with smaller values of *df*. For instance, it is better for a test comparing two models M_0 and M_1 when M_1 has at most a few more parameters than M_0 than it is for testing fit by comparing a model to the saturated model. In the latter case one is testing that *every* parameter equals 0 that could be in the model but is not, and *df* can be large when there are many cells. For models with only main effects, the adequacy of model-comparison tests depends more on the two-way marginal totals relating Y to each predictor than on cell counts. Cell counts can be small as long as most totals in these marginal tables exceed about 5.

When cell counts are so small that chi-squared approximations may be inadequate, one could combine categories of variables to obtain larger counts. This is usually not advisable unless there is a natural way to combine them and little information loss in defining the variable more crudely. In any case, poor sparse-data performance of chi-squared tests is becoming less problematic because of the development of small-sample methods. The following section discusses this for parameters in logistic regression.

5.4 CONDITIONAL LOGISTIC REGRESSION AND EXACT INFERENCE

For inference about logistic regression parameters, the ordinary sampling distributions are *approximately* normal or chi-squared. The approximation improves as the sample size increases. For small samples, it is better to use the *exact* sampling distributions. Methods that find and use the exact distribution are now feasible due to recent advances in computer power and software. For example, StatXact (Cytel Software) conducts many exact inferences for two-way and three-way contingency tables. LogXact (Cytel software) and PROC LOGISTIC (SAS) handle exact inference for logistic regression parameters.

5.4.1 Conditional Maximum Likelihood Inference

The exact inference approach deals with the primary parameters of interest using a *conditional likelihood* function that eliminates the other parameters. The technique uses a conditional probability distribution defined for potential samples that provide the same information about the other parameters that occurs in the observed sample. The distribution and the related conditional likelihood function depend only on the parameters of interest.

The *conditional maximum likelihood estimate* of a parameter is the value at which the conditional likelihood function achieves it maximum. Ordinary ML estimators of parameters work best when the sample size is large compared with the number of model parameters. When the sample size is small, or when there are many parameters relative to the sample size, conditional ML estimates of parameters work better than ordinary ML estimators.

Exact inference for a parameter uses the conditional likelihood function that eliminates *all* the other parameters. Since that conditional likelihood does not involve unknown parameters, probabilities such as P-values use exact distributions rather than approximations. When the sample size is small, conditional likelihood-based exact inference in logistic regression is more reliable than the ordinary large-sample inferences.

5.4.2 Small-Sample Tests for Contingency Tables

Consider first logistic regression with a single explanatory variable,

$$\text{logit}[\pi(x)] = \alpha + \beta x$$

When x takes only two values, the model applies to 2×2 tables of counts $\{n_{ij}\}$ for which the two columns are the levels of Y. The usual sampling model treats the responses on Y in the two rows as independent binomial variates. The row totals, which are the numbers of trials for those binomial variates, are naturally fixed.

For this model, the hypothesis of independence is H_0: $\beta = 0$. The unknown parameter α refers to the relative number of outcomes of $y = 1$ and $y = 0$, which are the column totals. Software eliminates α from the likelihood by conditioning also on the column totals, which are the information in the data about α. Fixing both sets of marginal totals yields a hypergeometric distribution for n_{11}, for which the probabilities do not depend on unknown parameters. The resulting exact test of H_0: $\beta = 0$ is the same as Fisher's exact test (Section 2.6.1).

Next, suppose the model also has a second explanatory factor, Z, with K levels. If Z is nominal-scale, a relevant model is

$$\text{logit}(\pi) = \alpha + \beta x + \beta_k^Z$$

Section 4.3.4 presented this model for $2 \times 2 \times K$ contingency tables. The test of H_0: $\beta = 0$ refers to the effect of X on Y, controlling for Z. The exact test eliminates the other parameters by conditioning on the marginal totals in each partial table. This gives an exact test of conditional independence between X and Y, controlling for Z.

For $2 \times 2 \times K$ tables $\{n_{ijk}\}$, conditional on the marginal totals in each partial table, the Cochran–Mantel–Haenszel test of conditional independence (Section 4.3.4) is a large-sample approximate method that compares $\sum_k n_{11k}$ to its null expected value. Exact tests use $\sum_k n_{11k}$ in the way they use n_{11} in Fisher's exact test for 2×2 tables. Hypergeometric distributions in each partial table determine the probability distribution of $\sum_k n_{11k}$. The P-value for H_a: $\beta > 0$ equals the right-tail probability that $\sum_k n_{11k}$ is at least as large as observed, for the fixed marginal totals. Two-sided alternatives can use a two-tail probability of those outcomes that are no more likely than the observed one.

5.4.3 Example: Promotion Discrimination

Table 5.8 refers to US Government computer specialists of similar seniority considered for promotion from classification level GS-13 to level GS-14. The table cross-classifies promotion decision, considered for three separate months, by employee's race. We test conditional independence of promotion decision and race. The table contains several small counts. The overall sample size is not small ($n = 74$), but one marginal count (collapsing over month of decision) equals zero, so we might be wary of using the *CMH* test.

Table 5.8. Promotion Decisions by Race and by Month

	July Promotions		August Promotions		September Promotions	
Race	Yes	No	Yes	No	Yes	No
Black	0	7	0	7	0	8
White	4	16	4	13	2	13

Source: J. Gastwirth, *Statistical Reasoning in Law and Public Policy*, Academic Press, New York (1988), p. 266.

Let $x = 1$ for black and 0 for white. We first use H_a: $\beta < 0$. This corresponds to potential discrimination against black employees, their probability of promotion being lower than for white employees. Fixing the row and column marginal totals in each partial table, the test uses n_{11k}, the first cell count in each. For the margins of the partial tables in Table 5.8, n_{111} can range between 0 and 4, n_{112} can range between 0 and 4, and n_{113} can range between 0 and 2. The total $\sum_k n_{11k}$ can take values between 0 and 10. The sample data represent the smallest possible count for blacks being promoted in each of the three cases. The observed $\sum_k n_{11k} = 0$.

Because the sample result is the most extreme possible, the conditional ML estimator of the effect of race in the logistic regression model is $\hat{\beta} = -\infty$. Exact conditional methods are still useful when ordinary ML or conditional ML methods report an infinite parameter estimate. The P-value is the null probability of this most extreme outcome, which software (StatXact) reveals to equal 0.026.

A two-sided P-value, based on summing the probabilities of all tables having probabilities no greater than the observed table, equals 0.056. There is some evidence, but not strong, that promotion is associated with race.

5.4.4 Small-Sample Confidence Intervals for Logistic Parameters and Odds Ratios

Software can also construct confidence intervals using exact conditional distributions. The 95% confidence interval for β consists of all values β_0 for which the P-value for H_0: $\beta = \beta_0$ is larger than 0.05 in the exact test.

Consider again Table 5.8 on promotion decisions and race. When $\hat{\beta}$ is infinite, a confidence interval is still useful because it reports a finite bound in the other direction. For these data, StatXact reports an exact 95% confidence interval for β of $(-\infty, 0.01)$. This corresponds to the interval $(e^{-\infty}, e^{0.01}) = (0, 1.01)$ for the true conditional odds ratio in each partial table.

5.4.5 Limitations of Small-Sample Exact Methods*

Although the use of exact distributions is appealing, the conditioning on certain margins can make that distribution highly discrete. Because of this, as Sections 1.4.4 and 2.6.3 discussed, inferences are conservative. When a null hypothesis is true, for instance, the P-value falls below 0.05 no more than 5% of the time, but possibly much less than 5%. For a 95% confidence interval, the true confidence level is at least 0.95, but is unknown.

To alleviate conservativeness, we recommend inference based on the mid P-value. Section 1.4.5 defined the mid P-value to be half the probability of the observed result plus the probability of more extreme results. A 95% confidence interval contains the null hypothesis parameter values having mid P-values exceeding 0.05 in two-sided tests. With mid P-based inference, the actual error probabilities usually more closely match the nominal level than either the exact or large-sample intervals.

Consider the promotion decisions and race example above in Section 5.4.3. For H_a: $\beta < 0$, the ordinary exact P-value of 0.026 was the null probability of the observed value of 0 for $\sum_k n_{11k}$, as this was the most extreme value. The mid P-value is half this, 0.013. Software reports that the mid P 95% confidence interval for the conditional odds ratio is (0, 0.78). This does not contain 1.0, which differs from the interval (0, 1.01) based on the ordinary exact P-value. In summary, we can be *at least* 95% confident that the conditional odds ratio falls in (0, 1.01) and *approximately* 95% confident that it falls in (0, 0.78).

When any predictor is continuous, the discreteness can be so extreme that the exact conditional distribution is degenerate – it is completely concentrated at the observed result. Then, the P-value is 1.0 and the confidence interval contains all possible parameter values. Such results are uninformative. The exact conditional approach is not then useful. Generally, small-sample exact conditional inference works with contingency tables but not with continuous predictors.

5.5 SAMPLE SIZE AND POWER FOR LOGISTIC REGRESSION

The major aim of many studies is to determine whether a particular variable has an effect on a binary response. The study design should determine the sample size needed to provide a good chance of detecting an effect of a given size.

5.5.1 Sample Size for Comparing Two Proportions

Many studies are designed to compare two groups. Consider the hypothesis that the group "success" probabilities π_1 and π_2 are identical. We could conduct a test for the 2×2 table that cross-classifies group by response, rejecting H_0 if the P-value $\leq \alpha$ for some fixed α. To determine sample size, we must specify the probability β of failing to detect a difference between π_1 and π_2 of some fixed size considered to be practically important. For this size of effect, β is the probability of failing to reject H_0 at the α level. Then, $\alpha = P(\text{type I error})$ and $\beta = P(\text{type II error})$. The *power* of the test equals $1 - \beta$.

A study using equal group sample sizes requires approximately

$$n_1 = n_2 = (z_{\alpha/2} + z_\beta)^2[\pi_1(1-\pi_1) + \pi_2(1-\pi_2)]/(\pi_1 - \pi_2)^2$$

This formula requires values for π_1, π_2, α, and β. For testing H_0: $\pi_1 = \pi_2$ at the 0.05 level, suppose we would like $P(\text{type II error}) = 0.10$ if π_1 and π_2 are truly about 0.20 and 0.30. Then $\alpha = 0.05$, $\beta = 0.10$, $z_{0.025} = 1.96$, $z_{0.10} = 1.28$, and we require

$$n_1 = n_2 = (1.96 + 1.28)^2[(0.2)(0.8) + (0.3)(0.7)]/(0.2 - 0.3)^2 = 389$$

This formula also provides the sample sizes needed for a comparable confidence interval for $\pi_1 - \pi_2$. Then, α is the error probability for the interval and β equals the probability that the confidence interval indicates a plausible lack of effect, in the sense that it contains the value zero. Based on the above calculation with $\alpha = 0.05$ and $\beta = 0.10$, we need about 400 subjects in each group for a 95% confidence interval to have only a 0.10 chance of containing 0 when actually $\pi_1 = 0.20$ and $\pi_2 = 0.30$.

This sample-size formula is approximate and tends to underestimate slightly the actual required values. It is adequate for most practical work, because normally conjectures for π_1 and π_2 are only rough. Fleiss et al. (2003, Chapter 4) provided more precise formulas.

The null hypothesis H_0: $\pi_1 = \pi_2$ in a 2×2 table corresponds to one for a parameter in a logistic regression model having the form

$$\text{logit}(\pi) = \beta_0 + \beta_1 x \tag{5.1}$$

where $x = 1$ for group 1 and $x = 0$ for group 2. (We use the β_0 and β_1 notation so as not to confuse these with the error probabilities.) H_0 corresponds to a log odds ratio of 0, or $\beta_1 = 0$. Thus, this example relates to sample size determination for a simple logistic regression model.

5.5.2 Sample Size in Logistic Regression*

For models of form (5.1) in which x is quantitative, the sample size needed for testing H_0: $\beta_1 = 0$ depends on the distribution of the x values. One needs to guess the probability of success $\bar{\pi}$ at the mean of x. The size of the effect is the odds ratio θ comparing $\bar{\pi}$ to the probability of success one standard deviation above the mean of x. Let $\lambda = \log(\theta)$. An approximate sample-size formula for a one-sided test (due to F. Y. Hsieh, *Statist. Med.*, **8**: 795–802, 1989) is

$$n = [z_\alpha + z_\beta \exp(-\lambda^2/4)]^2(1 + 2\bar{\pi}\delta)/(\bar{\pi}\lambda^2)$$

where

$$\delta = [1 + (1 + \lambda^2)\exp(5\lambda^2/4)]/[1 + \exp(-\lambda^2/4)]$$

We illustrate for modeling the dependence of the probability of severe heart disease on x = cholesterol level for a middle-aged population. Consider the test of H_0: $\beta_1 = 0$ against H_a: $\beta_1 > 0$. Suppose previous studies have suggested that $\bar{\pi}$ is about 0.08, and we want the test to be sensitive to a 50% increase (i.e., to 0.12), for a standard deviation increase in cholesterol.

The odds of severe heart disease at the mean cholesterol level equal $0.08/0.92 = 0.087$, and the odds one standard deviation above the mean equal $0.12/0.88 = 0.136$. The odds ratio equals $\theta = 0.136/0.087 = 1.57$, and $\lambda = \log(1.57) = 0.450$, $\lambda^2 = 0.202$. For $\beta = P(\text{type II error}) = 0.10$ in an $\alpha = 0.05$-level test, $z_\alpha = z_{0.05} = 1.645$, $z_\beta = z_{0.10} = 1.28$. Thus,

$$\delta = [1 + (1.202)\exp(5 \times 0.202/4)]/[1 + \exp(-0.202/4)] = 2.548/1.951 = 1.306$$

and

$$n = [1.645 + 1.28\exp(-0.202/4)]^2(1 + 2(.08)(1.306))/(.08)(0.202) = 612$$

The value n decreases as $\bar{\pi}$ gets closer to 0.50 and as $|\lambda|$ gets farther from the null value of 0. Its derivation assumes that X has approximately a normal distribution.

5.5.3 Sample Size in Multiple Logistic Regression*

A multiple logistic regression model requires larger sample sizes to detect partial effects. Let R denote the multiple correlation between the predictor X of interest and the others in the model. One divides the above formula for n by $(1 - R^2)$. In that formula, $\bar{\pi}$ denotes the probability at the mean value of all the explanatory variables, and the odds ratio refers to the effect of the predictor of interest at the mean level of the others.

We illustrate by continuing the previous example. Consider a test for the effect of cholesterol on severe heart disease, while controlling for blood pressure level. If the correlation between cholesterol and blood pressure levels is 0.40, we need $n \approx 612/[1 - (0.40)^2] = 729$ for detecting the stated partial effect of cholesterol.

These formulas provide, at best, rough ballpark indications of sample size. In most applications, one has only a crude guess for $\bar{\pi}$, θ, and R, and the explanatory variable may be far from normally distributed.

PROBLEMS

5.1 For the horseshoe crab data (available at www.stat.ufl.edu/~aa/intro-cda/appendix.html), fit a model using weight and width as predictors.

a. Report the prediction equation.

b. Conduct a likelihood-ratio test of H_0: $\beta_1 = \beta_2 = 0$. Interpret.

c. Conduct separate likelihood-ratio tests for the partial effects of each variable. Why does neither test show evidence of an effect when the test in (**b**) shows very strong evidence?

5.2 For the horseshoe crab data, use a stepwise procedure to select a model for the probability of a satellite when weight, spine condition, and color (nominal scale) are the predictors. Explain each step of the process.

5.3 For the horseshoe crab data with width, color, and spine as predictors, suppose you start a backward elimination process with the most complex model possible. Denoted by $C * S * W$, it uses main effects for each term as well as the three two-factor interactions and the three-factor interaction. Table 5.9 shows the fit for this model and various simpler models.

a. Conduct a likelihood-ratio test comparing this model to the simpler model that removes the three-factor interaction term but has all the two-factor interactions. Does this suggest that the three-factor term can be removed from the model?

b. At the next stage, if we were to drop one term, explain why we would select model $C * S + C * W$.

c. For the model at this stage, comparing to the model $S + C * W$ results in an increased deviance of 8.0 on $df = 6$ ($P = 0.24$); comparing to the model $W + C * S$ has an increased deviance of 3.9 on $df = 3$ ($P = 0.27$). Which term would you take out?

Table 5.9. Logistic Regression Models for Horseshoe Crab Data

Model	Predictors	Deviance	*df*	AIC
1	$C * S * W$	170.44	152	212.4
2	$C * S + C * W + S * W$	173.68	155	209.7
3a	$C * S + S * W$	177.34	158	207.3
3b	$C * W + S * W$	181.56	161	205.6
3c	$C * S + C * W$	173.69	157	205.7
4a	$S + C * W$	181.64	163	201.6
4b	$W + C * S$	177.61	160	203.6
5	$C + S + W$	186.61	166	200.6

d. Finally, compare the working model at this stage to the main-effects model $C + S + W$. Is it permissible to simplify to this model?

e. Of the models shown in the table, which is preferred according to the AIC?

5.4 Refer to Problem 4.16 on the four scales of the Myers–Briggs (MBTI) personality test. Table 5.10 shows the result of fitting a model using the four scales as predictors of whether a subject drinks alcohol frequently.

a. Conduct a model goodness-of-fit test, and interpret.

b. If you were to simplify the model by removing a predictor, which would you remove? Why?

c. When six interaction terms are added, the deviance decreases to 3.74. Show how to test the hypothesis that none of the interaction terms are needed, and interpret.

Table 5.10. Output for Fitting Model to Myers–Briggs Personality Scales Data of Table 4.13

Criteria For Assessing Goodness Of Fit

Criterion	DF	Value
Deviance	11	11.1491
Pearson Chi-Square	11	10.9756

Analysis of Parameter Estimates

Parameter		Estimate	Standard Error	Like-ratio 95% Conf. Limits		Chi-Square
Intercept		−2.4668	0.2429	−2.9617	−2.0078	103.10
EI	e	0.5550	0.2170	0.1314	0.9843	6.54
SN	s	−0.4292	0.2340	−0.8843	0.0353	3.36
TF	t	0.6873	0.2206	0.2549	1.1219	9.71
JP	j	−0.2022	0.2266	−0.6477	0.2426	0.80

5.5 Refer to the previous exercise. PROC LOGISTIC in SAS reports AIC values of 642.1 for the model with the four main effects and the six interaction terms, 637.5 for the model with only the four binary main effect terms, 644.0 for the model with only TF as a predictor, and 648.8 for the model with no predictors. According to this criterion, which model is preferred? Why?

5.6 Refer to the previous two exercises about MBTI and drinking.

a. The sample proportion who reported drinking alcohol frequently was 0.092. When this is the cutpoint for forming a classification table, sensitivity = 0.53 and specificity = 0.66. Explain what these mean.

b. Using (**a**), show that the sample proportion of correct classifications was 0.65.

c. The concordance index c equals 0.658 for the model with the four main effects and the six interaction terms, 0.640 for the model with only the four main effect terms, and 0.568 for the model with only T/F as a predictor. According to this criterion, which model would you choose (i) if you want to maximize sample predictive power (ii) if you think model parsimony is important?

5.7 From the same survey referred to in Problem 4.16, Table 5.11 cross-classifies whether a person smokes frequently with the four scales of the MBTI personality test. SAS reports model -2 log likelihood values of 1130.23 with only an intercept term, 1124.86 with also the main effect predictors, 1119.87 with also all the two-factor interactions, and 1116.47 with also all the three-factor interactions.

a. Write the model for each case, and show that the numbers of parameters are 1, 5, 11, and 15.

b. According to AIC, which of these four models is preferable?

c. When a classification table for the model containing the four main effect terms uses the sample proportion of frequent smokers of 0.23 as the cutoff, sensitivity $= 0.48$ and specificity $= 0.55$. The area under the ROC curve is $c = 0.55$. Does knowledge of personality type help you predict well whether someone is a frequent smoker? Explain.

Table 5.11. Data on Smoking Frequently and Four Scales of Myers–Briggs Personality Test

Extroversion/Introversion		E				I			
Sensing/iNtuitive		S		N		S		N	
		Smoking Frequently							
Thinking/ Feeling	Judging/ Perceiving	Yes	No	Yes	No	Yes	No	Yes	No
T	J	13	64	6	17	32	108	4	9
	P	11	31	4	14	9	43	9	26
F	J	16	89	6	25	34	104	4	27
	P	19	60	23	57	29	76	22	57

Source: Reproduced with special permission of CPP Inc., Mountain View, CA 94043.

5.8 Refer to the classification table in Table 5.3 with $\pi_0 = 0.50$.

a. Explain how this table was constructed.

b. Estimate the sensitivity and specificity, and interpret.

5.9 Problem 4.1 with Table 4.8 used a labeling index (LI) to predict $\pi =$ the probability of remission in cancer patients.

a. When the data for the 27 subjects are 14 binomial observations (for the 14 distinct levels of LI), the deviance for this model is 15.7 with $df = 12$. Is it appropriate to use this to check the fit of the model? Why or why not?

b. The model that also has a quadratic term for LI has deviance = 11.8. Conduct a test comparing the two models.

c. The model in (**b**) has fit, $\text{logit}(\hat{\pi}) = -13.096 + 0.9625(LI) - 0.0160(LI)^2$, with $SE = 0.0095$ for $\hat{\beta}_2 = -0.0160$. If you know basic calculus, explain why $\hat{\pi}$ is increasing for LI between 0 and 30. Since LI varies between 8 and 38 in this sample, the estimated effect of LI is positive over most of its observed values.

d. For the model with only the linear term, the Hosmer–Lemeshow test statistic = 6.6 with $df = 6$. Interpret.

5.10 For the horseshoe crab data, fit the logistic regression model with x = weight as the sole predictor of the presence of satellites.

a. For a classification table using the sample proportion of 0.642 as the cutoff, report the sensitivity and specificity. Interpret.

b. Form a ROC curve, and report and interpret the area under it.

c. Investigate the model goodness-of-fit using the Hosmer–Lemeshow statistic or some other model-checking approach. Interpret.

d. Inferentially compare the model to the model with x and x^2 as predictors. Interpret.

e. Compare the models in (**d**) using the AIC. Interpret.

5.11 Here is an alternative to the Hosmer–Lemeshow goodness-of-fit test when at least one predictor is continuous: Partition values for the explanatory variables into a set of regions. Add these regions as a predictor in the model by setting up dummy variables for the regions. The test statistic compares the fit of this model to the simpler one, testing that the extra parameters are not needed. Doing this for model (4.11) by partitioning according to the eight width regions in Table 4.11, the likelihood-ratio statistic for testing that the extra parameters are unneeded equals 7.5, based on $df = 7$. Interpret.

5.12 Refer to Table 7.27 in Chapter 7 with opinion about premarital sex as the response variable. Use a process (such as backward elimination) or criterion (such as AIC) to select a model. Interpret the parameter estimates for that model.

5.13 Logistic regression is often applied to large financial databases. For example, *credit scoring* is a method of modeling the influence of predictors on the probability that a consumer is credit worthy. The data archive found under the index at www.stat.uni-muenchen.de for a textbook by L. Fahrmeir and G. Tutz (*Multivariate Statistical Modelling Based on Generalized Linear Models*,

2001) contains such a data set that includes 20 predictors for 1000 observations. Build a model for credit-worthiness, using the predictors running account, duration of credit, payment of previous credits, intended use, gender, and marital status, explaining how you chose a final model.

5.14 Refer to the following artificial data:

x	Number of trials	Number of successes
0	4	1
1	4	2
2	4	4

Denote by M_0 the logistic model with only an intercept term and by M_1 the model that also has x as a linear predictor. Denote the maximized log likelihood values by L_0 for M_0, L_1 for M_1, and L_s for the saturated model. Recall that $G^2(M_i) = -2(L_i - L_s)$, $i = 0, 1$. Create a data file in two ways, entering the data as (i) ungrouped data: 12 individual binary observations, (ii) grouped data: three summary binomial observations each with sample size = 4. The saturated model has 12 parameters for data file (i) but three parameters for data file (ii).

a. Fit M_0 and M_1 for each data file. Report L_0 and L_1 (or $-2L_0$ and $-2L_1$) in each case. Note that they do not depend on the form of data entry.

b. Show that the deviances $G^2(M_0)$ and $G^2(M_1)$ depend on the form of data entry. Why is this? (Hint: They depend on L_s. Would L_s depend on the form of data entry? Why? Thus, you should group the data to use the deviance to check the fit of a model.)

c. Show that the difference between the deviances, $G^2(M_0 \mid M_1)$, does not depend on the form of data entry (because L_s cancels in the difference). Thus, for testing the effect of a predictor, it does not matter how you enter the data.

5.15 According to the *Independent* newspaper (London, March 8, 1994), the Metropolitan Police in London reported 30,475 people as missing in the year ending March 1993. For those of age 13 or less, 33 of 3271 missing males and 38 of 2486 missing females were still missing a year later. For ages 14–18, the values were 63 of 7256 males and 108 of 8877 females; for ages 19 and above, the values were 157 of 5065 males and 159 of 3520 females. Analyze these data, including checking model fit and interpreting parameter estimates. (Thanks to Dr. P. M. E. Altham for showing me these data.)

5.16 In Chapter 4, exercises 4.29, 4.30, 4.31, and 4.32 asked for a data analysis and report. Select one of those analyses, and conduct a goodness-of-fit test for the model you used. Interpret.

5.17 Refer to Table 2.10 on death penalty decisions. Fit a logistic model with the two race predictors.

a. Test the model goodness of fit. Interpret.

b. Report the standardized residuals. Interpret.

c. Interpret the parameter estimates.

5.18 Table 5.12 summarizes eight studies in China about smoking and lung cancer.

a. Fit a logistic model with smoking and study as predictors. Interpret the smoking effect.

b. Conduct a Pearson test of goodness of fit. Interpret.

c. Check residuals to analyze further the quality of fit. Interpret.

Table 5.12. Data for Problem 5.18 on Smoking and Lung Cancer

		Lung Cancer				Lung Cancer	
City	Smoking	Yes	No	City	Smoking	Yes	No
Beijing	Yes	126	100	Harbin	Yes	402	308
	No	35	61		No	121	215
Shanghai	Yes	908	688	Zhengzhou	Yes	182	156
	No	497	807		No	72	98
Shenyang	Yes	913	747	Taiyuan	Yes	60	99
	No	336	598		No	11	43
Nanjing	Yes	235	172	Nanchang	Yes	104	89
	No	58	121		No	21	36

Source: Based on data in Z. Liu, *Int. J. Epidemiol.*, **21**: 197–201, 1992. Reprinted by permission of Oxford University Press.

5.19 Problem 7.9 shows a $2 \times 2 \times 6$ table for $Y =$ whether admitted to graduate school at the University of California, Berkeley.

a. Set up indicator variables and specify the logit model that has department as a predictor (with no gender effect) for $Y =$ whether admitted ($1 =$ yes, $0 =$ no).

b. For the model in (**a**), the deviance equals 21.7 with $df = 6$. What does this suggest about the quality of the model fit?

c. For the model in (**a**), the standardized residuals for the number of females who were admitted are (4.15, 0.50, −0.87, 0.55, −1.00, 0.62) for Departments (1,2,3,4,5,6). Interpret.

d. Refer to (**c**). What would the standardized residual equal for the number of males who were admitted into Department 1? Interpret.

e. When we add a gender effect, the estimated conditional odds ratio between admissions and gender ($1 =$ male, $0 =$ female) is 0.90. The marginal table, collapsed over department, has odds ratio 1.84. Explain how these associations differ so much for these data.

5.20 Refer to Table 2.7 on mother's drinking and infant malformations.

a. Fit the logistic regression model using scores $\{0, 0.5, 1.5, 4, 7\}$ for alcohol consumption. Check goodness of fit.

b. Test independence using the likelihood-ratio test for the model in (**a**). (The trend test of Section 2.5.1 is the score test for this model.)

c. The sample proportion of malformations is much higher in the highest alcohol category because, although it has only one malformation, its sample size is only 38. Are the results sensitive to this single observation? Re-fit the model without it, entering 0 malformations for 37 observations, and compare the results of the likelihood-ratio test. (Because results are sensitive to a single observation, it is hazardous to make conclusions, even though n was extremely large.)

d. Fit the model and conduct the test of independence for all the data using scores $\{1, 2, 3, 4, 5\}$. Compare the results with (**b**). (Results for highly unbalanced data can be sensitive to the choice of scores.)

5.21 In the previous exercise, the table has some small counts, and exact methods have greater validity than large-sample ones. Conduct an exact test of independence using the scores in (**a**).

5.22 For the example in Section 5.3.1, $y = 0$ at $x = 10, 20, 30, 40$, and $y = 1$ at $x = 60, 70, 80, 90$.

a. Explain intuitively why $\hat{\beta} = \infty$ for the model, $\text{logit}(\pi) = \alpha + \beta x$.

b. Report $\hat{\beta}$ and its SE for the software you use.

c. Add two observations at $x = 50$, $y = 1$ for one and $y = 0$ for the other. Report $\hat{\beta}$ and its *SE*. Do you think these are correct? Why?

d. Replace the two observations in (c) by $y = 1$ at $x = 49.9$ and $y = 0$ at $x = 50.1$. Report $\hat{\beta}$ and its *SE*. Do you think these are correct? Why?

5.23 Table 5.13 refers to the effectiveness of immediately injected or $1\frac{1}{2}$-hour-delayed penicillin in protecting rabbits against lethal injection with β-hemolytic streptococci.

a. Let X = delay, Y = whether cured, and Z = penicillin level. Fit the model, $\text{logit}[P(Y = 1)] = \beta x + \beta_k^Z$, deleting an intercept term so each level of Z has its own parameter. Argue that the pattern of 0 cell counts suggests that $\hat{\beta}_1^Z = -\infty$ and $\hat{\beta}_5^Z = \infty$. What does your software report?

b. Using the logit model, conduct the likelihood-ratio test of XY conditional independence. Interpret.

c. Estimate the XY conditional odds ratio. Interpret.

Table 5.13. Data for Problem 5.23 on Penicillin in Rabbits

Penicillin Level	Delay	Response	
		Cured	Died
1/8	None	0	6
	$1\frac{1}{2}$ h	0	5
1/4	None	3	3
	$1\frac{1}{2}$ h	0	6
1/2	None	6	0
	$1\frac{1}{2}$ h	2	4
1	None	5	1
	$1\frac{1}{2}$ h	6	0
4	None	2	0
	$1\frac{1}{2}$ h	5	0

Source: Reprinted with permission from article by N. Mantel, *J. Am. Statist. Assoc.*, **58**: 690–700, 1963.

5.24 In the previous exercise, the small cell counts make large-sample analyses questionnable. Conduct small-sample inference, and interpret.

5.25 Table 5.14 is from a study of nonmetastatic osteosarcoma described in the *LogXact 7* manual (Cytel Software, 2005, p. 171). The response is whether the subject achieved a three-year disease-free interval.

Table 5.14. Data for Problem 5.25

Lymphocytic Infiltration	Sex	Osteoblastic Pathology	Disease-Free	
			Yes	No
High	Female	No	3	0
High	Female	Yes	2	0
High	Male	No	4	0
High	Male	Yes	1	0
Low	Female	No	5	0
Low	Female	Yes	3	2
Low	Male	No	5	4
Low	Male	Yes	6	11

a. Show that each predictor has a significant effect when it is used individually without the other predictors.

b. Try to fit a main-effects logistic regression model containing all three predictors. Explain why the ML estimate for the effect of lymphocytic infiltration is infinite.

c. Using conditional logistic regression, conduct an exact test of the hypothesis of no effect of lymphocytic infiltration, controlling for the other variables. Interpret.

d. Using conditional logistic regression, find a 95% confidence interval for the effect in (**c**). Interpret.

5.26 Table 5.15 describes results from a study in which subjects received a drug and the outcome measures whether the subject became incontinent ($y = 1$, yes; $y = 0$, no). The three explanatory variables are lower urinary tract variables that represent drug-induced physiological changes.

a. Report the prediction equations when each predictor is used separately in logistic regressions.

b. Try to fit a main-effects logistic regression model containing all three predictors. What does your software report for the effects and their standard errors? (The ML estimates are actually $-\infty$ for x_1 and x_2 and ∞ for x_3.)

c. Use conditional logistic regression to find an exact P-value for testing H_0: $\beta_3 = 0$. [The exact distribution is degenerate, and neither ordinary ML or exact conditional ML works with these data. For alternative approaches, see articles by D. M. Potter (*Statist. Med.*, **24**: 693–708, 2005) and G. Heinze and M. Schemper (*Statist. Med.*, **22**: 1409–1419, 2002).]

Table 5.15. Data from Incontinence Study of Problem 5.26

y	x_1	x_2	x_3	y	x_1	x_2	x_3
0	−1.9	−5.3	−43	0	−1.5	3.9	−15
0	−0.1	−5.2	−32	0	0.5	27.5	8
0	0.8	−3.0	−12	0	0.8	−1.6	−2
0	0.9	3.4	1	0	2.3	23.4	14
1	−5.6	−13.1	−1	1	−5.3	−19.8	−33
1	−2.4	1.8	−9	1	−2.3	−7.4	4
1	−2.0	−5.7	−7	1	−1.7	−3.9	13
1	−0.6	−2.4	−7	1	−0.5	−14.5	−12
1	−0.1	−10.2	−5	1	−0.1	−9.9	−11
1	0.4	−17.2	−9	1	0.7	−10.7	−10
1	1.1	−4.5	−15				

Source: D. M. Potter, *Statist. Med.*, **24**: 693–708, 2005.

5.27 About how large a sample is needed to test the hypothesis of equal probabilities so that $P(\text{type II error}) = 0.05$ when $\pi_1 = 0.40$ and $\pi_2 = 0.60$, if the hypothesis is rejected when the P-value is less than 0.01?

5.28 We expect two proportions to be about 0.20 and 0.30, and we want an 80% chance of detecting a difference between them using a 90% confidence interval.

a. Assuming equal sample sizes, how large should they be?

b. Compare the results with the sample sizes required for (i) a 90% interval with power 90%, (ii) a 95% interval with power 80%, and (iii) a 95% interval with power 90%.

5.29 The horseshoe crab x = width values in Table 3.2 have a mean of 26.3 and standard deviation of 2.1. If the true relationship were similar to the fitted equation reported in Section 4.1.3, namely, $\hat{\pi} = \exp(-12.351 + 0.497x)/[1 + \exp(-12.351 + 0.497x)]$, how large a sample yields $P(\text{type II error}) = 0.10$ in an $\alpha = 0.05$-level test of independence? Use the alternative of a positive effect of width on the probability of a satellite.

5.30 The following are true–false questions.

a. A model for a binary response has a continuous predictor. If the model truly holds, the deviance statistic for the model has an asymptotic chi-squared distribution as the sample size increases. It can be used to test model goodness of fit.

b. For the horseshoe crab data, when width or weight is the sole predictor for the probability of a satellite, the likelihood-ratio test of the predictor effect has P-value <0.0001. When both weight and width are in the model, it is possible that the likelihood-ratio tests for the partial effects of width and weight could both have P-values larger than 0.05.

c. For the model, $\text{logit}[\pi(x)] = \alpha + \beta x$, suppose $y = 1$ for all $x \leq 50$ and $y = 0$ for all $x > 50$. Then, the ML estimate $\hat{\beta} = -\infty$.

CHAPTER 6

Multicategory Logit Models

Logistic regression is used to model binary response variables. Generalizations of it model categorical responses with more than two categories. We will now study models for *nominal* response variables in Section 6.1 and for *ordinal* response variables in Sections 6.2 and 6.3. As in ordinary logistic regression, explanatory variables can be categorical and/or quantitative.

At each setting of the explanatory variables, the multicategory models assume that the counts in the categories of Y have a *multinomial* distribution. This generalization of the binomial distribution applies when the number of categories exceeds two (see Section 1.2.2).

6.1 LOGIT MODELS FOR NOMINAL RESPONSES

Let J denote the number of categories for Y. Let $\{\pi_1, \ldots, \pi_J\}$ denote the response probabilities, satisfying $\sum_j \pi_j = 1$. With n independent observations, the probability distribution for the number of outcomes of the J types is the multinomial. It specifies the probability for each possible way the n observations can fall in the J categories. Here, we will not need to calculate such probabilities.

Multicategory logit models simultaneously use all pairs of categories by specifying the odds of outcome in one category instead of another. For models of this section, the order of listing the categories is irrelevant, because the model treats the response scale as *nominal* (unordered categories).

6.1.1 Baseline-Category Logits

Logit models for nominal response variables pair each category with a baseline category. When the last category (J) is the baseline, the *baseline-category logits*

An Introduction to Categorical Data Analysis, Second Edition. By Alan Agresti

are

$$\log\left(\frac{\pi_j}{\pi_J}\right), \quad j = 1, \ldots, J-1$$

Given that the response falls in category j or category J, this is the log odds that the response is j. For $J = 3$, for instance, the model uses $\log(\pi_1/\pi_3)$ and $\log(\pi_2/\pi_3)$.

The baseline-category logit model with a predictor x is

$$\log\left(\frac{\pi_j}{\pi_J}\right) = \alpha_j + \beta_j x, \quad j = 1, \ldots, J-1 \tag{6.1}$$

The model has $J - 1$ equations, with separate parameters for each. The effects vary according to the category paired with the baseline. When $J = 2$, this model simplifies to a single equation for $\log(\pi_1/\pi_2) = \text{logit}(\pi_1)$, resulting in ordinary logistic regression for binary responses.

The equations (6.1) for these pairs of categories determine equations for all other pairs of categories. For example, for an arbitrary pair of categories a and b,

$$\begin{aligned}\log\left(\frac{\pi_a}{\pi_b}\right) &= \log\left(\frac{\pi_a/\pi_J}{\pi_b/\pi_J}\right) = \log\left(\frac{\pi_a}{\pi_J}\right) - \log\left(\frac{\pi_b}{\pi_J}\right)\\ &= (\alpha_a + \beta_a x) - (\alpha_b + \beta_b x)\\ &= (\alpha_a - \alpha_b) + (\beta_a - \beta_b)x \end{aligned} \tag{6.2}$$

So, the equation for categories a and b has the form $\alpha + \beta x$ with intercept parameter $\alpha = (\alpha_a - \alpha_b)$ and with slope parameter $\beta = (\beta_a - \beta_b)$.

Software for multicategory logit models fits all the equations (6.1) *simultaneously*. Estimates of the model parameters have smaller standard errors than when binary logistic regression software fits each component equation in (6.1) separately. For simultaneous fitting, the same parameter estimates occur for a pair of categories no matter which category is the baseline. The choice of the baseline category is arbitrary.

6.1.2 Example: Alligator Food Choice

Table 6.1 comes from a study by the Florida Game and Fresh Water Fish Commission of the foods that alligators in the wild choose to eat. For 59 alligators sampled in Lake George, Florida, Table 6.1 shows the primary food type (in volume) found in the alligator's stomach. Primary food type has three categories: Fish, Invertebrate, and Other. The invertebrates were primarily apple snails, aquatic insects, and crayfish. The "other" category included amphibian, mammal, plant material, stones or other debris, and reptiles (primarily turtles, although one stomach contained the tags of 23 baby alligators that had been released in the lake during the previous year!). The table also shows the alligator length, which varied between 1.24 and 3.89 meters.

Table 6.1. Alligator Size (Meters) and Primary Food Choice,[a] for 59 Florida Alligators

1.24 I	1.30 I	1.30 I	1.32 F	1.32 F	1.40 F	1.42 I	1.42 F
1.45 I	1.45 O	1.47 I	1.47 F	1.50 I	1.52 I	1.55 I	1.60 I
1.63 I	1.65 O	1.65 I	1.65 F	1.65 F	1.68 F	1.70 I	1.73 O
1.78 I	1.78 I	1.78 O	1.80 I	1.80 F	1.85 F	1.88 I	1.93 I
1.98 I	2.03 F	2.03 F	2.16 F	2.26 F	2.31 F	2.31 F	2.36 F
2.36 F	2.39 F	2.41 F	2.44 F	2.46 F	2.56 O	2.67 F	2.72 I
2.79 F	2.84 F	3.25 O	3.28 O	3.33 F	3.56 F	3.58 F	3.66 F
3.68 O	3.71 F	3.89 F					

[a] F = Fish, I = Invertebrates, O = Other.

Source: Thanks to M. F. Delany and Clint T. Moore for these data.

Let Y = primary food choice and x = alligator length. For model (6.1) with $J = 3$, Table 6.2 shows some output (from PROC LOGISTIC in SAS), with "other" as the baseline category. The ML prediction equations are

$$\log(\hat{\pi}_1/\hat{\pi}_3) = 1.618 - 0.110x$$

and

$$\log(\hat{\pi}_2/\hat{\pi}_3) = 5.697 - 2.465x$$

Table 6.2. Computer Output for Baseline-Category Logit Model with Alligator Data

```
          Testing Global Null Hypothesis: BETA = 0

        Test                 Chi-Square   DF   Pf > ChiSq

        Likelihood Ratio       16.8006     2       0.0002
        Score                  12.5702     2       0.0019
        Wald                    8.9360     2       0.0115

          Analysis of Maximum Likelihood Estimates

                                           Standard     Wald
Parameter  choice  DF  Estimate    Error    Chi-Square  Pr > ChiSq

Intercept     F     1    1.6177   1.3073      1.5314      0.2159
Intercept     I     1    5.6974   1.7938     10.0881      0.0015
length        F     1   -0.1101   0.5171      0.0453      0.8314
length        I     1   -2.4654   0.8997      7.5101      0.0061

                    Odds Ratio Estimates

                          Point           95% Wald
        Effect  choice  Estimate   Confidence Limits

        length    F       0.896      0.325      2.468
        length    I       0.085      0.015      0.496
```

By equation (6.2), the estimated log odds that the response is "fish" rather than "invertebrate" equals

$$\log(\hat{\pi}_1/\hat{\pi}_2) = (1.618 - 5.697) + [-0.110 - (-2.465)]x = -4.08 + 2.355x$$

Larger alligators seem relatively more likely to select fish rather than invertebrates.

The estimates for a particular equation are interpreted as in binary logistic regression, conditional on the event that the outcome was one of those two categories. For instance, given that the primary food type is fish or invertebrate, the estimated probability that it is fish increases in length x according to an S-shaped curve. For alligators of length $x + 1$ meters, the estimated odds that primary food type is "fish" rather than "invertebrate" equal $\exp(2.355) = 10.5$ times the estimated odds at length x meters.

The hypothesis that primary food choice is independent of alligator length is H_0: $\beta_1 = \beta_2 = 0$ for model (6.1). The likelihood-ratio test takes twice the difference in log likelihoods between this model and the simpler one without length as a predictor. As Table 6.2 shows, the test statistic equals 16.8, with $df = 2$. The P-value of 0.0002 provides strong evidence of a length effect.

6.1.3 Estimating Response Probabilities

The multicategory logit model has an alternative expression in terms of the response probabilities. This is

$$\pi_j = \frac{e^{\alpha_j + \beta_j x}}{\sum_h e^{\alpha_h + \beta_h x}}, \quad j = 1, \ldots, J \tag{6.3}$$

The denominator is the same for each probability, and the numerators for various j sum to the denominator. So, $\sum_j \pi_j = 1$. The parameters equal zero in equation (6.3) for whichever category is the baseline in the logit expressions.

The estimates in Table 6.3 contrast "fish" and "invertebrate" to "other" as the baseline category. The estimated probabilities (6.3) of the outcomes (Fish, Invertebrate, Other) equal

$$\hat{\pi}_1 = \frac{e^{1.62 - 0.11x}}{1 + e^{1.62 - 0.11x} + e^{5.70 - 2.47x}}$$

$$\hat{\pi}_2 = \frac{e^{5.70 - 2.47x}}{1 + e^{1.62 - 0.11x} + e^{5.70 - 2.47x}}$$

$$\hat{\pi}_3 = \frac{1}{1 + e^{1.62 - 0.11x} + e^{5.70 - 2.47x}}$$

The "1" term in each denominator and in the numerator of $\hat{\pi}_3$ represents $e^{\hat{\alpha}_3 + \hat{\beta}_3 x}$ for $\hat{\alpha}_3 = \hat{\beta}_3 = 0$ with the baseline category.

Table 6.3. Parameter Estimates and Standard Errors (in parentheses) for Baseline-category Logit Model Fitted to Table 6.1

	Food Choice Categories for Logit	
Parameter	(Fish/Other)	(Invertebrate/Other)
Intercept	1.618	5.697
Length	−0.110 (0.517)	−2.465 (0.900)

For example, for an alligator of the maximum observed length of $x = 3.89$ meters, the estimated probability that primary food choice is "other" equals

$$\hat{\pi}_3 = 1/\{1 + e^{1.62-0.11(3.89)} + e^{5.70-2.47(3.89)}\} = 0.23.$$

Likewise, you can check that $\hat{\pi}_1 = 0.76$ and $\hat{\pi}_2 = 0.005$. Very large alligators apparently prefer to eat fish. Figure 6.1 shows the three estimated response probabilities as a function of alligator length.

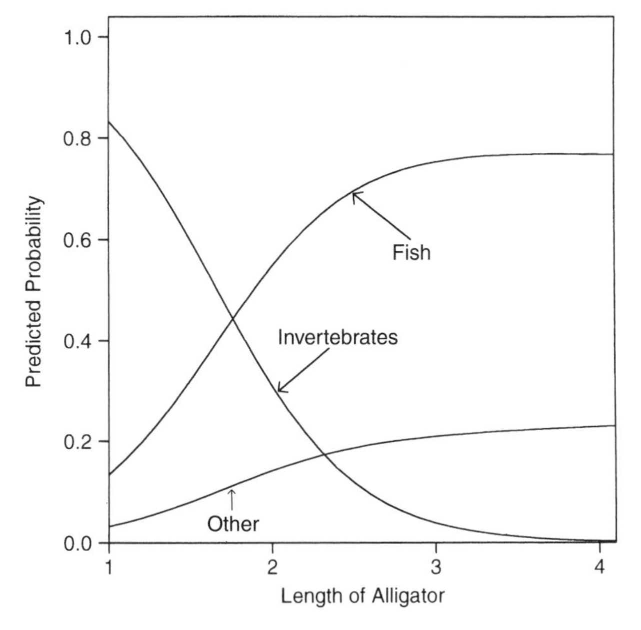

Figure 6.1. Estimated probabilities for primary food choice.

6.1.4 Example: Belief in Afterlife

When explanatory variables are entirely categorical, a contingency table can summarize the data. If the data are not sparse, one can test model goodness of fit using the X^2 or G^2 statistics of Section 5.2.2.

To illustrate, Table 6.4, from a General Social Survey, has Y = belief in life after death, with categories (Yes, Undecided, No), and explanatory variables x_1 = gender and x_2 = race. Let $x_1 = 1$ for females and 0 for males, and $x_2 = 1$ for whites and 0 for blacks. With "no" as the baseline category for Y, the model is

$$\log\left(\frac{\pi_j}{\pi_3}\right) = \alpha_j + \beta_j^{G} x_1 + \beta_j^{R} x_2, \quad j = 1, 2$$

where G and R superscripts identify the gender and race parameters.

Table 6.4. Belief in Afterlife by Gender and Race

Race	Gender	Belief in Afterlife		
		Yes	Undecided	No
White	Female	371	49	74
	Male	250	45	71
Black	Female	64	9	15
	Male	25	5	13

Source: General Social Survey.

For these data, the goodness-of-fit statistics are $X^2 = 0.9$ and $G^2 = 0.8$ (the "deviance"). The sample has two logits at each of four gender–race combinations, for a total of eight logits. The model, considered for $j = 1$ and 2, contains six parameters. Thus, the residual $df = 8 - 6 = 2$. The model fits well.

The model assumes a lack of interaction between gender and race in their effects on belief in life after death. Table 6.5 shows the parameter estimates. The effect parameters represent log odds ratios with the baseline category. For instance, β_1^{G} is

Table 6.5. Parameter Estimates and Standard Errors (in parentheses) for Baseline-category Logit Model Fitted to Table 6.4

Parameter	Belief Categories for logit	
	(Yes/No)	(Undecided/No)
Intercept	0.883 (0.243)	−0.758 (0.361)
Gender ($F = 1$)	0.419 (0.171)	0.105 (0.246)
Race ($W = 1$)	0.342 (0.237)	0.271 (0.354)

Table 6.6. Estimated Probabilities for Belief in Afterlife

		Belief in Afterlife		
Race	Gender	Yes	Undecided	No
White	Female	0.76	0.10	0.15
	Male	0.68	0.12	0.20
Black	Female	0.71	0.10	0.19
	Male	0.62	0.12	0.26

the conditional log odds ratio between gender and response categories 1 and 3 (yes and no), given race. Since $\hat{\beta}_1^G = 0.419$, for females the estimated odds of response "yes" rather than "no" on life after death are $\exp(0.419) = 1.5$ times those for males, controlling for race. For whites, the estimated odds of response "yes" rather than "no" on life after death are $\exp(0.342) = 1.4$ times those for blacks, controlling for gender.

The test of the gender effect has H_0: $\beta_1^G = \beta_2^G = 0$. The likelihood-ratio test compares $G^2 = 0.8$ $(df = 2)$ to $G^2 = 8.0$ $(df = 4)$ obtained by dropping gender from the model. The difference of deviances of $8.0 - 0.8 = 7.2$ has $df = 4 - 2 = 2$. The P-value of 0.03 shows evidence of a gender effect. By contrast, the effect of race is not significant: The model deleting race has $G^2 = 2.8$ $(df = 4)$, which is an increase in G^2 of 2.0 on $df = 2$. This partly reflects the larger standard errors that the effects of race have, due to a much greater imbalance between sample sizes in the race categories than in the gender categories.

Table 6.6 displays estimated probabilities for the three response categories. To illustrate, for white females ($x_1 = x_2 = 1$), the estimated probability of response 1 ("yes") on life after death equals

$$\frac{e^{0.883+0.419(1)+0.342(1)}}{1 + e^{0.883+0.419(1)+0.342(1)} + e^{-0.758+0.105(1)+0.271(1)}} = 0.76$$

6.1.5 Discrete Choice Models

The multicategory logit model is an important tool in marketing research for analyzing how subjects choose among a discrete set of options. For example, for subjects who recently bought an automobile, we could model how their choice of brand depends on the subject's annual income, size of family, level of education, and whether he or she lives in a rural or urban environment.

A generalization of model (6.1) allows the explanatory variables to take different values for different Y categories. For example, the choice of brand of auto would likely depend on price, which varies among the brand options. The generalized model is called a *discrete choice model*.[1]

[1] See Agresti (2002, Section 7.6) and Hensher et al. (2005) for details.

6.2 CUMULATIVE LOGIT MODELS FOR ORDINAL RESPONSES

When response categories are ordered, the logits can utilize the ordering. This results in models that have simpler interpretations and potentially greater power than baseline-category logit models.

A *cumulative probability* for Y is the probability that Y falls at or below a particular point. For outcome category j, the cumulative probability is

$$P(Y \leq j) = \pi_1 + \cdots + \pi_j, \quad j = 1, \ldots, J$$

The cumulative probabilities reflect the ordering, with $P(Y \leq 1) \leq P(Y \leq 2) \leq \cdots \leq P(Y \leq J) = 1$. Models for cumulative probabilities do not use the final one, $P(Y \leq J)$, since it necessarily equals 1.

The logits of the cumulative probabilities are

$$\text{logit}[P(Y \leq j)] = \log\left[\frac{P(Y \leq j)}{1 - P(Y \leq j)}\right] = \log\left[\frac{\pi_1 + \cdots + \pi_j}{\pi_{j+1} + \cdots + \pi_J}\right],$$
$$j = 1, \ldots, J - 1$$

These are called *cumulative logits*. For $J = 3$, for example, models use both $\text{logit}[P(Y \leq 1)] = \log[\pi_1/(\pi_2 + \pi_3)]$ and $\text{logit}[P(Y \leq 2)] = \log[(\pi_1 + \pi_2)/\pi_3]$. Each cumulative logit uses all the response categories.

6.2.1 Cumulative Logit Models with Proportional Odds Property

A model for cumulative logit j looks like a binary logistic regression model in which categories 1–j combine to form a single category and categories $j + 1$ to J form a second category. For an explanatory variable x, the model

$$\text{logit}[P(Y \leq j)] = \alpha_j + \beta x, \quad j = 1, \ldots, J - 1 \tag{6.4}$$

has parameter β describing the effect of x on the log odds of response in category j or below. In this formula, β does not have a j subscript. Therefore, the model assumes that the effect of x is identical for all $J - 1$ cumulative logits. When this model fits well, it requires a single parameter rather than $J - 1$ parameters to describe the effect of x.

Figure 6.2 depicts this model for a four category response and quantitative x. Each cumulative probability has it own curve, describing its change as a function of x. The curve for $P(Y \leq j)$ looks like a logistic regression curve for a binary response with pair of outcomes $(Y \leq j)$ and $(Y > j)$. The common effect β for each j implies that the three curves have the same shape. Any one curve is identical to any of the

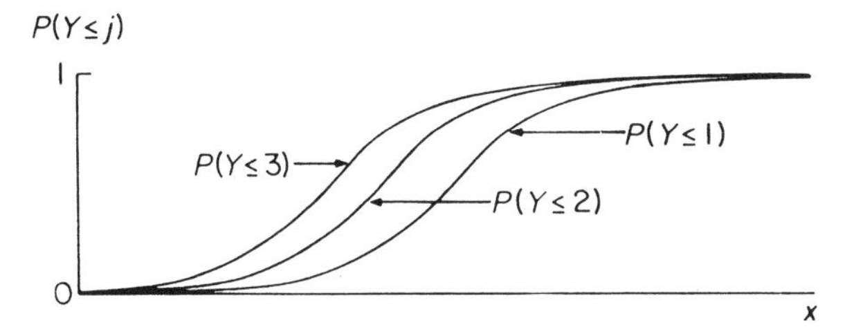

Figure 6.2. Depiction of cumulative probabilities in proportional odds model.

others shifted to the right or shifted to the left. As in logistic regression, the size of $|\beta|$ determines how quickly the curves climb or drop. At any fixed x value, the curves have the same ordering as the cumulative probabilities, the one for $P(Y \leq 1)$ being lowest.

Figure 6.2 has $\beta > 0$. Figure 6.3 shows corresponding curves for the category probabilities, $P(Y = j) = P(Y \leq j) - P(Y \leq j - 1)$. As x increases, the response on Y is more likely to fall at the low end of the ordinal scale. When $\beta < 0$, the curves in Figure 6.2 descend rather than ascend, and the labels in Figure 6.3 reverse order.

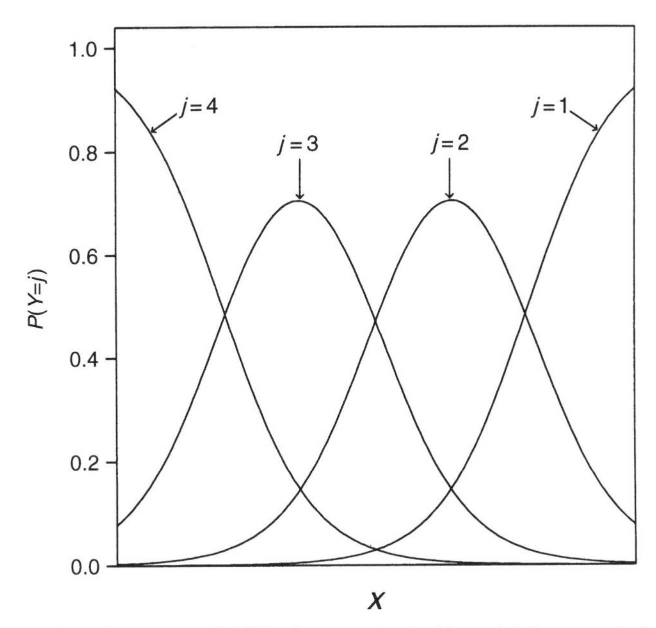

Figure 6.3. Depiction of category probabilities in proportional odds model. At any particular x value, the four probabilities sum to 1.

Then, as x increases, Y is more likely to fall at the high end of the scale.[2] When the model holds with $\beta = 0$, the graph has a horizontal line for each cumulative probability. Then, X and Y are statistically independent.

Model interpretations can use odds ratios for the cumulative probabilities and their complements. For two values x_1 and x_2 of x, an odds ratio comparing the cumulative probabilities is

$$\frac{P(Y \leq j \mid X = x_2)/P(Y > j \mid X = x_2)}{P(Y \leq j \mid X = x_1)/P(Y > j \mid X = x_1)}$$

The log of this odds ratio is the difference between the cumulative logits at those two values of x. This equals $\beta(x_2 - x_1)$, proportional to the distance between the x values. In particular, for $x_2 - x_1 = 1$, the odds of response below any given category multiply by e^{β} for each unit increase in x.

For this log odds ratio $\beta(x_2 - x_1)$, the same proportionality constant (β) applies for each cumulative probability. This property is called the *proportional odds* assumption of model (6.4).

Explanatory variables in cumulative logit models can be quantitative, categorical (with indicator variables), or of both types. The ML fitting process uses an iterative algorithm simultaneously for all j. When the categories are reversed in order, the same fit results but the sign of $\hat{\beta}$ reverses.

6.2.2 Example: Political Ideology and Party Affiliation

Table 6.7, from a General Social Survey, relates political ideology to political party affiliation. Political ideology has a five-point ordinal scale, ranging from very liberal to very conservative. Let x be an indicator variable for political party, with $x = 1$ for Democrats and $x = 0$ for Republicans.

Table 6.7. Political Ideology by Gender and Political Party

		Political Ideology				
Gender	Political Party	Very Liberal	Slightly Liberal	Moderate	Slightly Conservative	Very Conservative
Female	Democratic	44	47	118	23	32
	Republican	18	28	86	39	48
Male	Democratic	36	34	53	18	23
	Republican	12	18	62	45	51

Source: General Social Survey.

Table 6.8 shows output (from PROC LOGISTIC in SAS) for the ML fit of model (6.4). With $J = 5$ response categories, the model has four $\{\alpha_j\}$ intercepts. Usually

[2]The model is sometimes written instead as $\text{logit}[P(Y \leq j)] = \alpha_j - \beta x$, so that $\beta > 0$ corresponds to Y being more likely to fall at the high end of the scale as x increases.

Table 6.8. Computer Output (SAS) for Cumulative Logit Model with Political Ideology Data

Analysis of Maximum Likelihood Estimates

Parameter	DF	Estimate	Standard Error	Wald Chi-Square	Pr > ChiSq
Intercept 1	1	−2.4690	0.1318	350.8122	<.0001
Intercept 2	1	−1.4745	0.1091	182.7151	<.0001
Intercept 3	1	0.2371	0.0948	6.2497	.0124
Intercept 4	1	1.0695	0.1046	104.6082	<.0001
party	1	0.9745	0.1291	57.0182	<.0001

Odds Ratio Estimates

Effect	Point Estimate	95% Wald Confidence Limits	
party	2.650	2.058	3.412

Testing Global Null Hypothesis: BETA = 0

Test	Chi-Square	DF	Pr > ChiSq
Likelihood Ratio	58.6451	1	<.0001
Score	57.2448	1	<.0001
Wald	57.0182	1	<.0001

Deviance and Pearson Goodness-of-Fit Statistics

Criterion	Value	DF	Value/DF	Pr > ChiSq
Deviance	3.6877	3	1.2292	0.2972
Pearson	3.6629	3	1.2210	0.3002

these are not of interest except for estimating response probabilities. The estimated effect of political party is $\hat{\beta} = 0.975$ ($SE = 0.129$). For any fixed j, the estimated odds that a Democrat's response is in the liberal direction rather than the conservative direction (i.e., $Y \leq j$ rather than $Y > j$) equal $\exp(0.975) = 2.65$ times the estimated odds for Republicans. A fairly substantial association exists, with Democrats tending to be more liberal than Republicans.

The model expression for the cumulative probabilities themselves is

$$P(Y \leq j) = \exp(\alpha_j + \beta x)/[1 + \exp(\alpha_j + \beta x)]$$

For example, $\hat{\alpha}_1 = -2.469$, so the first estimated cumulative probability for Democrats ($x = 1$) is

$$\hat{P}(Y \leq 1) = \frac{\exp[-2.469 + 0.975(1)]}{1 + \exp[-2.469 + 0.975(1)]} = 0.18$$

Likewise, substituting $\hat{\alpha}_2$, $\hat{\alpha}_3$, and $\hat{\alpha}_4$ for Democrats yields $\hat{P}(Y \leq 2) = 0.38$, $\hat{P}(Y \leq 3) = 0.77$, and $\hat{P}(Y \leq 4) = 0.89$. Category probabilities are differences of cumulative probabilities. For example, the estimated probability that a Democrat is moderate (category 3) is

$$\hat{\pi}_3 = \hat{P}(Y = 3) = \hat{P}(Y \leq 3) - \hat{P}(Y \leq 2) = 0.39$$

6.2.3 Inference about Model Parameters

For testing independence (H_0: $\beta = 0$), Table 6.8 reports that the likelihood-ratio statistic is 58.6 with $df = 1$. This gives extremely strong evidence of an association ($P < 0.0001$). The test statistic equals the difference between the deviance value for the independence model (which is 62.3, with $df = 4$) and the model allowing a party effect on ideology (which is 3.7, with $df = 3$).

Since it is based on an ordinal model, this test of independence uses the ordering of the response categories. When the model fits well, it is more powerful than the tests of independence presented in Section 2.4 based on $df = (I - 1)(J - 1)$, because it focuses on a restricted alternative and has only a single degree of freedom. The df value is 1 because the hypothesis of independence (H_0: $\beta = 0$) has a single parameter. Capturing an effect with a smaller df value yields a test with greater power (Sections 2.5.3 and 4.4.3). Similar strong evidence results from the Wald test, using $z^2 = (\hat{\beta}/SE)^2 = (0.975/0.129)^2 = 57.1$.

A 95% confidence interval for β is $0.975 \pm 1.96 \times 0.129$, or $(0.72, 1.23)$. The confidence interval for the odds ratio of cumulative probabilities equals $[\exp(0.72), \exp(1.23)]$, or $(2.1, 3.4)$. The odds of being at the liberal end of the political ideology scale is at least twice as high for Democrats as for Republicans. The effect is practically significant as well as statistically significant.

6.2.4 Checking Model Fit

As usual, one way to check a model compares it with models that contain additional effects. For example, the likelihood-ratio test compares the working model to models containing additional predictors or interaction terms.

For a global test of fit, the Pearson X^2 and deviance G^2 statistics compare ML fitted cell counts that satisfy the model to the observed cell counts. When there are at most a few explanatory variables that are all categorical and nearly all the cell counts are at least about 5, these test statistics have approximate chi-squared distributions. For the political ideology data, Table 6.8 shows that $X^2 = 3.7$ and $G^2 = 3.7$, based on $df = 3$. The model fits adequately.

Some software also presents a score test of the proportional odds assumption that the effects are the same for each cumulative probability. This compares model (6.4), which has the same β for each j, to the more complex model having a separate β_j for each j. For these data, this statistic equals 3.9 with $df = 3$, again not showing evidence of lack of fit.

The model with proportional odds form implies that the distribution of Y at one predictor value tends to be higher, or tends to be lower, or tends to be similar, than the distribution of Y at another predictor value. Here, for example, Republicans tend to be higher than Democrats in degree of conservative political ideology. When x refers to two groups, as in Table 6.7, the model does *not* fit well when the response distributions differ in their variability, so such a tendency does not occur. If Democrats tended to be primarily moderate in ideology, while Republicans tended to be both very conservative and very liberal, then the Republicans' responses would show greater variability than the Democrats'. The two ideology distributions would be quite different, but the model would not detect this.

When the model does not fit well, one could use the more general model with separate effects for the different cumulative probabilities. This model replaces β in equation (6.4) with β_j. It implies that curves for different cumulative probabilities climb or fall at different rates, but then those curves cross at certain predictor values. This is inappropriate, because this violates the order that cumulative probabilities must have [such as $P(Y \leq 2) \leq P(Y \leq 3)$ for all x]. Therefore, such a model can fit adequately only over a narrow range of predictor values. Using the proportional odds form of model ensures that the cumulative probabilities have the proper order for all predictor values.

When the model fit is inadequate, another alternative is to fit baseline-category logit models [recall equation (6.1)] and use the ordinality in an informal way in interpreting the associations. A disadvantage this approach shares with the one just mentioned is the increase in the number of parameters. Even though the model itself may have less bias, estimates of measures of interest such as odds ratios or category probabilities may be poorer because of the lack of model parsimony. We do not recommend this approach unless the lack of fit of the ordinal model is severe in a practical sense.

Some researchers collapse ordinal responses to binary so they can use ordinary logistic regression. However, a loss of efficiency occurs in collapsing ordinal scales, in the sense that larger standard errors result. In practice, when observations are spread fairly evenly among the categories, the efficiency loss is minor when you collapse a large number of categories to about four categories. However, it can be severe when you collapse to a binary response. It is usually inadvisable to do this.

6.2.5 Example: Modeling Mental Health

Table 6.9 comes from a study of mental health for a random sample of adult residents of Alachua County, Florida. Mental impairment is ordinal, with categories (well, mild symptom formation, moderate symptom formation, impaired). The study related Y = mental impairment to two explanatory variables. The life events index x_1 is a composite measure of the number and severity of important life events such as birth of child, new job, divorce, or death in family that occurred to the subject within the past three years. In this sample it has a mean of 4.3 and standard deviation of 2.7. Socioeconomic status ($x_2 = SES$) is measured here as binary (1 = high, 0 = low).

Table 6.9. Mental Impairment by SES and Life Events

Subject	Mental Impairment	SES	Life Events	Subject	Mental Impairment	SES	Life Events
1	Well	1	1	21	Mild	1	9
2	Well	1	9	22	Mild	0	3
3	Well	1	4	23	Mild	1	3
4	Well	1	3	24	Mild	1	1
5	Well	0	2	25	Moderate	0	0
6	Well	1	0	26	Moderate	1	4
7	Well	0	1	27	Moderate	0	3
8	Well	1	3	28	Moderate	0	9
9	Well	1	3	29	Moderate	1	6
10	Well	1	7	30	Moderate	0	4
11	Well	0	1	31	Moderate	0	3
12	Well	0	2	32	Impaired	1	8
13	Mild	1	5	33	Impaired	1	2
14	Mild	0	6	34	Impaired	1	7
15	Mild	1	3	35	Impaired	0	5
16	Mild	0	1	36	Impaired	0	4
17	Mild	1	8	37	Impaired	0	4
18	Mild	1	2	38	Impaired	1	8
19	Mild	0	5	39	Impaired	0	8
20	Mild	1	5	40	Impaired	0	9

The main effects model of proportional odds form is

$$\text{logit}[P(Y \leq j)] = \alpha_j + \beta_1 x_1 + \beta_2 x_2$$

Table 6.10 shows SAS output. The estimates $\hat{\beta}_1 = -0.319$ and $\hat{\beta}_2 = 1.111$ suggest that the cumulative probability starting at the "well" end of the scale decreases as life

Table 6.10. Output for Fitting Cumulative Logit Model to Table 6.9

```
          Score Test for the Proportional Odds Assumption
        Chi-Square              DF              Pr > ChiSq
        2.3255                  4                  0.6761

                         Std      Like Ratio 95%      Chi-     Pr >
Parameter    Estimate   Error      Conf Limits       Square   ChiSq
Intercept1   -0.2819   0.6423   -1.5615    0.9839    0.19    0.6607
Intercept2    1.2128   0.6607   -0.0507    2.5656    3.37    0.0664
Intercept3    2.2094   0.7210    0.8590    3.7123    9.39    0.0022
life         -0.3189   0.1210   -0.5718   -0.0920    6.95    0.0084
ses           1.1112   0.6109   -0.0641    2.3471    3.31    0.0689
```

events increases and increases at the higher level of SES. Given the life events score, at the high SES level the estimated odds of mental impairment below any fixed level are $e^{1.111} = 3.0$ times the estimated odds at the low SES level.

For checking fit, the Pearson X^2 and deviance G^2 statistics are valid only for non-sparse contingency tables. They are inappropriate here. Instead, we can check the fit by comparing the model to more complex models. Permitting interaction yields a model with ML fit

$$\text{logit}[\hat{P}(Y \leq j)] = \hat{\alpha}_j - 0.420x_1 + 0.371x_2 + 0.181x_1x_2$$

The coefficient 0.181 of x_1x_2 has $SE = 0.238$. The estimated effect of life events is -0.420 for the low SES group ($x_2 = 0$) and $(-0.420 + 0.181) = -0.239$ for the high SES group ($x_2 = 1$). The impact of life events seems more severe for the low SES group, but the difference in effects is not significant.

An alternative test of fit, presented in Table 6.10, is the score test of the proportional odds assumption. This tests the hypothesis that the effects are the same for each cumulative logit. It compares the model with one parameter for x_1 and one for x_2 to the more complex model with three parameters for each, allowing different effects for logit$[P(Y \leq 1)]$, logit$[P(Y \leq 2)]$, and logit$[P(Y \leq 3)]$. Here, the score statistic equals 2.33. It has $df = 4$, because the more complex model has four additional parameters. The more complex model does not fit significantly better ($P = 0.68$).

6.2.6 Interpretations Comparing Cumulative Probabilities

Section 6.2.1 presented an odds ratio interpretation for the model. An alternative way of summarizing effects uses the cumulative probabilities for Y directly. To describe effects of quantitative variables, we compare cumulative probabilities at their quartiles. To describe effects of categorical variables, we compare cumulative probabilities for different categories. We control for quantitative variables by setting them at their mean. We control for qualitative variables by fixing the category, unless there are several in which case we can set them at the means of their indicator variables. In the binary case, Section 4.5.1 used these interpretations for ordinary logistic regression.

We illustrate with $P(Y \leq 1) = P(Y = 1)$, the *well* outcome, for the mental health data. First, consider the SES effect. At the mean life events of 4.3, $\hat{P}(Y = 1) = 0.37$ at high SES (i.e., $x_2 = 1$) and $\hat{P}(Y = 1) = 0.16$ at low SES ($x_2 = 0$). Next, consider the life events effect. The lower and upper quartiles for life events are 2.0 and 6.5. For high SES, $\hat{P}(Y = 1)$ changes from 0.55 to 0.22 between these quartiles; for low SES, it changes from 0.28 to 0.09. (Comparing 0.55 with 0.28 at the lower quartile and 0.22 with 0.09 at the upper quartile provides further information about the SES effect.) The sample effect is substantial for each predictor.

6.2.7 Latent Variable Motivation*

With the proportional odds form of cumulative logit model, a predictor's effect is the same in the equations for the different cumulative logits. Because each predictor has only a single parameter, it is simpler to summarize and interpret effects than in the baseline-category logit model (6.1).

One motivation for the proportional odds structure relates to a model for an assumed underlying continuous variable. With many ordinal variables, the category labels relate to a subjective assessment. It is often realistic to conceive that the observed response is a crude measurement of an underlying continuous variable. The example in Section 6.2.2 measured political ideology with five categories (very liberal, slightly liberal, moderate, slightly conservative, very conservative). In practice, there are differences in political ideology among people who classify themselves in the same category. With a precise enough way to measure political ideology, it is possible to imagine a continuous measurement. For example, if the underlying political ideology scale has a normal distribution, then a person whose score is 1.96 standard deviations above the mean is more conservative than 97.5% of the population. In statistics, an unobserved variable assumed to underlie what we actually observe is called a *latent variable*.

Let Y^* denote a latent variable. Suppose $-\infty = \alpha_0 < \alpha_1 < \cdots < \alpha_J = \infty$ are *cutpoints* of the continuous scale for Y^* such that the observed response Y satisfies

$$Y = j \quad \text{if } \alpha_{j-1} < Y^* \leq \alpha_j$$

In other words, we observe Y in category j when the latent variable falls in the jth interval of values. Figure 6.4 depicts this. Now, suppose the latent variable Y^* satisfies an ordinary regression model relating its mean to the predictor values. Then,

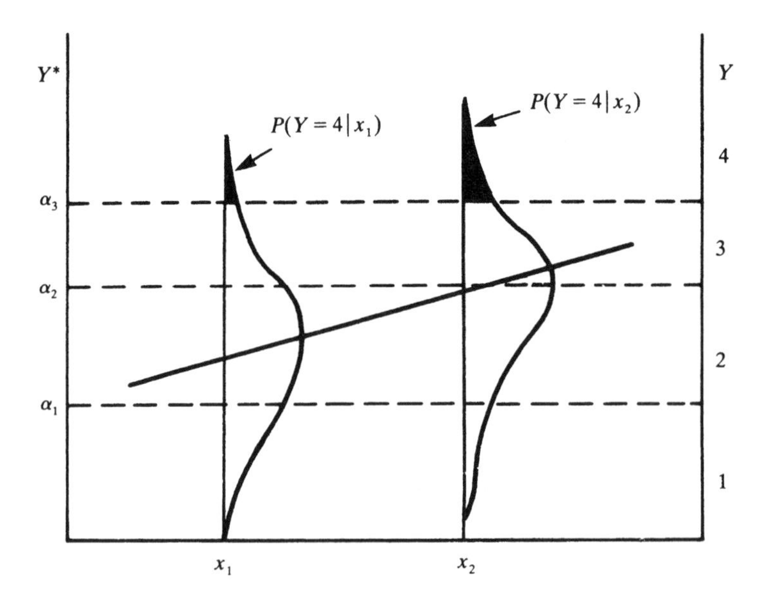

Figure 6.4. Ordinal measurement, and underlying regression model for a latent variable.

one can show[3] that the categorical variable we actually observe satisfies a model with the same linear predictor. Also, the predictor effects are the same for each cumulative probability. Moreover, the shape of the curve for each of the $J - 1$ cumulative probabilities is the same as the shape of the *cdf* of the distribution of Y^*.

At given values of the predictors, suppose Y^* has a normal distribution, with constant variance. Then a probit model holds for the cumulative probabilities. If the distribution of Y^* is the *logistic distribution*, which is bell-shaped and symmetric and nearly identical to the normal, then the cumulative logit model holds with the proportional odds form.

Here is the practical implication of this latent variable connection: If it is plausible to imagine that an ordinary regression model with the chosen predictors describes well the effects for an underlying latent variable, then it is sensible to fit the cumulative logit model with the proportional odds form.

6.2.8 Invariance to Choice of Response Categories

In the connection just mentioned between the model for Y and a model for a latent variable Y^*, the same parameters occur for the effects regardless of how the cutpoints $\{\alpha_j\}$ discretize the real line to form the scale for Y. The effect parameters are *invariant* to the choice of categories for Y.

For example, if a continuous variable measuring political ideology has a linear regression with some predictor variables, then the same effect parameters apply to a discrete version of political ideology with the categories (liberal, moderate, conservative) or (very liberal, slightly liberal, moderate, slightly conservative, very conservative). An implication is this: Two researchers who use different response categories in studying a predictor's effect should reach similar conclusions. If one models political ideology using (very liberal, slightly liberal, moderate, slightly conservative, very conservative) and the other uses (liberal, moderate, conservative), the parameters for the effect of a predictor are roughly the same. Their estimates should be similar, apart from sampling error. This nice feature of the model makes it possible to compare estimates from studies using different response scales.

To illustrate, we collapse Table 6.7 to a three-category response, combining the two liberal categories and combining the two conservative categories. Then, the estimated party affiliation effect changes only from 0.975 ($SE = 0.129$) to 1.006 ($SE = 0.132$). Interpretations are unchanged.

6.3 PAIRED-CATEGORY ORDINAL LOGITS

Cumulative logit models for ordinal responses use the entire response scale in forming each logit. Alternative logits for ordered categories use *pairs* of categories.

[3]For details, see Agresti 2002, pp. 277–279.

6.3.1 Adjacent-Categories Logits

One approach forms logits for all pairs of adjacent categories. The *adjacent-categories logits* are

$$\log\left(\frac{\pi_{j+1}}{\pi_j}\right), \quad j = 1, \ldots, J-1$$

For $J = 3$, these logits are $\log(\pi_2/\pi_1)$ and $\log(\pi_3/\pi_2)$.

With a predictor x, the adjacent-categories logit model has form

$$\log\left(\frac{\pi_{j+1}}{\pi_j}\right) = \alpha_j + \beta_j x, \quad j = 1, \ldots, J-1 \tag{6.5}$$

A simpler proportional odds version of the model is

$$\log\left(\frac{\pi_{j+1}}{\pi_j}\right) = \alpha_j + \beta x, \quad j = 1, \ldots, J-1 \tag{6.6}$$

For it, the effects $\{\beta_j = \beta\}$ of x on the odds of making the higher instead of the lower response are identical for each pair of adjacent response categories. Like the cumulative logit model (6.4) of proportional odds form, this model has a single parameter rather than $J - 1$ parameters for the effect of x. This makes it simpler to summarize an effect.

The adjacent-categories logits, like the baseline-category logits, determine the logits for all pairs of response categories. For the simpler model (6.6), the coefficient of x for the logit, $\log(\pi_a/\pi_b)$, equals $\beta(a - b)$. The effect depends on the distance between categories, so this model recognizes the ordering of the response scale.

6.3.2 Example: Political Ideology Revisited

Let's return to Table 6.7 and model political ideology using the adjacent-categories logit model (6.6) of proportional odds form. Let $x = 0$ for Democrats and $x = 1$ for Republicans.

Software reports that the party affiliation effect is $\hat{\beta} = 0.435$. The estimated odds that a Republican's ideology classification is in category $j + 1$ instead of j are $\exp(\hat{\beta}) = 1.54$ times the estimated odds for Democrats. This is the estimated odds ratio for each of the four 2×2 tables consisting of a pair of adjacent columns of Table 6.7. For instance, the estimated odds of "slightly conservative" instead of "moderate" ideology are 54% higher for Republicans than for Democrats. The estimated odds ratio for an arbitrary pair of columns a and b equals $\exp[\hat{\beta}(a - b)]$. The estimated odds that a Republican's ideology is "very conservative" (category 5) instead of "very liberal" (category 1) are $\exp[0.435(5 - 1)] = (1.54)^4 = 5.7$ times those for Democrats.

The model fit has deviance $G^2 = 5.5$ with $df = 3$, a reasonably good fit. The likelihood-ratio test statistic for the hypothesis that party affiliation has no effect on ideology (H_0: $\beta = 0$) equals the difference between the deviance values for the two models, $62.3 - 5.5 = 56.8$ with $df = 4 - 3 = 1$. There is very strong evidence of an association ($P < 0.0001$). Results are similar to those for the cumulative-logit analysis in Section 6.2.2.

6.3.3 Continuation-Ratio Logits

Another approach forms logits for ordered response categories in a sequential manner. The models apply simultaneously to

$$\log\left(\frac{\pi_1}{\pi_2}\right), \log\left(\frac{\pi_1 + \pi_2}{\pi_3}\right), \ldots, \log\left(\frac{\pi_1 + \cdots + \pi_{J-1}}{\pi_J}\right)$$

These are called *continuation-ratio logits*. They refer to a binary response that contrasts each category with a grouping of categories from *lower* levels of the response scale.

A second type of continuation-ratio logit contrasts each category with a grouping of categories from *higher* levels of the response scale; that is,

$$\log\left(\frac{\pi_1}{\pi_2 + \cdots + \pi_J}\right), \log\left(\frac{\pi_2}{\pi_3 + \cdots + \pi_J}\right), \ldots, \log\left(\frac{\pi_{J-1}}{\pi_J}\right)$$

Models using these logits have different parameter estimates and goodness-of-fit statistics than models using the other continuation-ratio logits.

6.3.4 Example: A Developmental Toxicity Study

Table 6.11 comes from a developmental toxicity study. Rodent studies are commonly used to test and regulate substances posing potential danger to developing fetuses. This study administered diethylene glycol dimethyl ether, an industrial solvent used in the manufacture of protective coatings, to pregnant mice. Each mouse was exposed to one of five concentration levels for 10 days early in the pregnancy. Two days later, the uterine contents of the pregnant mice were examined for defects. Each fetus had the three possible outcomes (Dead, Malformation, Normal). The outcomes are ordered.

We apply continuation-ratio logits to model the probability of a dead fetus, using $\log[\pi_1/(\pi_2 + \pi_3)]$, and the conditional probability of a malformed fetus, given that the fetus was live, using $\log(\pi_2/\pi_3)$. We used scores $\{0, 62.5, 125, 250, 500\}$ for concentration level. The two models are ordinary logistic regression models in which the responses are column 1 and columns 2–3 combined for one fit and column 2 and column 3 for the second fit. The estimated effect of concentration level is 0.0064 ($SE = 0.0004$) for the first logit and 0.0174 ($SE = 0.0012$) for the second logit. In each case, the less desirable outcome is more likely as concentration level increases.

Table 6.11. Outcomes for Pregnant Mice in Developmental Toxicity Study[a]

Concentration (mg/kg per day)	Response		
	Non-live	Malformation	Normal
0 (controls)	15	1	281
62.5	17	0	225
125	22	7	283
250	38	59	202
500	144	132	9

[a]Based on results in C. J. Price et al., *Fund. Appl. Toxicol.*, **8**: 115–126, 1987. I thank Dr. Louise Ryan for showing me these data.

For instance, given that a fetus was live, for every 100-unit increase in concentration level, the estimated odds that it was malformed rather than normal changes by a multiplicative factor of $\exp(100 \times 0.0174) = 5.7$.

When models for different continuation-ratio logits have separate parameters, as in this example, separate fitting of ordinary binary logistic regression models for different logits gives the same results as simultaneous fitting. The sum of the separate deviance statistics is an overall goodness-of-fit statistic pertaining to the simultaneous fitting. For Table 6.11, the deviance G^2 values are 5.8 for the first logit and 6.1 for the second, each based on $df = 3$. We summarize the fit by their sum, $G^2 = 11.8$, based on $df = 6$ ($P = 0.07$).

6.3.5 Overdispersion in Clustered Data

The above analysis treats pregnancy outcomes for different fetuses as independent observations. In fact, each pregnant mouse had a litter of fetuses, and statistical dependence may exist among different fetuses from the same litter. The model also treats fetuses from different litters at a given concentration level as having the same response probabilities. Heterogeneity of various types among the litters (for instance, due to different physical conditions of different pregnant mice) would usually cause these probabilities to vary somewhat among litters. Either statistical dependence or heterogeneous probabilities violates the binomial assumption. They typically cause *overdispersion* – greater variation than the binomial model implies (recall Section 3.3.3).

For example, consider mice having litters of size 10 at a fixed, low concentration level. Suppose the average probability of fetus death is low, but some mice have genetic defects that cause fetuses in their litter to have a high probability of death. Then, the number of fetuses that die in a litter may vary among pregnant mice to a greater degree than if the counts were based on identical probabilities. We might observe death counts (out of 10 fetuses in a litter) such as 1, 0, 1, 10, 0, 2, 10, 1, 0, 10; this is more variability than we expect with binomial variates.

The total G^2 for testing the continuation-ratio model shows some evidence of lack of fit. The structural form chosen for the model may be incorrect. The lack of fit may mainly, however, reflects overdispersion caused by dependence within litters or heterogeneity among litters. Both factors are common in developmental toxicity studies. Chapters 9 and 10 present ways of handling correlated and/or heterogeneous observations and the overdispersion that occurs because of these factors.

6.4 TESTS OF CONDITIONAL INDEPENDENCE

It is often useful to check whether one variable has an effect on another after we control for a third variable. Sections 4.3.1 and 4.3.4 discussed this for a binary response, using logit models and the Cochran–Mantel–Haenszel test. This section shows ways to test the hypothesis of conditional independence in three-way tables when the response variable is multicategory.

6.4.1 Example: Job Satisfaction and Income

Table 6.12, from the 1991 General Social Survey, refers to the relationship between $Y =$ job satisfaction and income, stratified by gender, for black Americans. Let us test the hypothesis of conditional independence using a cumulative logit model. Let $x_1 =$ gender and $x_2 =$ income. We will treat income as quantitative, by assigning scores to its categories. The likelihood-ratio test compares the model

$$\text{logit}[P(Y \leq j)] = \alpha_j + \beta_1 x_1 + \beta_2 x_2, \quad j = 1, 2, 3$$

to the simpler model without an income effect,

$$\text{logit}[P(Y \leq j)] = \alpha_j + \beta_1 x_1, \quad j = 1, 2, 3$$

Table 6.12. Job Satisfaction and Income, Controlling for Gender

		Job Satisfaction			
Gender	Income	Very Dissatisfied	A Little Satisfied	Moderately Satisfied	Very Satisfied
Female	<5000	1	3	11	2
	5000–15,000	2	3	17	3
	15,000–25,000	0	1	8	5
	>25,000	0	2	4	2
Male	<5000	1	1	2	1
	5000–15,000	0	3	5	1
	15,000–25,000	0	0	7	3
	>25,000	0	1	9	6

Source: General Social Survey, 1991.

With grouped continuous variables, it is sensible to use scores that are midpoints of the class intervals. For income, we use scores {3, 10, 20, 35}, which use midpoints of the middle two categories, in thousands of dollars. The model with an income effect has deviance 13.9 ($df = 20$), and the model with no income effect has deviance 19.6 ($df = 19$). The difference between the deviances is 5.7, based on $df = 20 - 19 = 1$. This gives $P = 0.017$ and provides evidence of an association.

This test works well when the association is similar in each partial table, because the model does not have an interaction term. In this sense, it is directed toward an alternative of homogeneous association as that model characterizes the association (i.e., with odds ratios that use the entire response scale, dichotomized by response below vs above any particular point).

A baseline-category logit model treats the response variable as nominal rather than ordinal. It would also treat income as nominal if we used indicator variables for its categories rather than assumed a linear trend. For example let $x_2 = 1$ if income is in the first category, $x_3 = 1$ if income is in the second category, and $x_4 = 1$ if income is in the third category, in each case 0 otherwise. The model is

$$\log\left(\frac{\pi_j}{\pi_4}\right) = \alpha_j + \beta_{j1}x_1 + \beta_{j2}x_2 + \beta_{j3}x_3 + \beta_{j4}x_4$$

For this model, conditional independence of job satisfaction and income is

$$H_0\colon \beta_{j2} = \beta_{j3} = \beta_{j4} = 0, \quad j = 1, 2, 3$$

equating nine parameters equal to 0.

In this case, the difference of deviances equals 12.3, based on $df = 9$, for which the P-value is 0.20. This test has the advantage of not assuming as much about the model structure. A disadvantage is that it often has low power, because the null hypothesis has so many (nine) parameters. If there truly is a trend in the relationship, we are more likely to capture it with the ordinal analysis. In testing that the single association parameter equals 0, that chi-squared test focuses the analysis on $df = 1$.

Alternatively, a model could treat one variable as ordinal and one as nominal.[4] For example, a cumulative logit model could treat a predictor as nominal by using indicator variables for its categories. This would be appropriate if the response variable did not tend to increase regularly or to decrease regularly as the predictor value changed. For these data, since the parsimonious model that treats both variables as ordinal fits well (deviance = 13.9 with $df = 20$), we prefer it to these other models.

6.4.2 Generalized Cochran–Mantel–Haenszel Tests*

Alternative tests of conditional independence generalize the Cochran–Mantel–Haenszel (CMH) statistic (4.9) to $I \times J \times K$ tables. Like the CMH statistic and the

[4] See Section 7.5 of Agresti (2002) for details.

model-based statistics without interaction terms, these statistics perform well when the conditional association is similar in each partial table. There are three versions, according to whether both, one, or neither of Y and the predictor are treated as ordinal.

When both variables are ordinal, the test statistic generalizes the correlation statistic (2.10) for two-way tables. It is designed to detect a linear trend in the association that has the same direction in each partial table. The generalized correlation statistic has approximately a chi-squared distribution with $df = 1$. Its formula is complex and we omit computational details. It is available in standard software (e.g., PROC FREQ in SAS).

For Table 6.12 with the scores $\{3, 10, 20, 35\}$ for income and $\{1, 3, 4, 5\}$ for satisfaction, the sample correlation between income and job satisfaction equals 0.16 for females and 0.37 for males. The generalized correlation statistic equals 6.2 with $df = 1$ $(P = 0.01)$. This gives the same conclusion as the ordinal-model-based likelihood-ratio test of the previous subsection.

When in doubt about scoring, perform a sensitivity analysis by using a few different choices that seem sensible. Unless the categories exhibit severe imbalance in their totals, the choice of scores usually has little impact on the results. With the row and column numbers as the scores, the sample correlation equals 0.17 for females and 0.38 for males, and the generalized correlation statistic equals 6.6 with $df = 1$ $(P = 0.01)$, giving the same conclusion.

6.4.3 Detecting Nominal–Ordinal Conditional Association*

When the predictor is nominal and Y is ordinal, scores are relevant only for levels of Y. We summarize the responses of subjects within a given row by the mean of their scores on Y, and then average this row-wise mean information across the K strata. The test of conditional independence compares the I rows using a statistic based on the variation among the I averaged row mean responses. This statistic is designed to detect differences among their true mean values. It has a large-sample chi-squared distribution with $df = (I - 1)$.

For Table 6.12, this test treats job satisfaction as ordinal and income as nominal. The test searches for differences among the four income levels in their mean job satisfaction. Using scores $\{1, 2, 3, 4\}$, the mean job satisfaction at the four levels of income equal (2.82, 2.84, 3.29, 3.00) for females and (2.60, 2.78, 3.30, 3.31) for males. For instance, the mean for the 17 females with income <5000 equals $[1(1) + 2(3) + 3(11) + 4(2)]/17 = 2.82$. The pattern of means is similar for each gender, roughly increasing as income increases. The generalized CMH statistic for testing whether the true row mean scores differ equals 9.2 with $df = 3$ $(P = 0.03)$. The evidence is not quite as strong as with the fully ordinal analyses above based on $df = 1$.

Unlike this statistic, the correlation statistic of the previous subsection also treats the rows as ordinal. It detects a linear trend across rows in the row mean scores, and it utilizes the approximate increase in mean satisfaction as income increases. One can use the nominal–ordinal statistic when both variables are ordinal but such a linear trend may not occur. For instance, one might expect responses on Y to tend to be higher in some rows than in others, without the mean of Y increasing consistently or decreasing consistently as income increases.

6.4.4 Detecting Nominal–Nominal Conditional Association*

Another *CMH*-type statistic, based on $df = (I-1)(J-1)$, provides a "general association" test. It is designed to detect *any* type of association that is similar in each partial table. It treats the variables as nominal, so it does not require category scores.

For Table 6.12, the general association statistic equals 10.2, with $df = 9$ ($P = 0.34$). We pay a price for ignoring the ordinality of job satisfaction and income. For ordinal variables, the general association test is usually not as powerful as narrower tests with smaller df values that use the ordinality.

Table 6.13 summarizes results of the three generalized *CMH* tests applied to Table 6.12. The format is similar to that used by SAS with the CMH option in PROC FREQ. Normally, we would prefer a model-based approach to a CHM-type test. A model, besides providing significance tests, provides estimates of *sizes* of the effects.

Table 6.13. Summary of Generalized Cochran–Mantel–Haenszel Tests of Conditional Independence for Table 6.12

Alternative Hypothesis	Statistic	*df*	*P*-value
General association	10.2	9	0.34
Row mean scores differ	9.2	3	0.03
Nonzero correlation	6.6	1	0.01

PROBLEMS

6.1 A model fit predicting preference for President (Democrat, Republican, Independent) using x = annual income (in \$10,000 dollars) is $\log(\hat{\pi}_D/\hat{\pi}_I) = 3.3 - 0.2x$ and $\log(\hat{\pi}_R/\hat{\pi}_I) = 1.0 + 0.3x$.

a. State the prediction equation for $\log(\hat{\pi}_R/\hat{\pi}_D)$. Interpret its slope.

b. Find the range of x for which $\hat{\pi}_R > \hat{\pi}_D$.

c. State the prediction equation for $\hat{\pi}_I$.

6.2 Refer to the alligator food choice example in Section 6.1.2.

a. Using the model fit, estimate an odds ratio that describes the effect of length on primary food choice being either "invertebrate" or "other."

b. Estimate the probability that food choice is invertebrate, for an alligator of length 3.9 meters.

c. Estimate the length at which the outcomes "invertebrate" and "other" are equally likely.

6.3 Table 6.14 displays primary food choice for a sample of alligators, classified by length (≤ 2.3 meters, >2.3 meters) and by the lake in Florida in which they were caught.

a. Fit a model to describe effects of length and lake on primary food choice. Report the prediction equations.

b. Using the fit of your model, estimate the probability that the primary food choice is "fish," for each length in Lake Oklawaha. Interpret the effect of length.

Table 6.14. Data on Alligators for Exercise 6.3

		Primary Food Choice				
Lake	Size	Fish	Invertebrate	Reptile	Bird	Other
Hancock	≤ 2.3	23	4	2	2	8
	>2.3	7	0	1	3	5
Oklawaha	≤ 2.3	5	11	1	0	3
	>2.3	13	8	6	1	0
Trafford	≤ 2.3	5	11	2	1	5
	>2.3	8	7	6	3	5
George	≤ 2.3	16	19	1	2	3
	>2.3	17	1	0	1	3

Source: Wildlife Research Laboratory, Florida Game and Fresh Water Fish Commission.

6.4 Refer to the belief in afterlife example in Section 6.1.4.

a. Estimate the probability of response "yes" for black females.

b. Describe the gender effect by reporting and interpreting the estimated conditional odds ratio for the (i) "undecided" and "no" pair of response categories, (ii) "yes" and "undecided" pair.

6.5 For a recent General Social Survey, a prediction equation relating Y = job satisfaction (four ordered categories; 1 = the least satisfied) to the subject's report of x_1 = earnings compared with others with similar positions (four ordered categories; 1 = much less, 4 = much more), x_2 = freedom to make decisions about how to do job (four ordered categories; 1 = very true, 4 = not at all true), and x_3 = work environment allows productivity (four ordered categories; 1 = strongly agree, 4 = strongly disagree), was $\text{logit}[\hat{P}(Y \leq j)] = \hat{\alpha}_j - 0.54x_1 + 0.60x_2 + 1.19x_3$.

a. Summarize each partial effect by indicating whether subjects tend to be more satisfied, or less satisfied, as (i) x_1, (ii) x_2, (iii) x_3, increases.

b. Report the settings for x_1, x_2, x_3 at which a subject is most likely to have highest job satisfaction.

6.6 Does marital happiness depend on family income? For the 2002 General Social Survey, counts in the happiness categories (not, pretty, very) were (6, 43, 75) for below average income, (6, 113, 178) for average income, and (6, 57, 117) for above average income. Table 6.15 shows output for a baseline-category logit model with very happy as the baseline category and scores {1, 2, 3} for the income categories.

a. Report the prediction equations from this table.

b. Interpret the income effect in the first equation.

c. Report the Wald test statistic and P-value for testing that marital happiness is independent of family income. Interpret.

d. Does the model fit adequately? Justify your answer.

e. Estimate the probability that a person with average family income reports a very happy marriage.

Table 6.15. Output on Modeling Happiness for Problem 6.6

Deviance and Pearson Goodness-of-Fit Statistics

Criterion	Value	DF	Value/DF	Pr > ChiSq
Deviance	3.1909	2	1.5954	0.2028
Pearson	3.1510	2	1.5755	0.2069

Testing Global Null Hypothesis: BETA = 0

Test	Chi-Square	DF	Pr > ChiSq
Likelihood Ratio	0.9439	2	0.6238
Wald	0.9432	2	0.6240

Analysis of Maximum Likelihood Estimates

Parameter	happy	DF	Estimate	Standard Error	Wald Chi-Square	Pr > ChiSq
Intercept	1	1	−2.5551	0.7256	12.4009	0.0004
Intercept	2	1	−0.3513	0.2684	1.7133	0.1906
income	1	1	−0.2275	0.3412	0.4446	0.5049
income	2	1	−0.0962	0.1220	0.6210	0.4307

6.7 Refer to the previous exercise. Table 6.16 shows output for a cumulative logit model with scores {1, 2, 3} for the income categories.

a. Explain why the output reports two intercepts but one income effect.

b. Interpret the income effect.

c. Report a test statistic and P-value for testing that marital happiness is independent of family income. Interpret.

d. Does the model fit adequately? Justify your answer.

e. Estimate the probability that a person with average family income reports a very happy marriage.

Table 6.16. Output on Modeling Happiness for Problem 6.7

Deviance and Pearson Goodness-of-Fit Statistics

Criterion	Value	DF	Value/DF	Pr > ChiSq
Deviance	3.2472	3	1.0824	0.3551
Pearson	3.2292	3	1.0764	0.3576

Testing Global Null Hypothesis: BETA = 0

Test	Chi-Square	DF	Pr > ChiSq
Likelihood Ratio	0.8876	1	0.3461
Wald	0.8976	1	0.3434

Analysis of Maximum Likelihood Estimates

Parameter	DF	Estimate	Standard Error	Wald Chi-Square	Pr > ChiSq
Intercept 1	1	−3.2467	0.3404	90.9640	<.0001
Intercept 2	1	−0.2378	0.2592	0.8414	0.3590
income	1	−0.1117	0.1179	0.8976	0.3434

6.8 Table 6.17 results from a clinical trial for the treatment of small-cell lung cancer. Patients were randomly assigned to two treatment groups. The sequential therapy administered the same combination of chemotherapeutic agents in each treatment cycle. The alternating therapy used three different combinations, alternating from cycle to cycle.

a. Fit a cumulative logit model with main effects for treatment and gender. Interpret the estimated treatment effect.

b. Fit the model that also contains an interaction term between treatment and gender. Interpret the interaction term by showing how the estimated treatment effect varies by gender.

c. Does the interaction model give a significantly better fit?

Table 6.17. Data for Problem 6.8 on Lung Cancer Treatment

		Response to Chemotherapy			
Therapy	Gender	Progressive Disease	No Change	Partial Remission	Complete Remission
Sequential	Male	28	45	29	26
	Female	4	12	5	2
Alternating	Male	41	44	20	20
	Female	12	7	3	1

Source: Holtbrugge, W. and Schumacher, M., *Appl. Statist.*, **40**: 249–259, 1991.

6.9 A cumulative logit model is fitted to data from the 2004 General Social Survey, with Y = political ideology (extremely liberal or liberal, slightly liberal, moderate, slightly conservative, extremely conservative or conservative) and predictor religious preference (Protestant, Catholic, Jewish, None). With indicator variables for the first three religion categories, the ML fit has $\hat{\alpha}_1 = -1.03$, $\hat{\alpha}_2 = -0.13$, $\hat{\alpha}_3 = 1.57$, $\hat{\alpha}_4 = 2.41$, $\hat{\beta}_1 = -1.27$, $\hat{\beta}_2 = -1.22$, $\hat{\beta}_3 = -0.44$.

a. Why are there four $\{\hat{\alpha}_j\}$? Why is $\hat{\alpha}_1 < \hat{\alpha}_2 < \hat{\alpha}_3 < \hat{\alpha}_4$?

b. Which group is estimated to be the (i) most liberal, (ii) most conservative? Why?

c. Estimate the probability of the most liberal response for the Protestant and None groups.

d. Use an estimated odds ratio to compare political ideology for the (i) Protestant and None groups, (ii) Protestant and Catholic groups.

6.10 Refer to the interpretations in Section 6.2.6 for the mental health data. Summarize the SES effect by finding $P(Y \leq 2)$ for high SES and for low SES, at the mean life events of 4.3.

6.11 Refer to Table 6.12. Treating job satisfaction as the response, analyze the data using a cumulative logit model.

a. Describe the effect of income, using scores $\{3, 10, 20, 35\}$.

b. Compare the estimated income effect to the estimate obtained after combining categories "Very dissatisfied" and "A little satisfied." What property of the model does this reflect?

c. Can you drop gender from the model in (**a**)?

6.12 Table 6.18 shows results from the 2000 General Social Survey relating happiness and religious attendance (1 = at most several times a year, 2 = once a month to several times a year, 3 = every week to several times a week).

a. Fit a multinomial model. Conduct descriptive and inferential analyses about the association.

b. Analyze the model goodness of fit.

Table 6.18. GSS Data for Exercise 6.12 on Happiness

	Happiness		
Religion	Not Too Happy	Pretty Happy	Very Happy
1	189	908	382
2	53	311	180
3	46	335	294

6.13 Fit an adjacent-categories logit model with main effects to the job satisfaction data in Table 6.12, using scores $\{1, 2, 3, 4\}$ for income.

a. Use proportional odds structure. Interpret the estimated effect of income.

b. Fit the model allowing different effects for each logit, which is equivalent to a baseline-category logit model. Interpret the income effect.

c. What is the difference between the two models in terms of how they treat job satisfaction?

6.14 Consider Table 6.4 on belief in an afterlife. Fit a model using (a) adjacent-categories logits, (b) alternative ordinal logits. In each case, prepare a one-page report, summarizing your analyses and interpreting results.

6.15 Analyze the job satisfaction data of Table 6.12 using continuation-ratio logits. Prepare a one-page summary.

6.16 Table 6.19 refers to a study that randomly assigned subjects to a control group or a treatment group. Daily during the study, treatment subjects ate cereal containing psyllium. The purpose of the study was to analyze whether this resulted in lowering LDL cholesterol.

a. Model the ending cholesterol level as a function of treatment, using the beginning level as a covariate. Analyze the treatment effect, and interpret.

b. Repeat the analysis in (**a**), treating the beginning level as a categorical control variable. Compare results.

c. An alternative approach to (**b**) uses a generalized Cochran–Mantel–Haenszel test with 2×4 tables relating treatment to the ending response for four partial tables based on beginning cholesterol level. Apply such a test, taking into account the ordering of the response. Interpret, and compare results with (**b**).

Table 6.19. Data for Problem 6.16 on Cholesterol Study

	Ending LDL Cholesterol Level							
	Control				Treatment			
Beginning	≤3.4	3.4–4.1	4.1–4.9	>4.9	≤3.4	3.4–4.1	4.1–4.9	>4.9
≤3.4	18	8	0	0	21	4	2	0
3.4–4.1	16	30	13	2	17	25	6	0
4.1–4.9	0	14	28	7	11	35	36	6
>4.9	0	2	15	22	1	5	14	12

Source: Dr. Sallee Anderson, Kellogg Co.

6.17 Table 6.20 is an expanded version of a data set Section 7.2.6 presents about a sample of auto accidents, with predictors gender, location of accident, and

Table 6.20. Data for Problem 6.17 on Auto Accidents

			Severity of Injury				
Gender	Location	Seat-belt	1	2	3	4	5
Female	Urban	No	7,287	175	720	91	10
		Yes	11,587	126	577	48	8
	Rural	No	3,246	73	710	159	31
		Yes	6,134	94	564	82	17
Male	Urban	No	10,381	136	566	96	14
		Yes	10,969	83	259	37	1
	Rural	No	6,123	141	710	188	45
		Yes	6,693	74	353	74	12

Source: Dr. Cristanna Cook, Medical Care Development, Augusta, ME.

whether the subject used a seat belt. The response categories are (1) not injured, (2) injured but not transported by emergency medical services, (3) injured and transported by emergency medical services but not hospitalized, (4) injured and hospitalized but did not die, (5) injured and died. Analyze these data. Prepare a two-page report, summarizing your descriptive and inferential analyses.

6.18 A response scale has the categories (strongly agree, mildly agree, mildly disagree, strongly disagree, do not know). How might you model this response? (*Hint*: One approach handles the ordered categories in one model and combines them and models the "do not know" response in another model.)

6.19 The sample in Table 6.12 consists of 104 black Americans. A similar table relating income and job satisfaction for white subjects in the same General Social Survey had counts (by row) of (3, 10, 30, 27/7, 8, 45, 39/8, 7, 46, 51/4, 2, 28, 47) for females and (1, 4, 9, 9/1, 2, 37, 29/0, 10, 35, 39/7, 14, 69, 109) for males. Test the hypothesis of conditional independence between income and job satisfaction, (**a**) using a model that treats income and job satisfaction as nominal, (**b**) using a model that incorporates the category orderings, (**c**) with a generalized CMH test for the alternative that the mean job satisfaction varies by level of income, controlling for gender, (**d**) with a generalized CMH test not designed to detect any particular pattern of association. Interpret, and compare results, indicating the extent to which conclusions suffer when you do not use the ordinality.

6.20 For $K = 1$, the generalized *CMH* correlation statistic equals formula (2.10). When there truly is a trend, Section 2.5.3 noted that this test is more powerful than the X^2 and G^2 tests of Section 2.4.3. To illustrate, for Table 6.12 on job satisfaction and income, construct the marginal 4×4 table.

a. Show that the Pearson $X^2 = 11.5$ with $df = 9$ $(P = 0.24)$. Show that the correlation statistic with equally-spaced scores is $M^2 = 7.6$ based on $df = 1$ $(P = 0.006)$. Interpret.

b. Conduct an analysis with a model for which you would expect the test of the income effect also to be powerful.

6.21 For the 2000 GSS, counts in the happiness categories (not too, pretty, very) were $(67, 650, 555)$ for those who were married and $(65, 276, 93)$ for those who were divorced. Analyze these data, preparing a one-page report summarizing your descriptive and inferential analyses.

6.22 True, or false?

a. One reason it is usually wise to treat an ordinal variable with methods that use the ordering is that in tests about effects, chi-squared statistics have smaller df values, so it is easier for them to be farther out in the tail and give small P-values; that is, the ordinal tests tend to be more powerful.

b. The cumulative logit model assumes that the response variable Y is ordinal; it should not be used with nominal variables. By contrast, the baseline-category logit model treats Y as nominal. It can be used with ordinal Y, but it then ignores the ordering information.

c. If political ideology tends to be mainly in the moderate category in New Zealand and mainly in the liberal and conservative categories in Australia, then the cumulative logit model with proportional odds assumption should fit well for comparing these countries.

d. Logistic regression for binary Y is a special case of the baseline-category logit and cumulative logit model with $J = 2$.

CHAPTER 7

Loglinear Models for Contingency Tables

Section 3.3.1 introduced loglinear models as generalized linear models (GLMs) for count data. One use of them is modeling cell counts in contingency tables. The models specify how the size of a cell count depends on the levels of the categorical variables for that cell. They help to describe association patterns among a set of categorical response variables.

Section 7.1 introduces loglinear models. Section 7.2 discusses statistical inference for model parameters and model checking. When one variable is a binary response variable, logistic models for that response are equivalent to certain loglinear models. Section 7.3 presents the connection. We shall see that loglinear models are mainly of use when at least two variables in a contingency table are response variables. Section 7.4 introduces graphical representations that portray a model's association patterns and indicate when conditional odds ratios are identical to marginal odds ratios. The loglinear models of Sections 7.1–7.4 treat all variables as nominal. Section 7.5 presents a loglinear model that describes association between ordinal variables.

7.1 LOGLINEAR MODELS FOR TWO-WAY AND THREE-WAY TABLES

Consider an $I \times J$ contingency table that cross-classifies n subjects. When the responses are statistically independent, the joint cell probabilities $\{\pi_{ij}\}$ are determined by the row and column marginal totals,

$$\pi_{ij} = \pi_{i+}\pi_{+j}, \quad i = 1, \dots, I, \quad j = 1, \dots, J$$

An Introduction to Categorical Data Analysis, Second Edition. By Alan Agresti

The cell probabilities $\{\pi_{ij}\}$ are parameters for a *multinomial* distribution. Loglinear model formulas use expected frequencies $\{\mu_{ij} = n\pi_{ij}\}$ rather than $\{\pi_{ij}\}$. Then they apply also to the *Poisson* distribution for cell counts with expected values $\{\mu_{ij}\}$. Under independence, $\mu_{ij} = n\pi_{i+}\pi_{+j}$ for all i and j.

7.1.1 Loglinear Model of Independence for Two-Way Table

Denote the row variable by X and the column variable by Y. The condition of independence, $\mu_{ij} = n\pi_{i+}\pi_{+j}$, is multiplicative. Taking the log of both sides of the equation yields an additive relation. Namely, $\log \mu_{ij}$ depends on a term based on the sample size, a term based on the probability in row i, and a term based on the probability in column j. Thus, independence has the form

$$\log \mu_{ij} = \lambda + \lambda_i^X + \lambda_j^Y \tag{7.1}$$

for a *row effect* λ_i^X and a *column effect* λ_j^Y. (The X and Y superscripts are labels, not "power" exponents.) This model is called the *loglinear model of independence*. The parameter λ_i^X represents the effect of classification in row i. The larger the value of λ_i^X, the larger each expected frequency is in row i. Similarly, λ_j^Y represents the effect of classification in column j.

The null hypothesis of independence is equivalently the hypothesis that this loglinear model holds. The fitted values that satisfy the model are $\{\hat{\mu}_{ij} = n_{i+}n_{+j}/n\}$. These are the estimated expected frequencies for the X^2 and G^2 tests of independence (Section 2.4). Those tests are also goodness-of-fit tests of this loglinear model.

7.1.2 Interpretation of Parameters in Independence Model

As formula (7.1) illustrates, loglinear models for contingency tables do not distinguish between response and explanatory classification variables. Model (7.1) treats both X and Y as responses, modeling the cell counts. Loglinear models are examples of generalized linear models. The GLM treats the cell counts as independent observations from some distribution, typically the Poisson. The model regards the observations to be the cell counts rather than the individual classifications of the subjects.

Parameter interpretation is simplest when we view one response as a function of the others. For instance, consider the independence model (7.1) for $I \times 2$ tables. In row i, the logit for the probability that $Y = 1$ equals

$$\begin{aligned} \log[P(Y = 1)/(1 - P(Y = 1))] &= \log(\mu_{i1}/\mu_{i2}) = \log \mu_{i1} - \log \mu_{i2} \\ &= (\lambda + \lambda_i^X + \lambda_1^Y) - (\lambda + \lambda_i^X + \lambda_2^Y) = \lambda_1^Y - \lambda_2^Y \end{aligned}$$

This logit does not depend on i. That is, the logit for Y does not depend on the level of X. The loglinear model corresponds to the simple model of form, $\text{logit}[P(Y = 1)] = \alpha$, whereby the logit takes the same value in every row i. In each row, the odds of response in column 1 equal $\exp(\alpha) = \exp(\lambda_1^Y - \lambda_2^Y)$. In model (7.1), differences between two parameters for a given variable relate to the log odds of making one response, relative to another, on that variable.

For the independence model, one of $\{\lambda_i^X\}$ is redundant, and one of $\{\lambda_j^Y\}$ is redundant. This is analogous to ANOVA and multiple regression models with factors, which require one fewer indicator variable than the number of factor levels. Most software sets the parameter for the last category equal to 0. Another approach lets the parameters for each factor sum to 0. The choice of constraints is arbitrary. What *is* unique is the *difference* between two main effect parameters of a particular type. As just noted, that is what determines odds and odds ratios.

For example, in the 2000 General Social Survey, subjects were asked whether they believed in life after death. The number who answered "yes" was 1339 of the 1639 whites, 260 of the 315 blacks and 88 of the 110 classified as "other" on race. Table 7.1 shows results of fitting the independence loglinear model to the 3 × 2 table. The model fits well. For the constraints used, $\lambda_1^Y = 1.50$ and $\lambda_2^Y = 0$. Therefore, the estimated odds of belief in life after death was $\exp(1.50) = 4.5$ for each race.

Table 7.1. Results of Fitting Independence Loglinear Model to Cross-Classification of Race by Belief in Life after Death

Criteria For Assessing Goodness Of Fit

Criterion	DF	Value
Deviance	2	0.3565
Pearson Chi-Square	2	0.3601

Parameter		DF	Estimate	Standard Error
Intercept		1	3.0003	0.1061
race	white	1	2.7014	0.0985
race	black	1	1.0521	0.1107
race	other	0	0.0000	0.0000
belief	yes	1	1.4985	0.0570
belief	no	0	0.0000	0.0000

7.1.3 Saturated Model for Two-Way Tables

Variables that are statistically dependent rather than independent satisfy the more complex loglinear model,

$$\log \mu_{ij} = \lambda + \lambda_i^X + \lambda_j^Y + \lambda_{ij}^{XY} \tag{7.2}$$

The $\{\lambda_{ij}^{XY}\}$ parameters are association terms that reflect deviations from independence. The parameters represent interactions between X and Y, whereby the effect of one variable on the expected cell count depends on the level of the other variable. The independence model (7.1) is the special case in which all $\lambda_{ij}^{XY} = 0$.

Direct relationships exist between log odds ratios and the $\{\lambda_{ij}^{XY}\}$ association parameters. For example, the model for 2×2 tables has log odds ratio

$$\begin{aligned}
\log \theta &= \log\left(\frac{\mu_{11}\mu_{22}}{\mu_{12}\mu_{21}}\right) = \log \mu_{11} + \log \mu_{22} - \log \mu_{12} - \log \mu_{21} \\
&= (\lambda + \lambda_1^X + \lambda_1^Y + \lambda_{11}^{XY}) + (\lambda + \lambda_2^X + \lambda_2^Y + \lambda_{22}^{XY}) \\
&\quad - (\lambda + \lambda_1^X + \lambda_2^Y + \lambda_{12}^{XY}) - (\lambda + \lambda_2^X + \lambda_1^Y + \lambda_{21}^{XY}) \\
&= \lambda_{11}^{XY} + \lambda_{22}^{XY} - \lambda_{12}^{XY} - \lambda_{21}^{XY} \qquad (7.3)
\end{aligned}$$

Thus, $\{\lambda_{ij}^{XY}\}$ determine the log odds ratio. When these parameters equal zero, the log odds ratio is zero, and X and Y are independent.

In $I \times J$ tables, only $(I - 1)(J - 1)$ association parameters are nonredundant. One can specify the parameters so that the ones in the last row and in the last column are zero. These parameters are coefficients of cross-products of $(I - 1)$ indicator variables for X with $(J - 1)$ indicator variables for Y. Tests of independence analyze whether these $(I - 1)(J - 1)$ parameters equal zero, so the tests have residual $df = (I - 1)(J - 1)$.

Table 7.2 shows estimates for fitting model (7.2) to the 3×2 table mentioned above on X = gender and Y = belief in afterlife. The estimated odds ratios between belief and race are exp(0.1096) = 1.12 for white and other, exp(0.1671) = 1.18 for black and other, and exp(0.1096 − 0.1671) = 0.94 for white and black. For example, the estimated odds of belief in an life after death for whites are 0.94 times the estimated odds for blacks. Since the independence model fitted well, none of these estimated odds ratios differ significantly from 1.0.

Table 7.2. Estimates for Fitting Saturated Loglinear Model to Cross-Classification of Race by Belief in Life after Death

Parameter			DF	Estimate	Standard error
Intercept			1	3.0910	0.2132
race	white		1	2.6127	0.2209
race	black		1	0.9163	0.2523
race	other		0	0.0000	0.0000
belief	yes		1	1.3863	0.2384
belief	no		0	0.0000	0.0000
race*belief	white	yes	1	0.1096	0.2468
race*belief	white	no	0	0.0000	0.0000
race*belief	black	yes	1	0.1671	0.2808
race*belief	black	no	0	0.0000	0.0000
race*belief	other	yes	0	0.0000	0.0000
race*belief	other	no	0	0.0000	0.0000

Model (7.2) has a single constant parameter (λ), $(I-1)$ nonredundant λ_i^X parameters, $(J-1)$ nonredundant λ_j^Y parameters, and $(I-1)(J-1)$ nonredundant λ_{ij}^{XY} parameters. The total number of parameters equals $1+(I-1)+(J-1)+(I-1)(J-1)=IJ$. The model has as many parameters as observed cell counts. It is the *saturated* loglinear model, having the maximum possible number of parameters. Because of this, it is the most general model for two-way tables. It describes perfectly any set of expected frequencies. It gives a perfect fit to the data. The estimated odds ratios just reported are the same as the sample odds ratios. In practice, unsaturated models are preferred, because their fit smooths the sample data and has simpler interpretations.

When a model has two-factor terms, be cautious in interpreting the single-factor terms. By analogy with two-way ANOVA, when there is two-factor interaction, it can be misleading to report main effects. The estimates of the main effect terms depend on the coding scheme used for the higher-order effects, and the interpretation also depends on that scheme. Normally, we restrict our attention to the highest-order terms for a variable.

7.1.4 Loglinear Models for Three-Way Tables

With three-way contingency tables, loglinear models can represent various independence and association patterns. Two-factor association terms describe the conditional odds ratios between variables.

For cell expected frequencies $\{\mu_{ijk}\}$, consider loglinear model

$$\log \mu_{ijk} = \lambda + \lambda_i^X + \lambda_j^Y + \lambda_k^Z + \lambda_{ik}^{XZ} + \lambda_{jk}^{YZ} \tag{7.4}$$

Since it contains an XZ term (λ_{ik}^{XZ}), it permits association between X and Z, controlling for Y. This model also permits a YZ association, controlling for X. It does not contain an XY association term. This loglinear model specifies conditional independence between X and Y, controlling for Z.

We symbolize this model by (XZ, YZ). The symbol lists the highest-order terms in the model for each variable. This model is an important one. It holds, for instance, if an association between two variables (X and Y) disappears when we control for a third variable (Z).

Models that delete additional association terms are too simple to fit most data sets well. For instance, the model that contains only single-factor terms, denoted by (X, Y, Z), is called the *mutual independence model*. It treats each pair of variables as independent, both conditionally and marginally. When variables are chosen wisely for a study, this model is rarely appropriate.

A model that permits all three pairs of variables to have conditional associations is

$$\log \mu_{ijk} = \lambda + \lambda_i^X + \lambda_j^Y + \lambda_k^Z + \lambda_{ij}^{XY} + \lambda_{ik}^{XZ} + \lambda_{jk}^{YZ} \tag{7.5}$$

For it, the next subsection shows that conditional odds ratios between any two variables are the same at each level of the third variable. This is the property of *homogeneous association* (Section 2.7.6). This loglinear model is called the *homogeneous association model* and symbolized by (XY, XZ, YZ).

The most general loglinear model for three-way tables is

$$\log \mu_{ijk} = \lambda + \lambda_i^X + \lambda_j^Y + \lambda_k^Z + \lambda_{ij}^{XY} + \lambda_{ik}^{XZ} + \lambda_{jk}^{YZ} + \lambda_{ijk}^{XYZ}$$

Denoted by (XYZ), it is the saturated model. It provides a perfect fit.

7.1.5 Two-Factor Parameters Describe Conditional Associations

Model interpretations refer to the highest-order parameters. For instance, consider the homogeneous association model (7.5). Its parameters relate directly to conditional odds ratios. We illustrate this for $2 \times 2 \times K$ tables. The XY conditional odds ratio $\theta_{XY(k)}$ describes association between X and Y in partial table k (recall Section 2.7.4). From an argument similar to that in Section 7.1.3,

$$\log \theta_{XY(k)} = \log\left(\frac{\mu_{11k}\mu_{22k}}{\mu_{12k}\mu_{21k}}\right) = \lambda_{11}^{XY} + \lambda_{22}^{XY} - \lambda_{12}^{XY} - \lambda_{21}^{XY} \tag{7.6}$$

The right-hand side of this expression does not depend on k, so the odds ratio is the same at every level of Z. Similarly, model (XY, XZ, YZ) also has equal XZ odds ratios at different levels of Y, and it has equal YZ odds ratios at different levels of X. Any model not having the three-factor term λ_{ijk}^{XYZ} satisfies homogeneous association.

7.1.6 Example: Alcohol, Cigarette, and Marijuana Use

Table 7.3 is from a survey conducted by the Wright State University School of Medicine and the United Health Services in Dayton, Ohio. The survey asked students in their final year of a high school near Dayton, Ohio whether they had ever

Table 7.3. Alcohol (A), Cigarette (C), and Marijuana (M) Use for High School Seniors

Alcohol Use	Cigarette Use	Marijuana Use: Yes	Marijuana Use: No
Yes	Yes	911	538
	No	44	456
No	Yes	3	43
	No	2	279

Source: I am grateful to Professor Harry Khamis, Wright State University, for supplying these data.

used alcohol, cigarettes, or marijuana. Denote the variables in this $2 \times 2 \times 2$ table by A for alcohol use, C for cigarette use, and M for marijuana use.

Loglinear models are simple to fit with software. Table 7.4 shows fitted values for several models. The fit for model (AC, AM, CM) is close to the observed data, which are the fitted values for the saturated model (ACM). The other models fit poorly.

Table 7.4. Fitted Values for Loglinear Models Applied to Table 7.3

Alcohol Use	Cigarette Use	Marijuana Use	Loglinear Model (A, C, M)	(AC, M)	(AM, CM)	(AC, AM, CM)	(ACM)
Yes	Yes	Yes	540.0	611.2	909.24	910.4	911
		No	740.2	837.8	438.84	538.6	538
	No	Yes	282.1	210.9	45.76	44.6	44
		No	386.7	289.1	555.16	455.4	456
No	Yes	Yes	90.6	19.4	4.76	3.6	3
		No	124.2	26.6	142.16	42.4	43
	No	Yes	47.3	118.5	0.24	1.4	2
		No	64.9	162.5	179.84	279.6	279

Table 7.5 illustrates association patterns for these models by presenting estimated marginal and conditional odds ratios. For example, the entry 1.0 for the AC conditional odds ratio for model (AM, CM) is the common value of the AC fitted odds ratios at the two levels of M,

$$1.0 = \frac{909.24 \times 0.24}{45.76 \times 4.76} = \frac{438.84 \times 179.84}{555.16 \times 142.16}$$

This model implies conditional independence between alcohol use and cigarette use, controlling for marijuana use. The entry 2.7 for the AC *marginal* association for this model is the odds ratio for the marginal AC fitted table,

$$2.7 = \frac{(909.24 + 438.84)(0.24 + 179.84)}{(45.76 + 555.16)(4.76 + 142.16)}$$

Table 7.5. Estimated Odds Ratios for Loglinear Models in Table 7.4

	Conditional Association			Marginal Association		
Model	AC	AM	CM	AC	AM	CM
(A, C, M)	1.0	1.0	1.0	1.0	1.0	1.0
(AC, M)	17.7	1.0	1.0	17.7	1.0	1.0
(AM, CM)	1.0	61.9	25.1	2.7	61.9	25.1
(AC, AM, CM)	7.8	19.8	17.3	17.7	61.9	25.1
(ACM) level 1	13.8	24.3	17.5	17.7	61.9	25.1
(ACM) level 2	7.7	13.5	9.7			

The odds ratios for the observed data are those reported for the saturated model (*ACM*).

From Table 7.5, conditional odds ratios equal 1.0 for each pairwise term not appearing in a model. An example is the *AC* association in model (*AM*, *CM*). For that model, the estimated marginal *AC* odds ratio differs from 1.0. Section 2.7.5 noted that conditional independence does not imply marginal independence. Some models have conditional odds ratios that equal the corresponding marginal odds ratios. Section 7.4.2 presents a condition that guarantees this. This equality does not normally happen for loglinear models containing all pairwise associations.

Model (*AC*, *AM*, *CM*) permits all pairwise associations but has homogeneous odds ratios. The *AC* fitted conditional odds ratios for this model equal 7.8. For each level of *M*, students who have smoked cigarettes have estimated odds of having drunk alcohol that are 7.8 times the estimated odds for students who have not smoked cigarettes. The *AC* marginal odds ratio of 17.7 ignores the third factor (*M*), whereas the conditional odds ratio of 7.8 controls for it.

For model (*AC*, *AM*, *CM*) (or simpler models), one can calculate an estimated conditional odds ratio using the model's fitted values at either level of the third variable. Or, one can calculate it from equation (7.6) using the parameter estimates. For example, the estimated conditional *AC* odds ratio is

$$\exp(\hat{\lambda}_{11}^{AC} + \hat{\lambda}_{22}^{AC} - \hat{\lambda}_{12}^{AC} - \hat{\lambda}_{21}^{AC})$$

Table 7.6. Output for Fitting Loglinear Model to Table 7.3

Criteria For Assessing Goodness Of Fit

Criterion	DF	Value
Deviance	1	0.3740
Pearson Chi-Square	1	0.4011

Parameter			Estimate	Standard Error	Wald Chi-Square	Pr > ChiSq
Intercept			5.6334	0.0597	8903.96	<.0001
a	1		0.4877	0.0758	41.44	<.0001
c	1		−1.8867	0.1627	134.47	<.0001
m	1		−5.3090	0.4752	124.82	<.0001
a*m	1	1	2.9860	0.4647	41.29	<.0001
a*c	1	1	2.0545	0.1741	139.32	<.0001
c*m	1	1	2.8479	0.1638	302.14	<.0001

LR Statistics

Source	DF	Chi-Square	Pr > ChiSq
a*m	1	91.64	<.0001
a*c	1	187.38	<.0001
c*m	1	497.00	<.0001

Table 7.6 shows software output, using constraints for which parameters at the second level of any variable equal 0. Thus, $\hat{\lambda}_{22}^{AC} = \hat{\lambda}_{12}^{AC} = \hat{\lambda}_{21}^{AC} = 0$, and the estimated conditional AC odds ratio is $\exp(\hat{\lambda}_{11}^{AC}) = \exp(2.05) = 7.8$.

7.2 INFERENCE FOR LOGLINEAR MODELS

Table 7.5 shows that estimates of conditional and marginal odds ratios are highly dependent on the model. This highlights the importance of good model selection. An estimate from this table is informative only if its model fits well. This section shows how to check goodness of fit, conduct inference, and extend loglinear models to higher dimensions.

7.2.1 Chi-Squared Goodness-of-Fit Tests

Consider the null hypothesis that a given loglinear model holds. As usual, large-sample chi-squared statistics assess goodness of fit by comparing the cell fitted values to the observed counts. In the three-way case, the likelihood-ratio and Pearson statistics are

$$G^2 = 2\sum n_{ijk} \log\left(\frac{n_{ijk}}{\hat{\mu}_{ijk}}\right), \quad X^2 = \sum \frac{(n_{ijk} - \hat{\mu}_{ijk})^2}{\hat{\mu}_{ijk}}$$

The G^2 statistic is the *deviance* for the model (recall Section 3.4.3). The degrees of freedom equal the number of cell counts minus the number of model parameters. The df value decreases as the model becomes more complex. The saturated model has $df = 0$.

For the student drug survey (Table 7.3), Table 7.7 presents goodness-of-fit tests for several models. For a given df, larger G^2 or X^2 values have smaller P-values

Table 7.7. Goodness-of-Fit Tests for Loglinear Models Relating Alcohol (A), Cigarette (C), and Marijuana (M) Use

Model	G^2	X^2	df	P-value*
(A, C, M)	1286.0	1411.4	4	<0.001
(A, CM)	534.2	505.6	3	<0.001
(C, AM)	939.6	824.2	3	<0.001
(M, AC)	843.8	704.9	3	<0.001
(AC, AM)	497.4	443.8	2	<0.001
(AC, CM)	92.0	80.8	2	<0.001
(AM, CM)	187.8	177.6	2	<0.001
(AC, AM, CM)	0.4	0.4	1	0.54
(ACM)	0.0	0.0	0	—

*P-value for G^2 statistic.

and indicate poorer fits. The models that lack any association term fit poorly, having P-values below 0.001. The model (AC, AM, CM) that permits all pairwise associations but assumes homogeneous association fits well ($P = 0.54$). Table 7.6 shows the way PROC GENMOD in SAS reports the goodness-of-fit statistics for this model.

7.2.2 Loglinear Cell Residuals

Cell residuals show the quality of fit cell-by-cell. They show where a model fits poorly. Sometimes they indicate that certain cells display lack of fit in an otherwise good-fitting model. When a table has many cells, some residuals may be large purely by chance.

Section 2.4.5 introduced standardized residuals for the independence model, and Section 3.4.5 discussed them generally for GLMs. They divide differences between observed and fitted counts by their standard errors. When the model holds, standardized residuals have approximately a standard normal distribution. Lack of fit is indicated by absolute values larger than about 2 when there are few cells or about 3 when there are many cells.

Table 7.8 shows standardized residuals for the model (AM, CM) of AC conditional independence with Table 7.3. This model has $df = 2$ for testing fit. The two nonredundant residuals refer to checking AC independence at each level of M. The large residuals reflect the overall poor fit. [In fact, X^2 relates to the two nonredundant residuals by $X^2 = (3.70)^2 + (12.80)^2 = 177.6$.]. Extremely large residuals occur for students who have not smoked marijuana. For them, the positive residuals occur when A and C are both "yes" or both "no." More of these students have used both or neither of alcohol and cigarettes than one would expect if their usage were conditionally independent. The same pattern persists for students who have smoked marijuana, but the differences between observed and fitted counts are then not as striking.

Table 7.8 also shows standardized residuals for model (AC, AM, CM). Since $df = 1$ for this model, only one residual is nonredundant. Both G^2 and X^2 are small, so

Table 7.8. Standardized Residuals for Two Loglinear Models

Drug Use				Model (AM, CM)		Model (AC, AM, CM)	
A	C	M	Observed Count	Fitted Count	Standardized Residual	Fitted Count	Standardized Residual
Yes	Yes	Yes	911	909.2	3.70	910.4	0.63
		No	538	438.8	12.80	538.6	−0.63
	No	Yes	44	45.8	−3.70	44.6	−0.63
		No	456	555.2	−12.80	455.4	0.63
No	Yes	Yes	3	4.8	−3.70	3.6	−0.63
		No	43	142.2	−12.80	42.4	0.63
	No	Yes	2	0.2	3.70	1.4	0.63
		No	279	179.8	12.80	279.6	−0.63

these residuals indicate a good fit. (In fact, when $df = 1$, X^2 equals the square of each standardized residual.)

7.2.3 Tests about Conditional Associations

To test a conditional association in a model, we compare the model to the simpler model not containing that association. For example, for model (AC, AM, CM) for the drug use survey, the null hypothesis of conditional independence between alcohol use and cigarette smoking states that the λ^{AC} term equals zero. The test analyzes whether the simpler model (AM, CM) of AC conditional independence holds, against the alternative that model (AC, AM, CM) holds.

As Section 3.4.4 discussed, the likelihood-ratio statistic for testing that a model term equals zero is identical to the difference between the deviances for the model without that term and the model with the term. The deviance for a model is also its G^2 goodness-of-fit statistic. The df for the test equal the difference between the corresponding residual df values.

Denote the test statistic for testing that $\lambda^{AC} = 0$ in model (AC, AM, CM) by $G^2[(AM, CM) \mid (AC, AM, CM)]$. It equals

$$G^2[(AM, CM) \mid (AC, AM, CM)] = G^2(AM, CM) - G^2(AC, AM, CM)$$

From Table 7.7, this test statistic equals $187.8 - 0.4 = 187.4$. It has $df = 2 - 1 = 1$ $(P < 0.0001)$. This is strong evidence of an AC conditional association. Also, the statistics comparing models (AC, CM) and (AC, AM) with model (AC, AM, CM) provide strong evidence of AM and CM conditional associations, as the bottom of Table 7.6 shows. Further analyses of the data should use model (AC, AM, CM) rather than any simpler model.

7.2.4 Confidence Intervals for Conditional Odds Ratios

ML estimators of loglinear model parameters have large-sample normal distributions. For models in which the highest-order terms are two-factor associations, the estimates refer to conditional log odds ratios. One can use the estimates and their standard errors to construct confidence intervals for true log odds ratios and then exponentiate them to form intervals for odds ratios.

Consider the association between alcohol and cigarettes for the student drug-use data, using model (AC, AM, CM). Software that sets redundant association parameters in the last row and the last column equal to zero (such as PROC GENMOD in SAS) reports $\hat{\lambda}_{11}^{AC} = 2.054$, with $SE = 0.174$. For that approach, the lone nonzero term equals the estimated conditional log odds ratio. A 95% confidence interval for the true conditional log odds ratio is $2.054 \pm 1.96(0.174)$ or $(1.71, 2.39)$, yielding $(e^{1.71}, e^{2.39}) = (5.5, 11.0)$ for the odds ratio. There is a strong positive association between cigarette use and alcohol use, both for users and nonusers of marijuana.

For model (AC, AM, CM), the 95% confidence intervals are (8.0, 49.2) for the AM conditional odds ratio and (12.5, 23.8) for the CM conditional odds ratio. The intervals are wide, but these associations also are strong. In summary, this model reveals strong conditional associations for each pair of drugs. There is a strong tendency for users of one drug to be users of a second drug, and this is true both for users and for nonusers of the third drug. Table 7.5 shows that estimated marginal associations are even stronger. Controlling for outcome on one drug moderates the association somewhat between the other two drugs.

The analyses in this section pertain to association structure. A different analysis pertains to comparing marginal distributions, for instance to determine if one drug has more usage than the others. Section 8.1 presents this type of analysis.

7.2.5 Loglinear Models for Higher Dimensions

Loglinear models are more complex for three-way tables than for two-way tables, because of the variety of potential association patterns. Basic concepts for models with three-way tables extend readily, however, to multiway tables.

We illustrate this for four-way tables, with variables W, X, Y, and Z. Interpretations are simplest when the model has no three-factor terms. Such models are special cases of (WX, WY, WZ, XY, XZ, YZ), which has homogenous associations. Each pair of variables is conditionally associated, with the same odds ratios at each combination of levels of the other two variables. An absence of a two-factor term implies conditional independence for those variables. Model (WX, WY, WZ, XZ, YZ) does not contain an XY term, so it treats X and Y as conditionally independent at each combination of levels of W and Z.

A variety of models have three-factor terms. A model could contain WXY, WXZ, WYZ, or XYZ terms. The XYZ term permits the association between any pair of those three variables to vary across levels of the third variable, at each fixed level of W. The saturated model contains all these terms plus a four-factor term.

7.2.6 Example: Automobile Accidents and Seat Belts

Table 7.9 shows results of accidents in the state of Maine for 68,694 passengers in autos and light trucks. The table classifies passengers by gender (G), location of accident (L), seat-belt use (S), and injury (I). The table reports the sample proportion of passengers who were injured. For each GL combination, the proportion of injuries was about halved for passengers wearing seat belts.

Table 7.10 displays tests of fit for several loglinear models. To investigate the complexity of model needed, we consider model (G, I, L, S) containing only single-factor terms, model (GI, GL, GS, IL, IS, LS) containing also all the two-factor terms, and model (GIL, GIS, GLS, ILS) containing also all the three-factor terms. Model (G, I, L, S) implies mutual independence of the four variables. It fits very poorly ($G^2 = 2792.8, df = 11$). Model (GI, GL, GS, IL, IS, LS) fits much better ($G^2 = 23.4, df = 5$) but still has lack of fit ($P < 0.001$). Model (GIL, GIS, GLS, ILS)

Table 7.9. Injury (I) by Gender (G), Location (L), and Seat Belt Use (S), with Fit of Models (GI, GL, GS, IL, IS, LS) and (GLS, GI, IL, IS)

		Seat	Injury		(GI, GL, GS, IL, IS, LS)		(GLS, GI, IL, IS)		Sample
Gender	Location	Belt	No	Yes	No	Yes	No	Yes	Proportion Yes
Female	Urban	No	7,287	996	7,166.4	993.0	7,273.2	1,009.8	0.12
		Yes	11,587	759	11,748.3	721.3	11,632.6	713.4	0.06
	Rural	No	3,246	973	3,353.8	988.8	3,254.7	964.3	0.23
		Yes	6,134	757	5,985.5	781.9	6,093.5	797.5	0.11
Male	Urban	No	10,381	812	10,471.5	845.1	10,358.9	834.1	0.07
		Yes	10,969	380	10,837.8	387.6	10,959.2	389.8	0.03
	Rural	No	6,123	1,084	6,045.3	1,038.1	6,150.2	1,056.8	0.15
		Yes	6,693	513	6,811.4	518.2	6,697.6	508.4	0.07

Source: I am grateful to Dr. Cristanna Cook, Medical Care Development, Augusta, ME, for supplying these data.

Table 7.10. Goodness-of-fit Tests for Loglinear Models Relating Injury (I), Gender (G), Location (L), and Seat-Belt Use (S)

Model	G^2	df	P-value
(G, I, L, S)	2792.8	11	<0.0001
(GI, GL, GS, IL, IS, LS)	23.4	5	<0.001
(GIL, GIS, GLS, ILS)	1.3	1	0.25
(GIL, GS, IS, LS)	18.6	4	0.001
(GIS, GL, IL, LS)	22.8	4	<0.001
(GLS, GI, IL, IS)	7.5	4	0.11
(ILS, GI, GL, GS)	20.6	4	<0.001

fits well ($G^2 = 1.3$, $df = 1$) but is difficult to interpret. This suggests studying models that are more complex than (GI, GL, GS, IL, IS, LS) but simpler than (GIL, GIS, GLS, ILS). We do this in the next subsection, but first we analyze model (GI, GL, GS, IL, IS, LS).

Table 7.9 shows the fitted values for (GI, GL, GS, IL, IS, LS), which assumes homogeneous conditional odds ratios for each pair of variables. Table 7.11 reports the model-based estimated odds ratios. One can obtain them directly using the fitted values for partial tables relating two variables at any combination of levels of the other two. The log odds ratios also follow directly from loglinear parameter estimates. For instance, $\log(0.44) = -0.814 = \hat{\lambda}_{11}^{IS}$ when parameters at the second level of either factor are equated to 0.

Table 7.11. Estimated Conditional Odds Ratios for Two Loglinear Models

	Loglinear Model	
Odds Ratio	(GI, GL, GS, IL, IS, LS)	(GLS, GI, IL, IS)
GI	0.58	0.58
IL	2.13	2.13
IS	0.44	0.44
GL (S = no)	1.23	1.33
GL (S = yes)	1.23	1.17
GS (L = urban)	0.63	0.66
GS (L = rural)	0.63	0.58
LS (G = female)	1.09	1.17
LS (G = male)	1.09	1.03

Since the sample size is large, the estimates of odds ratios are precise. For example, the SE of the estimated IS conditional log odds ratio is 0.028. A 95% confidence interval for the true log odds ratio is $-0.814 \pm 1.96(0.028)$, or $(-0.868, -0.760)$, which translates to (0.42, 0.47) for the odds ratio. The odds of injury for passengers wearing seat belts were less than half the odds for passengers not wearing them, for each gender–location combination. The fitted odds ratios in Table 7.11 also suggest that, other factors being fixed, injury was more likely in rural than urban accidents and more likely for females than males. Also, the estimated odds that males used seat belts are only 0.63 times the estimated odds for females.

7.2.7 Three-Factor Interaction

Interpretations are more complicated when a model contains three-factor terms. Such terms refer to interactions, the association between two variables varying across levels of the third variable. Table 7.10 shows results of adding a single three-factor term to model (GI, GL, GS, IL, IS, LS). Of the four possible models, (GLS, GI, IL, IS) fits best. Table 7.9 also displays its fit.

For model (GLS, GI, IL, IS), each pair of variables is conditionally dependent, and at each level of I the association between G and L or between G and S or between L and S varies across the levels of the remaining variable. For this model, it is inappropriate to interpret the GL, GS, and LS two-factor terms on their own. For example, the presence of the GLS term implies that the GS odds ratio varies across the levels of L. Because I does not occur in a three-factor term, the conditional odds ratio between I and each variable is the same at each combination of levels of the other two variables. The first three lines of Table 7.11 report the fitted odds ratios for the GI, IL, and IS associations.

When a model has a three-factor term, to study the interaction, calculate fitted odds ratios between two variables at each level of the third. Do this at any levels of remaining variables not involved in the interaction. The bottom six lines of Table 7.11 illustrates this for model (GLS, GI, IL, IS). For example, the fitted GS odds ratio of 0.66 for (L = urban) refers to four fitted values for urban accidents, both the four with (injury = no) and the four with (injury = yes); that is,

$$0.66 = (7273.2)(10,959.2)/(11,632.6)(10,358.9)$$
$$= (1009.8)(389.8)/(713.4)(834.1)$$

7.2.8 Large Samples and Statistical Versus Practical Significance

The sample size can strongly influence results of any inferential procedure. We are more likely to detect an effect as the sample size increases. This suggests a cautionary remark. For small sample sizes, reality may be more complex than indicated by the simplest model that passes a goodness-of-fit test. By contrast, for large sample sizes, statistically significant effects can be weak and unimportant.

We saw above that model (GLS, GI, IL, IS) seems to fit much better than (GI, GL, GS, IL, IS, LS): The difference in G^2 values is $23.4 - 7.5 = 15.9$, based on $df = 5 - 4 = 1$ ($P = 0.0001$). The fitted odds ratios in Table 7.11, however, show that the three-factor interaction is weak. The fitted odds ratio between any two of G, L, and S is similar at both levels of the third variable. The significantly better fit of model (GLS, GI, IL, IS) mainly reflects the enormous sample size. Although the three-factor interaction is weak, it is significant because the large sample provides small standard errors. A comparison of fitted odds ratios for the two models suggests that the simpler model (GI, GL, GS, IL, IS, LS) is adequate for practical purposes. Simpler models are easier to summarize. A goodness-of-fit test should not be the sole criterion for selecting a model.

For large samples, it is helpful to summarize the closeness of a model fit to the sample data in a way that, unlike a test statistic, is not affected by the sample size. For a table of arbitrary dimension with cell counts $\{n_i = np_i\}$ and fitted values $\{\hat{\mu}_i = n\hat{\pi}_i\}$, one such measure is the *dissimilarity index*,

$$D = \sum |n_i - \hat{\mu}_i|/2n = \sum |p_i - \hat{\pi}_i|/2$$

This index takes values between 0 and 1. Smaller values represent a better fit. It represents the proportion of sample cases that must move to different cells for the model to achieve a perfect fit.

The dissimilarity index helps indicate whether the lack of fit is important in a practical sense. A very small D value suggests that the sample data follow the model pattern closely, even though the model is not perfect.

For Table 7.9, model (GI, GL, GS, IL, IS, LS) has $D = 0.008$, and model (GLS, GI, IL, IS) has $D = 0.003$. These values are very small. For either model, moving less than 1% of the data yields a perfect fit. The relatively large value of G^2 for model (GI, GL, GS, IL, IS, LS) indicated that the model does not truly hold. Nevertheless, the small value for D suggests that, in practical terms, the model provides a decent fit.

7.3 THE LOGLINEAR–LOGISTIC CONNECTION

Loglinear models for contingency tables focus on associations between categorical response variables. Logistic regression models, on the other hand, describe how a categorical response depends on a set of explanatory variables. Though the model types seem distinct, connections exist between them. For a loglinear model, one can construct logits for one response to help interpret the model. Moreover, logistic models with categorical explanatory variables have equivalent loglinear models.

7.3.1 Using Logistic Models to Interpret Loglinear Models

To understand implications of a loglinear model formula, it can help to form a logit for one of the variables. We illustrate with the homogeneous association model for three-way tables,

$$\log \mu_{ijk} = \lambda + \lambda_i^X + \lambda_j^Y + \lambda_k^Z + \lambda_{ij}^{XY} + \lambda_{ik}^{XZ} + \lambda_{jk}^{YZ}$$

Suppose Y is binary. We treat it as a response and X and Z as explanatory. When X is at level i and Z is at level k,

$$\begin{aligned}\text{logit}[P(Y=1)] &= \log\left[\frac{P(Y=1)}{1-P(Y=1)}\right] = \log\left[\frac{P(Y=1 \mid X=i, Z=k)}{P(Y=2 \mid X=i, Z=k)}\right] \\ &= \log\left(\frac{\mu_{i1k}}{\mu_{i2k}}\right) = \log(\mu_{i1k}) - \log(\mu_{i2k}) \\ &= (\lambda + \lambda_i^X + \lambda_1^Y + \lambda_k^Z + \lambda_{i1}^{XY} + \lambda_{ik}^{XZ} + \lambda_{1k}^{YZ}) \\ &\quad - (\lambda + \lambda_i^X + \lambda_2^Y + \lambda_k^Z + \lambda_{i2}^{XY} + \lambda_{ik}^{XZ} + \lambda_{2k}^{YZ}) \\ &= (\lambda_1^Y - \lambda_2^Y) + (\lambda_{i1}^{XY} - \lambda_{i2}^{XY}) + (\lambda_{1k}^{YZ} - \lambda_{2k}^{YZ})\end{aligned}$$

The first parenthetical term does not depend on i or k. The second parenthetical term depends on the level i of X. The third parenthetical term depends on the level k of Z. The logit has the additive form

$$\text{logit}[P(Y=1)] = \alpha + \beta_i^X + \beta_k^Z \tag{7.7}$$

Section 4.3.3 discussed this model, in which the logit depends on X and Z in an additive manner. Additivity on the logit scale is the standard definition of "no interaction" for categorical variables. When Y is binary, the loglinear model of homogeneous association is equivalent to this logistic regression model. When X is also binary, model (7.7) and loglinear model (XY, XZ, YZ) are characterized by equal odds ratios between X and Y at each of the K levels of Z.

7.3.2 Example: Auto Accident Data Revisited

For the data on Maine auto accidents (Table 7.9), Section 7.2.6 showed that loglinear model (GLS, GI, LI, IS) fits well. That model is

$$\begin{aligned}\log\mu_{gi\ell s} &= \lambda + \lambda_g^G + \lambda_i^I + \lambda_\ell^L + \lambda_s^S + \lambda_{gi}^{GI} + \lambda_{g\ell}^{GL} + \lambda_{gs}^{GS} + \lambda_{i\ell}^{IL} + \lambda_{is}^{IS} \\ &\quad + \lambda_{\ell s}^{LS} + \lambda_{g\ell s}^{GLS}\end{aligned} \tag{7.8}$$

We could treat injury (I) as a response variable, and gender (G), location (L), and seat-belt use (S) as explanatory variables. You can check that this loglinear model implies a logistic model of the form

$$\text{logit}[P(I=1)] = \alpha + \beta_g^G + \beta_\ell^L + \beta_s^S \tag{7.9}$$

Here, G, L, and S all affect I, but without interacting. The parameters in the two models are related by

$$\beta_g^G = \lambda_{g1}^{GI} - \lambda_{g2}^{GI},\ \beta_\ell^L = \lambda_{1\ell}^{IL} - \lambda_{2\ell}^{IL},\ \beta_s^S = \lambda_{1s}^{IS} - \lambda_{2s}^{IS}$$

In the logit calculation, all terms in the loglinear model not having the injury index i in the subscript cancel.

Odds ratios relate to two-factor loglinear parameters and main-effect logistic parameters. For instance, in model (7.9), the log odds ratio for the effect of S on I equals $\beta_1^S - \beta_2^S$. This equals $\lambda_{11}^{IS} + \lambda_{22}^{IS} - \lambda_{12}^{IS} - \lambda_{21}^{IS}$ in the loglinear model. These values are the same no matter how software sets up constraints for the parameters. For example, $\hat{\beta}_1^S - \hat{\beta}_2^S = -0.817$ for model (7.9), and $\hat{\lambda}_{11}^{IS} + \hat{\lambda}_{22}^{IS} - \hat{\lambda}_{12}^{IS} - \hat{\lambda}_{21}^{IS} = -0.817$ for model (*GLS*, *GI*, *LI*, *IS*). We obtain the same results whether we use software for logistic regression or software for the equivalent loglinear model. Fitted values, goodness-of-fit statistics, residual *df*, and standardized residuals for logistic model (7.9) are identical to those in Tables 7.9–7.11 for loglinear model (*GLS*, *GI*, *IL*, *IS*).

Loglinear models are GLMs that treat the 16 cell counts in Table 7.9 as outcomes of 16 Poisson variates. Logistic models are GLMs that treat the table as outcomes of eight binomial variates giving injury counts at the eight possible settings of (g, ℓ, s). Although the sampling models differ, the results from fits of corresponding models are identical.

7.3.3 Correspondence Between Loglinear and Logistic Models

Refer back to the derivation of logistic model (7.7) from loglinear model (*XY*, *XZ*, *YZ*). The λ_{ik}^{XZ} term in model (*XY*, *XZ*, *YZ*) cancels when we form the logit. It might seem as if the model (*XY*, *YZ*) omitting this term is also equivalent to that logit model. Indeed, forming the logit on Y for loglinear model (*XY*, *YZ*) results in a logistic model of the same form. The loglinear model that has the same fit, however, is the one that contains a general interaction term for relationships among the explanatory variables. The logistic model does not describe relationships among explanatory variables, so it assumes nothing about their association structure.

Table 7.12 summarizes equivalent logistic and loglinear models for three-way tables when Y is a binary response variable. The loglinear model (*Y*, *XZ*) states that Y is jointly independent of both X and Z. It is equivalent to the special case of logistic model (7.7) with $\{\beta_i^X\}$ and $\{\beta_k^Z\}$ terms equal to zero. In each pairing of models in Table 7.12, the loglinear model contains the *XZ* association term relating the variables that are explanatory in the logistic models.

Logistic model (7.9) for a four-way table contains main effect terms for the explanatory variables, but no interaction terms. This model corresponds to the loglinear model that contains the fullest interaction term among the explanatory variables, and associations between each explanatory variable and the response I, namely model (*GLS*, *GI*, *LI*, *IS*).

7.3.4 Strategies in Model Selection

When there is a single response variable and it is binary, relevant loglinear models correspond to logistic models for that response. When the response has more than two

Table 7.12. Equivalent Loglinear and Logistic Models for a Three-Way Table With Binary Response Variable Y

Loglinear Symbol	Logistic Model	Logistic Symbol
(Y, XZ)	α	(—)
(XY, XZ)	$\alpha + \beta_i^X$	(X)
(YZ, XZ)	$\alpha + \beta_k^Z$	(Z)
(XY, YZ, XZ)	$\alpha + \beta_i^X + \beta_k^Z$	$(X + Z)$
(XYZ)	$\alpha + \beta_i^X + \beta_k^Z + \beta_{ik}^{XZ}$	$(X{*}Z)$

categories, relevant loglinear models correspond to baseline-category logit models (Section 6.1). In such cases it is more sensible to fit logistic models directly, rather than loglinear models. Indeed, one can see by comparing equations (7.8) and (7.9) how much simpler the logistic structure is. The loglinear approach is better suited for cases with more than one response variable, as in studying association patterns for the drug use example in Section 7.1.6. In summary, loglinear models are most natural when at least two variables are response variables and we want to study their association structure. Otherwise, logistic models are more relevant.

Selecting a loglinear model becomes more difficult as the number of variables increases, because of the increase in possible associations and interactions. One exploratory approach first fits the model having only single-factor terms, the model having only two-factor and single-factor terms, the model having only three-factor and lower terms, and so forth, as Section 7.2.6 showed. Fitting such models often reveals a restricted range of good-fitting models.

When certain marginal totals are fixed by the sampling design or by the response–explanatory distinction, the model should contain the term for that margin. This is because the ML fit forces the corresponding fitted totals to be identical to those marginal totals. To illustrate, suppose one treats the counts $\{n_{g+\ell+}\}$ in Table 7.9 as fixed at each combination of levels of G = gender and L = location. Then a loglinear model should contain the GL two-factor term, because this ensures that $\{\hat{\mu}_{g+\ell+} = n_{g+\ell+}\}$. That is, the model should be at least as complex as model (GL, S, I). If 20,629 women had accidents in urban locations, then the fitted counts have 20,629 women in urban locations.

Related to this point, the modeling process should concentrate on terms linking response variables and terms linking explanatory variables to response variables. Allowing a general interaction term among the explanatory variables has the effect of fixing totals at combinations of their levels. If G and L are both explanatory variables, models assuming conditional independence between G and L are not of interest.

For Table 7.9, I is a response variable, and S might be treated either as a response or explanatory variable. If it is explanatory, we treat the $\{n_{g+\ell s}\}$ totals as fixed and fit logistic models for the I response. If S is also a response, we consider the $\{n_{g+\ell+}\}$ totals as fixed and consider loglinear models that are at least as complex as (GL, S, I).

Such models focus on the effects of G and L on S and on I as well as the association between S and I.

7.4 INDEPENDENCE GRAPHS AND COLLAPSIBILITY

We next present a graphical representation for conditional independences in loglinear models. The graph indicates which pairs of variables are conditionally independent, given the others. This representation is helpful for revealing implications of models, such as determining when marginal and conditional odds ratios are identical.

7.4.1 Independence Graphs

An *independence graph* for a loglinear model has a set of vertices, each vertex representing a variable. There are as many vertices as dimensions of the contingency table. Any two vertices either are or are not connected by an edge. A missing edge between two vertices represents a conditional independence between the corresponding two variables.

For example, for a four-way table, the loglinear model (WX, WY, WZ, YZ) lacks XY and XZ association terms. It assumes that X and Y are independent and that X and Z are independent, conditional on the other two variables. The independence graph portrays this model.

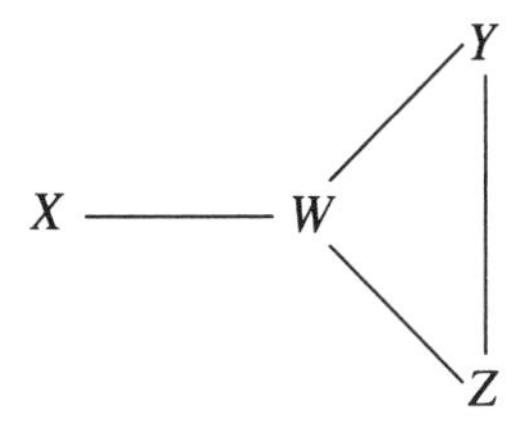

Edges connect W with X, W with Y, W with Z, and Y with Z. These represent pairwise conditional associations. Edges do not connect X with Y or X with Z, because those pairs are conditionally independent.

Two loglinear models that have the same conditional independences have the same independence graph. For instance, the independence graph just portrayed for model (WX, WY, WZ, YZ) is also the one for model (WX, WYZ) that also contains a three-factor WYZ term.

A *path* in an independence graph is a sequence of edges leading from one variable to another. Two variables X and Y are said to be *separated* by a subset of variables if all paths connecting X and Y intersect that subset. In the above graph, W separates X and Y, since any path connecting X with Y goes through W. The subset $\{W, Z\}$ also separates X and Y. A fundamental result states that two variables are conditionally independent given *any* subset of variables that separates them. Thus, not only are X and Y conditionally independent given W and Z, but also given W alone. Similarly, X and Z are conditionally independent given W alone.

The loglinear model (WX, XY, YZ) has independence graph

$$W \text{———} X \text{———} Y \text{———} Z$$

Here, W and Z are separated by X, by Y, and by X and Y. So, W and Z are independent given X alone or given Y alone or given both X and Y. Also, W and Y are independent, given X alone or given X and Z, and X and Z are independent, given Y alone or given Y and W.

7.4.2 Collapsibility Conditions for Three-Way Tables

Sometimes researchers collapse multiway contingency tables to make them simpler to describe and analyze. However, Section 2.7.5 showed that marginal associations may differ from conditional associations. For example, if X and Y are conditionally independent, given Z, they are not necessarily marginally independent. Under the following *collapsibility conditions*, a model's odds ratios *are* identical in partial tables as in the marginal table:

For three-way tables, XY marginal and conditional odds ratios are identical if either Z and X are conditionally independent or if Z and Y are conditionally independent.

The conditions state that the variable treated as the control (Z) is conditionally independent of X or Y, or both. These conditions correspond to loglinear models (XY, YZ) and (XY, XZ). That is, the XY association is identical in the partial tables and the marginal table for models with independence graphs

$$X \text{———} Y \text{———} Z \quad \text{and} \quad Y \text{———} X \text{———} Z$$

or even simpler models.

For Table 7.3 from Section 7.1.6 with A = alcohol use, C = cigarette use, and M = marijuana use, the model (AM, CM) of AC conditional independence has independence graph

$$A \text{———} M \text{———} C$$

Consider the AM association, identifying C with Z in the collapsibility conditions. In this model, since C is conditionally independent of A, the AM conditional odds ratios are the same as the AM marginal odds ratio collapsed over C. In fact, from Table 7.5, both the fitted marginal and conditional AM odds ratios equal 61.9. Similarly, the CM association is collapsible. The AC association is not, however. The collapsibility conditions are not satisfied, because M is conditionally dependent with both A and C in model (AM, CM). Thus, A and C may be marginally dependent, even though they are conditionally independent in this model. In fact, from Table 7.5, the model's fitted AC marginal odds ratio equals 2.7, not 1.0.

For the model (AC, AM, CM) of homogeneous association, no pair is conditionally independent. No collapsibility conditions are fulfilled. In fact, from Table 7.5, for this model each pair of variables has quite different fitted marginal and conditional associations. When a model contains all two-factor effects, collapsing over any variable may cause effects to change.

7.4.3 Collapsibility and Logistic Models

The collapsibility conditions apply also to logistic models. For example, consider a clinical trial to study the association between a binary response Y and a binary treatment variable X, using data from several centers (Z). The model

$$\text{logit}[P(Y = 1)] = \alpha + \beta x + \beta_k^Z \tag{7.10}$$

assumes that the treatment effect β is the same for each center. Since this model corresponds to loglinear model (XY, XZ, YZ), the estimated treatment effect may differ if we collapse the table over the center factor. The estimated XY conditional odds ratio, $\exp(\hat{\beta})$, differs from the sample XY odds ratio in the marginal 2×2 table.

The simpler model that lacks the center effects is

$$\text{logit}[P(Y = 1)] = \alpha + \beta x$$

For each treatment, this model states that the success probability is identical for each center. For it, the conditional and marginal treatment effects are identical, because the model states that Z is conditionally independent of Y. This model corresponds to loglinear model (XY, XZ) with independence graph Y —— X —— Z, for which the XY association is collapsible. In practice, this suggests that, when center effects are negligible, the estimated treatment effect is similar to the marginal XY odds ratio.

7.4.4 Collapsibility and Independence Graphs for Multiway Tables

The collapsibility conditions extend to multiway tables:

> Suppose that variables in a model for a multiway table partition into three mutually exclusive subsets, A, B, C, such that B separates A and C; that is, the model does not contain parameters linking variables from A with variables from C. When we collapse the table over the variables in C, model parameters relating variables in A and model parameters relating variables in A with variables in B are unchanged.

That is, when the subsets of variables have the form

$$A \text{ —— } B \text{ —— } C$$

collapsing over the variables in C, the same parameter values relate the variables in A, and the same parameter values relate variables in A to variables in B. It follows that the corresponding associations are unchanged, as described by odds ratios based on those parameters.

7.4.5 Example: Model Building for Student Drug Use

Sections 7.1.6 and 7.2 analyzed data on usage of alcohol (A), cigarettes (C), and marijuana (M) by high school students. When we classify the students also by gender (G) and race (R), the five-dimensional contingency table shown in Table 7.13 results. In selecting a model, we treat A, C, and M as response variables and G and R as explanatory variables. Since G and R are explanatory, it does not make sense to estimate association or assume conditional independence for that pair. From remarks in Section 7.3.4, a model should contain the GR term. Including this term forces the GR fitted marginal totals to be the same as the corresponding sample marginal totals.

Table 7.13. Alcohol, Cigarette, and Marijuana Use for High School Seniors, by Gender and Race

		Marijuana Use							
		White				Other			
		Female		Male		Female		Male	
Alcohol Use	Cigarette Use	Yes	No	Yes	No	Yes	No	Yes	No
Yes	Yes	405	268	453	228	23	23	30	19
	No	13	218	28	201	2	19	1	18
No	Yes	1	17	1	17	0	1	1	8
	No	1	117	1	133	0	12	0	17

Source: Professor Harry Khamis, Wright State University.

Table 7.14 summarizes goodness-of-fit tests for several models. Because many cell counts are small, the chi-squared approximation for G^2 may be poor. It is best not to take the G^2 values too seriously, but this index is useful for comparing models.

The first model listed in Table 7.14 contains only the GR association and assumes conditional independence for the other nine pairs of associations. It fits horribly. The homogeneous association model, on the other hand, seems to fit well. The only large standardized residual results from a fitted value of 3.1 in the cell having a count of 8. The model containing all the three-factor terms also fits well, but the improvement in fit is not great (difference in G^2 of $15.3 - 5.3 = 10.0$ based on $df = 16 - 6 = 10$). Thus, we consider models without three-factor terms. Beginning with the homogeneous association model, we eliminate two-factor terms that do not make significant contributions. We use a backward elimination process, sequentially taking out terms for which the resulting increase in G^2 is smallest, when refitting the model. However, we do not delete the GR term relating the explanatory variables.

Table 7.14. Goodness-of-Fit Tests for Models Relating Alcohol (A), Cigarette (C), and Marijuana (M) Use, by Gender (G) and Race (R)

Model	G^2	df
1. Mutual independence + GR	1325.1	25
2. Homogeneous association	15.3	16
3. All three-factor terms	5.3	6
4a. (2)–AC	201.2	17
4b. (2)–AM	107.0	17
4c. (2)–CM	513.5	17
4d. (2)–AG	18.7	17
4e. (2)–AR	20.3	17
4f. (2)–CG	16.3	17
4g. (2)–CR	15.8	17
4h. (2)–GM	25.2	17
4i. (2)–MR	18.9	17
5. $(AC, AM, CM, AG, AR, GM, GR, MR)$	16.7	18
6. $(AC, AM, CM, AG, AR, GM, GR)$	19.9	19
7. (AC, AM, CM, AG, AR, GR)	28.8	20

Table 7.14 shows the start of this process. Nine pairwise associations are candidates for removal from model (2), shown in models numbered (4a)–(4i). The smallest increase in G^2, compared with model (2), occurs in removing the CR term. The increase is $15.8 - 15.3 = 0.5$, based on $df = 17 - 16 = 1$, so this elimination seems reasonable. After removing the CR term (model 4g), the smallest additional increase results from removing the CG term (model 5). This results in $G^2 = 16.7$ with $df = 18$, an increase in G^2 of 0.9 based on $df = 1$. Removing next the MR term (model 6) yields $G^2 = 19.9$ with $df = 19$, a change in G^2 of 3.2 based on $df = 1$.

At this stage, the only large standardized residual occurs for a fitted value of 2.9 in the cell having a count of 8. Additional removals have a more severe effect. For instance, removing next the AG term increases G^2 by 5.3, based on $df = 1$, for a P-value of 0.02. We cannot take such P-values too literally, because these tests are suggested by the data. However, it seems safest not to drop additional terms. Model (6), denoted by $(AC, AM, CM, AG, AR, GM, GR)$, has independence graph

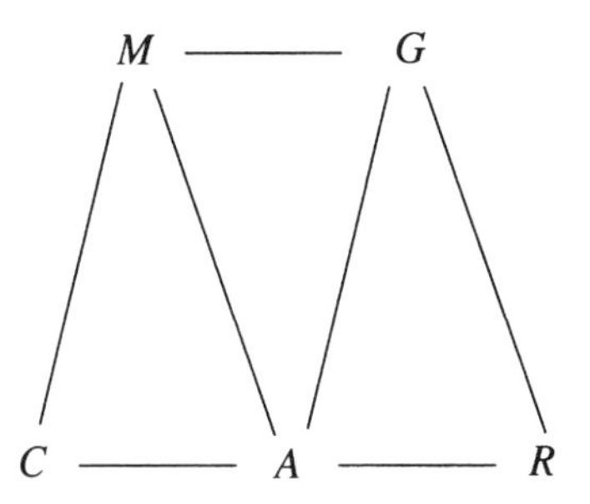

Consider the sets $\{C\}$, $\{A, M\}$, and $\{G, R\}$. For this model, every path between C and $\{G, R\}$ involves a variable in $\{A, M\}$. Given the outcome on alcohol use and marijuana use, the model states that cigarette use is independent of both gender and race. Collapsing over the explanatory variables race and gender, the conditional associations between C and A and between C and M are the same as with the model (AC, AM, CM) fitted in Section 7.1.6.

7.4.6 Graphical Models

The first independence graph shown in Section 7.4.1 lacked edges between X and Y and between X and Z. As noted there, that graph results both from loglinear model (WX, WY, WZ, YZ) and from loglinear model (WX, WYZ). A subclass of loglinear models, called *graphical models*, have a unique correspondence between the models and the independence graph representations. For any group of variables in an independence graph having no missing edges, the graphical model contains the highest-order term for those variables.

For example, the first independence graph shown in Section 7.4.1 has no missing edges for the group of variables $\{W, Y, Z\}$. Thus, the corresponding graphical model must contain the three-factor λ^{WYZ} term (as well as all its lower-order terms). Likewise, the group of variables $\{X, W\}$ has no missing edge. Therefore, the corresponding graphical model must contain the two-factor λ^{WX} term. The graphical model for this independence graph is the loglinear model (WX, WYZ).

The loglinear model (WX, WY, WZ, YZ) is not a graphical model. This is because the group of variables $\{W, Y, Z\}$ has no missing edges, yet the loglinear model does not contain the three-factor term for those variables.

A substantial theory has been developed for the subclass of loglinear models that are graphical models. This is beyond our scope here, but for a nontechnical introduction see Whittaker (1990).

7.5 MODELING ORDINAL ASSOCIATIONS

The loglinear models presented so far have a serious limitation: they treat all classifications as nominal. If we change the order of a variable's categories in any way, we get the same fit. For ordinal variables, these models ignore important information.

Table 7.15, from a General Social Survey, illustrates the inadequacy of ordinary loglinear models for analyzing ordinal data. Subjects were asked their opinion about a man and woman having sexual relations before marriage. They were also asked whether methods of birth control should be made available to teenagers between the ages of 14 and 16. Both classifications have ordered categories. The loglinear model of independence, denoted by (X, Y), has goodness-of-fit statistics $G^2(X, Y) = 127.6$ and $X^2(X, Y) = 128.7$, based on $df = 9$. These tests of fit are equivalently the tests of independence of Section 2.4. The model fits poorly, providing strong evidence of dependence. Yet, adding the ordinary association term makes the model saturated [see model (7.2)] and of little use.

Table 7.15. Opinions about Premarital Sex and Availability of Teenage Birth Control, Showing Independence Model Fit, Standardized Residuals for that Fit, and Linear-by-Linear Association Model Fit

	Teenage Birth Control			
Premarital Sex	Strongly Disagree	Disagree	Agree	Strongly Agree
Always wrong	81 (42.4) 7.6 (80.9)	68 (51.2) 3.1 (67.6)	60 (86.4) −4.1 (69.4)	38 (67.0) −4.8 (29.1)
Almost always wrong	24 (16.0) 2.3 (20.8)	26 (19.3) 1.8 (23.1)	29 (32.5) −0.8 (31.5)	14 (25.2) −2.8 (17.6)
Wrong only sometimes	18 (30.1) −2.7 (24.4)	41 (36.3) 1.0 (36.1)	74 (61.2) 2.2 (65.7)	42 (47.4) −1.0 (48.8)
Not wrong at all	36 (70.6) −6.1 (33.0)	57 (85.2) −4.6 (65.1)	161 (143.8) 2.4 (157.4)	157 (111.4) 6.8 (155.5)

Source: General Social Survey, 1991.

Table 7.15 also contains fitted values and standardized residuals (Section 2.4.5). The residuals in the corners of the table are large. Observed counts are much larger than the independence model predicts in the corners where both responses are the most negative possible ("always wrong" with "strongly disagree") or the most positive possible ("not wrong at all" with "strongly agree"). By contrast, observed counts are much smaller than fitted counts in the other two corners. Cross-classifications of ordinal variables often exhibit their greatest deviations from independence in the corner cells. This pattern suggests a positive trend. Subjects who feel more favorable to making birth control available to teenagers also tend to feel more tolerant about premarital sex.

The independence model is too simple to fit most data well. Models for ordinal variables use association terms that permit negative or positive trends. The models are more complex than the independence model yet simpler than the saturated model.

7.5.1 Linear-by-Linear Association Model

An ordinal loglinear model assigns scores $\{u_i\}$ to the I rows and $\{v_j\}$ to the J columns. To reflect category orderings, $u_1 \leq u_2 \leq \cdots \leq u_I$ and $v_1 \leq v_2 \leq \cdots \leq v_J$.

In practice, the most common choice is $\{u_i = i\}$ and $\{v_j = j\}$, the row and column numbers. The model is

$$\log \mu_{ij} = \lambda + \lambda_i^X + \lambda_j^Y + \beta u_i v_j \tag{7.11}$$

The independence model is the special case $\beta = 0$. Since the model has one more parameter (β) than the independence model, its residual *df* are 1 less, $df = (I-1)(J-1) - 1 = IJ - I - J$.

The final term in model (7.11) represents the deviation of $\log \mu_{ij}$ from independence. The deviation is linear in the Y scores at a fixed level of X and linear in the X scores at a fixed level of Y. In column j, for instance, the deviation is a linear function of X, having form (slope) × (score for X), with slope βv_j. Because of this property, equation (7.11) is called the *linear-by-linear association model* (abbreviated, $L \times L$). This linear-by-linear deviation implies that the model has its greatest departures from independence in the corners of the table.

The parameter β in model (7.11) specifies the direction and strength of association. When $\beta > 0$, there is a tendency for Y to increase as X increases. Expected frequencies are larger than expected, under independence, in cells of the table where X and Y are both high or both low. When $\beta < 0$, there is a tendency for Y to decrease as X increases. When we fit the model to data, the correlation between the row scores for X and the column scores for Y is the same for the observed counts as it is for the joint distribution given by the fitted counts. Thus, the fitted counts display the same positive or negative trend as the observed data.

For the 2×2 table created with the four cells intersecting rows a and c with columns b and d, the $L \times L$ model has the odds ratio

$$\frac{\mu_{ab}\mu_{cd}}{\mu_{ad}\mu_{cb}} = \exp[\beta(u_c - u_a)(v_d - v_b)] \tag{7.12}$$

The association is stronger as $|\beta|$ increases. For given β, pairs of categories that are farther apart have odds ratios farther from 1.

The odds ratios formed using adjacent rows and adjacent columns are called *local odds ratios*. Figure 7.1 portrays some local odds ratios. For unit-spaced scores such as $\{u_i = i\}$ and $\{v_j = j\}$, equation (7.12) simplifies so that the local odds ratios have the common value

$$\frac{\mu_{ab}\mu_{a+1,b+1}}{\mu_{a,b+1}\mu_{a+1,b}} = \exp(\beta)$$

Any set of equally spaced row and column scores has the property of uniform local odds ratios. This special case of the model is called *uniform association*.

7.5.2 Example: Sex Opinions

Table 7.15 also reports fitted values for the linear-by-linear association model applied to the opinions about premarital sex and availability of teen birth control, using row

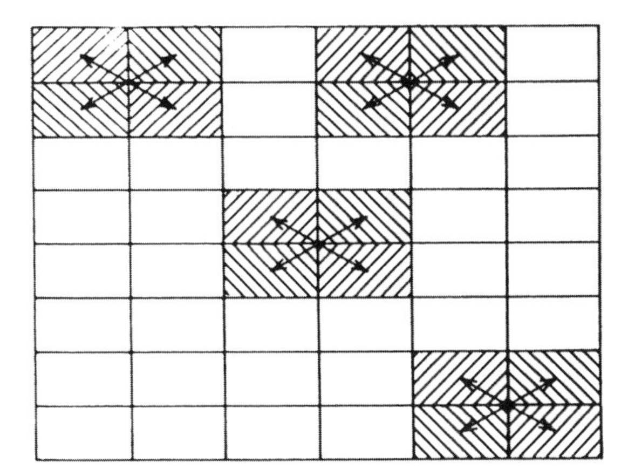

Figure 7.1. Constant local odds ratio implied by uniform association model.

scores {1, 2, 3, 4} and column scores {1, 2, 3, 4}. The goodness-of-fit statistics for this uniform association version of the model are $G^2(L \times L) = 11.5$ and $X^2(L \times L) = 11.5$, with $df = 8$. Compared with the independence model, the $L \times L$ model provides a dramatic improvement in fit, especially in the corners of the table.

The ML estimate of the association parameter is $\hat{\beta} = 0.286$, with $SE = 0.028$. The positive estimate suggests that subjects having more favorable attitudes about the availability of teen birth control also tend to have more tolerant attitudes about premarital sex. The estimated local odds ratio is $\exp(\hat{\beta}) = \exp(0.286) = 1.33$. The strength of association seems weak. From equation (7.12), however, nonlocal odds ratios are stronger. For example, the estimated odds ratio for the four corner cells equals

$$\exp[\hat{\beta}(u_4 - u_1)(v_4 - v_1)] = \exp[0.286(4 - 1)(4 - 1)] = \exp(2.57) = 13.1$$

Or, using the fitted values from Table 7.15, (80.9)(155.5)/(29.1)(33.0) = 13.1.

Two sets of scores having the same spacings yield the same $\hat{\beta}$ and the same fit. For instance, $\{u_1 = 1,\ u_2 = 2,\ u_3 = 3,\ u_4 = 4\}$ yields the same results as $\{u_1 = -1.5,\ u_2 = -0.5,\ u_3 = 0.5,\ u_4 = 1.5\}$. Other sets of equally spaced scores yield the same fit but an appropriately rescaled $\hat{\beta}$. For instance, the row scores {2, 4, 6, 8} with $\{v_j = j\}$ also yield $G^2 = 11.5$, but then $\hat{\beta} = 0.143$ with $SE = 0.014$ (both half as large).

To treat categories 2 and 3 as farther apart than categories 1 and 2 or categories 3 and 4, we could instead use scores such as {1, 2, 4, 5} for the rows and columns. The $L \times L$ model then has $G^2 = 8.8$. One need not, however, regard the model scores as approximate distances between categories. They simply imply a certain structure for the odds ratios. From equation (7.12), fitted odds ratios are stronger for pairs of categories having greater distances between scores.

7.5.3 Ordinal Tests of Independence

For the linear-by-linear association model, the hypothesis of independence is H_0: $\beta = 0$. The likelihood-ratio test statistic equals the reduction in G^2 goodness-of-fit statistics between the independence (X, Y) and $L \times L$ models,

$$G^2[(X, Y) \mid L \times L] = G^2(X, Y) - G^2(L \times L) \tag{7.13}$$

This statistic refers to a single parameter (β), and has $df = 1$. For Table 7.15, $G^2(X, Y) - G^2(L \times L) = 127.6 - 11.5 = 116.1$ has $P < 0.0001$, extremely strong evidence of an association.

The Wald statistic $z^2 = (\hat{\beta}/SE)^2$ provides an alternative chi-squared test statistic having $df = 1$. For these data, $z^2 = (0.286/0.0282)^2 = 102.4$ also shows strong evidence of a positive trend. The correlation statistic (2.10) of Section 2.5.1 for testing independence is usually similar to the likelihood-ratio and Wald statistics for H_0: $\beta = 0$ in this model. (In fact, it is the score statistic.) For Table 7.15, it equals 112.6, also with $df = 1$.

Generalizations of the linear-by-linear association model exist for multiway tables. See Problem 7.25. Recall that Sections 6.2 and 6.3 presented other ways of using ordinality, based on models that create ordinal logits. To distinguish between an ordinal response variable and explanatory variables, it is more sensible to apply an ordinal logit model than a loglinear model. Many loglinear models for ordinal variables have simple representations as adjacent-category logit models. See Problem 7.26.

PROBLEMS

7.1 For Table 2.1 on X = gender and Y = belief in an afterlife, Table 7.16 shows the results of fitting the independence loglinear model.

a. Report and interpret results of a goodness-of-fit test.

b. Report $\{\hat{\lambda}_j^Y\}$. Interpret $\hat{\lambda}_1^Y - \hat{\lambda}_2^Y$.

Table 7.16. Computer Output for Problem 7.1 on Belief in Afterlife

Criteria For Assessing Goodness Of Fit		
Criterion	DF	Value
Deviance	1	0.8224
Pearson Chi-Square	1	0.8246

Parameter		DF	Estimate	Std. Error
Intercept		1	4.5849	0.0752
gender	females	1	0.2192	0.0599
gender	males	0	0.0000	0.0000
belief	yes	1	1.4165	0.0752
belief	no	0	0.0000	0.0000

7.2 For the saturated model with Table 2.1, software reports the $\{\hat{\lambda}_{ij}^{XY}\}$ estimates:

```
Parameter                              DF    Estimate    Std Error
gender*belief    females    yes         1     0.1368       0.1507
gender*belief    females    no          0     0.0000       0.0000
gender*belief    males      yes         0     0.0000       0.0000
gender*belief    males      no          0     0.0000       0.0000
```

Show how to use these to estimate the odds ratio.

7.3 Table 7.17 is from a General Social Survey. White subjects in the sample were asked: (B) Do you favor busing (Negro/Black) and white school children from one school district to another?, (P) If your party nominated a (Negro/Black) for President, would you vote for him if he were qualified for the job?, (D) During the last few years, has anyone in your family brought a friend who was a (Negro/Black) home for dinner? The response scale for each item was (1 = Yes, 2 = No or Do not know). Table 7.18 shows output from fitting model (BD, BP, DP). Estimates equal 0 at the second category for any variable.

Table 7.17. Data for Problem 7.3

		Home	
President	Busing	1	2
1	1	41	65
	2	72	175
2	1	2	9
	2	4	55

Source: 1991 General Social Survey, with categories 1 = yes, 2 = no or do not know.

a. Analyze the model goodness of fit. Interpret.

b. Estimate the conditional odds ratios for each pair of variables. Interpret.

c. Show all steps of the likelihood-ratio test for the BP association, including explaining which loglinear model holds under the null hypothesis. Interpret.

d. Construct a 95% confidence interval for the BP conditional odds ratio. Interpret.

7.4 In a General Social Survey respondents were asked "Do you support or oppose the following measures to deal with AIDS? (1) Have the government pay all of the health care costs of AIDS patients; (2) develop a government information program to promote safe sex practices, such as the use of condoms." Table 7.19 shows responses on these two items, classified also by the respondent's gender.

Table 7.18. Output for Fitting Model to Table 7.17

Criteria For Assessing Goodness Of Fit

Criterion	DF	Value
Deviance	1	0.4794
Pearson Chi-Square	1	0.5196

Analysis Of Parameter Estimates

Parameter	DF	Estimate	Std Error
Intercept	1	3.9950	0.1346
president	1	1.1736	0.1536
busing	1	−1.7257	0.3300
home	1	−2.4533	0.4306
president*busing	1	0.7211	0.3539
president*home	1	1.5520	0.4436
busing*home	1	0.4672	0.2371

LR Statistics

Source	DF	Chi-Square	Pr > ChiSq
president*busing	1	4.64	0.0313
president*home	1	17.18	<.0001
busing*home	1	3.83	0.0503

Denote the variables by G for gender, H for opinion on health care costs, and I for opinion on an information program.

a. Fit the model (GH, GI, HI) and test its goodness of fit.

b. For this model, estimate the GI conditional odds ratio, construct a 95% confidence interval, and interpret.

c. Given the model, test whether G and I are conditionally independent. Do you think the GI term needs to be in the model?

Table 7.19. Data for Problem 7.4 on Measures to Deal with AIDS

		Health Opinion	
Gender	Information Opinion	Support	Oppose
Male	Support	76	160
	Oppose	6	25
Female	Support	114	181
	Oppose	11	48

Source: 1988 General Social Survey.

7.5 Refer to Table 2.10 on death penalty verdicts. Let D = defendant's race, V = victim's race, and P = death penalty verdict. Table 7.20 shows output for fitting model (DV, DP, PV). Estimates equal 0 at the second category for any variable.

a. Report the estimated conditional odds ratio between D and P at each level of V. Interpret.

b. The marginal odds ratio between D and P is 1.45. Contrast this odds ratio with that in (a), and remark on how Simpson's paradox occurs for these data.

c. Test the goodness of fit of this model. Interpret.

d. Specify the corresponding logistic model with P as the response.

Table 7.20. Computer Output for Problem 7.5 on Death Penalty

Criteria For Assessing Goodness Of Fit

Criterion	DF	Value
Deviance	1	0.3798
Pearson Chi-Square	1	0.1978

Parameter			DF	Estimate	Standard Error	LR 95% Confidence Limits	
Intercept			1	3.9668	0.1374	3.6850	4.2245
v	black		1	−5.6696	0.6459	−7.0608	−4.4854
d	black		1	−1.5525	0.3262	−2.2399	−0.9504
p	no		1	2.0595	0.1458	1.7836	2.3565
v*d	black	black	1	4.5950	0.3135	4.0080	5.2421
v*p	black	no	1	2.4044	0.6006	1.3068	3.7175
d*p	black	no	1	−0.8678	0.3671	−1.5633	−0.1140

LR Statistics

Source	DF	Chi-Square	Pr > ChiSq
v*d	1	384.05	<.0001
v*p	1	20.35	<.0001
d*p	1	5.01	0.0251

7.6 Table 7.21 shows the result of cross classifying a sample of people from the MBTI Step II National Sample, collected and compiled by CPP Inc., on the four scales of the Myers–Briggs personality test: Extroversion/Introversion (E/I), Sensing/iNtuitive (S/N), Thinking/Feeling (T/F) and Judging/Perceiving (J/P). The 16 cells in this table correspond to the 16 personality types: ESTJ, ESTP, ESFJ, ESFP, ENTJ, ENTP, ENFJ, ENFP, ISTJ, ISTP, ISFJ, ISFP, INTJ, INTP, INFJ, INFP.

a. Fit the loglinear model by which the variables are mutually independent. Report the results of the goodness-of-fit test.

Table 7.21. Data on Four Scales of the Myers–Briggs Personality Test

Extroversion/Introversion	E				I			
Sensing/iNtuitive	S		N		S		N	
	Thinking/Feeling							
Judging/Perceiving	T	F	T	F	T	F	T	F
J	77	106	23	31	140	138	13	31
P	42	79	18	80	52	106	35	79

Source: Reproduced with special permission of CPP, Inc., Mountain View, CA 94043.

b. Fit the loglinear model of homogeneous association. Based on the fit, show that the estimated conditional association is strongest between the S/N and J/P scales.

c. Using the model in (**b**), show that there is not strong evidence of conditional association between the E/I and T/F scale or between the E/I and J/P scales.

7.7 Refer to the previous exercise. Table 7.22 shows the fit of the model that assumes conditional independence between E/I and T/F and between E/I and J/P but has the other pairwise associations.

a. Compare this to the fit of the model containing all the pairwise associations, which has deviance 10.16 with $df = 5$. What do you conclude?

b. Show how to use the limits reported to construct a 95% likelihood-ratio confidence interval for the conditional odds ratio between the S/N and J/P scales. Interpret.

Table 7.22. Partial Output for Fitting Loglinear Model to Table 7.21

```
              Criteria For Assessing Goodness Of Fit
          Criterion                 DF          Value
          Deviance                   7         12.3687
          Pearson Chi-Square         7         12.1996

                   Analysis Of Parameter Estimates
                                     Standard  LR 95% confidence   Wald Chi-
Parameter          DF   Estimate     Error          limits          Square
EI*SN       e  n    1     0.3219     0.1360    0.0553    0.5886      5.60
SN*TF       n  f    1     0.4237     0.1520    0.1278    0.7242      7.77
SN*JP       n  j    1    -1.2202     0.1451   -1.5075   -0.9382     70.69
TF*JP       f  j    1    -0.5585     0.1350   -0.8242   -0.2948     17.12
```

c. The estimates shown use N for the first category of the S/N scale and J for the first category of the J/P scale. Suppose you instead use S for the first category of the S/N scale. Then, report the estimated conditional odds ratio and the 95% likelihood-ratio confidence interval, and interpret.

7.8 Refer to the previous two exercises. PROC GENMOD in SAS reports the maximized log likelihood as 3475.19 for the model of mutual independence ($df = 11$), 3538.05 for the model of homogeneous association ($df = 5$), and 3539.58 for the model containing all the three-factor interaction terms.

a. Write the loglinear model for each case, and show that the numbers of parameters are 5, 11, and 15.

b. According to AIC (see Section 5.1.5), which of these models seems best? Why?

7.9 Table 7.23 refers to applicants to graduate school at the University of California, Berkeley for the fall 1973 session. Admissions decisions are presented by gender of applicant, for the six largest graduate departments. Denote the three variables by A = whether admitted, G = gender, and D = department. Fit loglinear model (AD, AG, DG).

a. Report the estimated AG conditional odds ratio, and compare it with the AG marginal odds ratio. Why are they so different?

b. Report G^2 and df values, and comment on the quality of fit. Conduct a residual analysis. Describe the lack of fit.

c. Deleting the data for Department 1, re-fit the model. Interpret.

d. Deleting the data for Department 1 and treating A as the response variable, fit an equivalent logistic model for model (AD, AG, DG) in (**c**). Show how to use each model to obtain an odds ratio estimate of the effect of G on A, controlling for D.

Table 7.23. Data for Problem 7.9 on Admissions to Berkeley

	Whether Admitted			
	Male		Female	
Department	Yes	No	Yes	No
1	512	313	89	19
2	353	207	17	8
3	120	205	202	391
4	138	279	131	244
5	53	138	94	299
6	22	351	24	317
Total	1198	1493	557	1278

Note: For further details, see Bickel et al., *Science*, **187:** 398–403, 1975.

Table 7.24. Data for Problem 7.10

Safety Equipment in Use	Whether Ejected	Injury	
		Nonfatal	Fatal
Seat belt	Yes	1,105	14
	No	411,111	483
None	Yes	4,624	497
	No	157,342	1008

Source: Florida Department of Highway Safety and Motor Vehicles.

7.10 Table 7.24 is based on automobile accident records in 1988, supplied by the state of Florida Department of Highway Safety and Motor Vehicles. Subjects were classified by whether they were wearing a seat belt, whether ejected, and whether killed.

a. Find a loglinear model that describes the data well. Interpret the associations.

b. Treating whether killed as the response variable, fit an equivalent logistic model. Interpret the effects on the response.

c. Since the sample size is large, goodness-of-fit statistics are large unless the model fits very well. Calculate the dissimilarity index, and interpret.

7.11 Refer to the loglinear models in Section 7.2.6 for the auto accident injury data shown in Table 7.9. Explain why the fitted odds ratios in Table 7.11 for model (GI, GL, GS, IL, IS, LS) suggest that the most likely case for injury is accidents for females not wearing seat belts in rural locations.

7.12 Consider the following two-stage model for Table 7.9. The first stage is a logistic model with S as the response, for the three-way $G \times L \times S$ table. The second stage is a logistic model with these three variables as predictors for I in the four-way table. Explain why this composite model is sensible, fit the models, and interpret results.

7.13 Table 7.25 is from a General Social Survey. Subjects were asked about government spending on the environment (E), health (H), assistance to big cities (C), and law enforcement (L). The common response scale was $(1 =$ too little, $2 =$ about right, $3 =$ too much).

a. Table 7.26 shows some results, including the two-factor estimates, for the homogeneous association model. All estimates at category 3 of each variable equal 0. Test the model goodness of fit, and interpret.

b. Explain why the estimated conditional log odds ratio for the "too much" and "too little" categories of E and H equals

$$\hat{\lambda}_{11}^{EH} + \hat{\lambda}_{33}^{EH} - \hat{\lambda}_{13}^{EH} - \hat{\lambda}_{31}^{EH}$$

Table 7.25. Opinions about Government Spending

Cities		1			2			3		
Law Enforcement		1	2	3	1	2	3	1	2	3
Environment	Health									
1	1	62	17	5	90	42	3	74	31	11
	2	11	7	0	22	18	1	19	14	3
	3	2	3	1	2	0	1	1	3	1
2	1	11	3	0	21	13	2	20	8	3
	2	1	4	0	6	9	0	6	5	2
	3	1	0	1	2	1	1	4	3	1
3	1	3	0	0	2	1	0	9	2	1
	2	1	0	0	2	1	0	4	2	0
	3	1	0	0	0	0	0	1	2	3

Source: 1989 General Social Survey; 1 = too little, 2 = about right, 3 = too much.

which has estimated $SE = 0.523$. Show that a 95% confidence interval for the true odds ratio equals (3.1, 24.4). Interpret.

c. Estimate the conditional odds ratios using the "too much" and "too little" categories for each of the other pairs of variables. Summarize the associations. Based on these results, which term(s) might you consider dropping from the model? Why?

Table 7.26. Output for Fitting Model to Table 7.25

Criteria For Assessing Goodness Of Fit

Criterion	DF	Value
Deviance	48	31.6695
Pearson Chi-Square	48	26.5224

Parameter			DF	Estimate	Standard Error	Parameter			DF	Estimate	Standard Error
e*h	1	1	1	2.1425	0.5566	h*c	1	1	1	−0.1865	0.4547
e*h	1	2	1	1.4221	0.6034	h*c	1	2	1	0.7464	0.4808
e*h	2	1	1	0.7294	0.5667	h*c	2	1	1	−0.4675	0.4978
e*h	2	2	1	0.3183	0.6211	h*c	2	2	1	0.7293	0.5023
e*l	1	1	1	−0.1328	0.6378	h*l	1	1	1	1.8741	0.5079
e*l	1	2	1	0.3739	0.6975	h*l	1	2	1	1.0366	0.5262
e*l	2	1	1	−0.2630	0.6796	h*l	2	1	1	1.9371	0.6226
e*l	2	2	1	0.4250	0.7361	h*l	2	2	1	1.8230	0.6355
e*c	1	1	1	1.2000	0.5177	c*l	1	1	1	0.8735	0.4604
e*c	1	2	1	1.3896	0.4774	c*l	1	2	1	0.5707	0.4863
e*c	2	1	1	0.6917	0.5605	c*l	2	1	1	1.0793	0.4326
e*c	2	2	1	1.3767	0.5024	c*l	2	2	1	1.2058	0.4462

7.14 Table 7.27, from a General Social Survey, relates responses on R = religious service attendance (1 = at most a few times a year, 2 = at least several times a year), P = political views (1 = Liberal, 2 = Moderate, 3 = Conservative), B = birth control availability to teenagers between ages of 14 and 16 (1 = agree, 2 = disagree), S = sexual relations before marriage (1 = wrong only sometimes or not wrong at all, 2 = always or almost always wrong).

a. Find a loglinear model that fits these data well.

b. Interpret this model by estimating conditional odds ratios for each pair of variables.

c. Consider the logistic model predicting (S) using the other variables as main-effect predictors, without any interaction. Fit the corresponding loglinear model. Does it fit adequately? Interpret the effects of the predictors on the response, and compare to results from (**b**).

d. Draw the independence graph of the loglinear model selected in (**a**). Remark on conditional independence patterns. For each pair of variables, indicate whether the fitted marginal and conditional associations are identical.

Table 7.27. Data for Problem 7.14

		Premarital Sex							
		1				2			
	Religious Attendence	1		2		1		2	
	Birth Control	1	2	1	2	1	2	1	2
	1	99	15	73	25	8	4	24	22
Political views	2	73	20	87	37	20	13	50	60
	3	51	19	51	36	6	12	33	88

Source: 1991 General Social Survey.

7.15 Refer to Table 7.13 in Section 7.4.5 on the substance use survey, which also classified students by gender (G) and race (R).

a. Analyze these data using logistic models, treating marijuana use as the response variable. Select a model.

b. Which loglinear model is equivalent to your choice of logistic model?

7.16 For the Maine accident data modeled in Section 7.3.2:

a. Verify that logistic model (7.9) follows from loglinear model (GLS, GI, LI, IS).

b. Show that the conditional log odds ratio for the effect of S on I equals $\beta_1^S - \beta_2^S$ in the logistic model and $\lambda_{11}^{IS} + \lambda_{22}^{IS} - \lambda_{12}^{IS} - \lambda_{21}^{IS}$ in the loglinear model.

7.17 For a multiway contingency table, when is a logistic model more appropriate than a loglinear model, and when is a loglinear model more appropriate?

7.18 For a three-way table, consider the independence graph,

$$X \text{———} Z \qquad Y$$

a. Write the corresponding loglinear model.

b. Which, if any, pairs of variables are conditionally independent?

c. If Y is a binary response, what is the corresponding logistic model?

d. Which pairs of variables have the same marginal association as their conditional association?

7.19 Consider loglinear model (WXZ, WYZ).

a. Draw its independence graph, and identify variables that are conditionally independent.

b. Explain why this is the most general loglinear model for a four-way table for which X and Y are conditionally independent.

7.20 For a four-way table, are X and Y independent, given Z alone, for model (a) (WX, XZ, YZ, WZ), (b) (WX, XZ, YZ, WY)?

7.21 Refer to Problem 7.13 with Table 7.25.

a. Show that model (CE, CH, CL, EH, EL, HL) fits well. Show that model (CEH, CEL, CHL, EHL) also fits well but does not provide a significant improvement. Beginning with (CE, CH, CL, EH, EL, HL), show that backward elimination yields (CE, CL, EH, HL). Interpret its fit.

b. Based on the independence graph for (CE, CL, EH, HL), show that: (i) every path between C and H involves a variable in $\{E, L\}$; (ii) collapsing over H, one obtains the same associations between C and E and between C and L, and collapsing over C, one obtains the same associations between H and E and between H and L; (iii) the conditional independence patterns between C and H and between E and L are not collapsible.

7.22 Consider model $(AC, AM, CM, AG, AR, GM, GR)$ for the drug use data in Section 7.4.5.

a. Explain why the AM conditional odds ratio is unchanged by collapsing over race, but it is not unchanged by collapsing over gender.

b. Suppose we remove the GM term. Construct the independence graph, and show that $\{G, R\}$ are separated from $\{C, M\}$ by A.

c. For the model in (**b**), explain why all conditional associations among A, C, and M are identical to those in model (AC, AM, CM), collapsing over G and R.

7.23 Consider logit models for a four-way table in which X_1, X_2, and X_3 are predictors of Y. When the table is collapsed over X_3, indicate whether the association between X_1 and Y remains unchanged, for the model (a) that has main effects of all predictors, (b) that has main effects of X_1 and X_2 but assumes no effect for X_3.

7.24 Table 7.28 is from a General Social Survey. Subjects were asked whether methods of birth control should be available to teenagers between the ages of 14 and 16, and how often they attend religious services.

a. Fit the independence model, and use residuals to describe lack of fit.

b. Using equally spaced scores, fit the linear-by-linear association model. Describe the association.

c. Test goodness of fit of the model in (**b**); test independence in a way that uses the ordinality, and interpret.

d. Fit the $L \times L$ model using column scores $\{1, 2, 4, 5\}$. Repeat (**b**), and indicate whether results are substantively different with these scores.

e. Using formula (7.12) with the model in (**d**), explain why a fitted local log odds ratio using columns 2 and 3 is double a fitted local log odds ratio using columns 1 and 2 or columns 3 and 4. What is the relationship between the odds ratios?

Table 7.28. Data for Problem 7.24

	Teenage Birth Control			
Religious Attendance	Strongly Agree	Agree	Disagree	Strongly Disagree
Never	49	49	19	9
Less than once a year	31	27	11	11
Once or twice a year	46	55	25	8
Several times a year	34	37	19	7
About once a month	21	22	14	16
2–3 times a month	26	36	16	16
Nearly every week	8	16	15	11
Every week	32	65	57	61
Several times a week	4	17	16	20

Source: General Social Survey.

7.25 Generalizations of the linear-by-linear model (7.11) analyze association between ordinal variables X and Y while controlling for a categorical variable that may be nominal or ordinal. The model

$$\log \mu_{ijk} = \lambda + \lambda_i^X + \lambda_j^Y + \lambda_k^Z + \beta u_i v_j + \lambda_{ik}^{XZ} + \lambda_{jk}^{YZ}$$

with ordered scores $\{u_i\}$ and $\{v_j\}$ is a special case of model (XY, XZ, YZ) that replaces λ_{ij}^{XY} by a linear-by-linear term.

a. Show that the XY conditional independence model (XZ, YZ) is a special case of this model.

b. Assuming the ordinal model, explain how one could construct a test with $df = 1$ for testing XY conditional independence.

c. For this model, equation (7.12) applies for the cells at each fixed level of Z. With unit-spaced scores, explain why the model implies that every local odds ratio in each partial table equals $\exp(\beta)$.

d. If we replace β in this model by β_k, is there homogeneous association? Why or why not? (The fit of this model is equivalent to fitting the $L \times L$ association model separately for each partial table.)

7.26 For the linear-by-linear association model applied with column scores $\{v_j = j\}$, show that the adjacent-category logits within row i have form (6.6), identifying α_j with $(\lambda_{j+1}^{Y} - \lambda_j^{Y})$ and the row scores $\{u_i\}$ with the levels of x. In fact, the two models are equivalent. The logit representation (6.6) provides an interpretion for model (7.11).

7.27 True, or false?

a. When there is a single categorical response variable, logistic models are more appropriate than loglinear models.

b. When you want to model the association and interaction structure among several categorical response variables, logistic models are more appropriate than loglinear models.

c. A difference between logistic and loglinear models is that the logistic model is a GLM assuming a binomial random component whereas the loglinear model is a GLM assuming a Poisson random component. Hence, when both are fitted to a contingency table having 50 cells with a binary response, the logistic model treats the cell counts as 25 binomial observations whereas the loglinear model treats the cell counts as 50 Poisson observations.

CHAPTER 8

Models for Matched Pairs

This chapter introduces methods for comparing categorical responses for two samples that have a natural pairing between each subject in one sample and a subject in the other sample. Because each observation in one sample pairs with an observation in the other sample, the responses in the two samples are *matched pairs*. Because of the matching, the samples are statistically *dependent*. Methods that treat the two sets of observations as independent samples are inappropriate.

The most common way dependent samples occur is when each sample has the same subjects. Table 8.1 illustrates this for data from the 2000 General Social Survey. Subjects were asked whether, to help the environment, they would be willing to (1) pay higher taxes or (2) accept a cut in living standards. The rows of the table are the categories for opinion about paying higher taxes. The columns are the same categories for opinion about accepting a cut in living standards.

For matched-pairs observations, one way to summarize the data is with a two-way table that has the same categories for both classifications. The table is *square*, having the same number of rows and columns. The marginal counts display the frequencies for the outcomes for the two samples. In Table 8.1, the row marginal counts (359, 785) are the (yes, no) totals for paying higher taxes. The column marginal counts (334, 810) are the (yes, no) totals for accepting a cut in living standards.

Section 8.1 presents ways of comparing proportions from dependent samples. Section 8.2 introduces logistic regression models for matched-pairs binary data. Section 8.3 extends the methods to multicategory responses. Section 8.4 uses loglinear models to describe the structure of the table and to compare marginal distributions. The final two sections discuss two applications that yield matched-pairs data: measuring agreement between two observers who each evaluate the same subjects (Section 8.5), and evaluating preferences between pairs of outcome categories (Section 8.6).

An Introduction to Categorical Data Analysis, Second Edition. By Alan Agresti

Table 8.1. Opinions Relating to Environment

Pay Higher Taxes	Cut Living Standards Yes	No	Total
Yes	227	132	359
No	107	678	785
Total	334	810	1144

8.1 COMPARING DEPENDENT PROPORTIONS

For Table 8.1, how can we compare the probabilities of a "yes" outcome for the two environmental questions? Let n_{ij} denote the number of subjects who respond in category i for the first question and j for the second. In Table 8.1, $n_{1+} = n_{11} + n_{12} = 359$ subjects said "yes" for raising taxes, and $n_{+1} = n_{11} + n_{21} = 334$ subjects said "yes" for accepting cuts in living standards. The sample proportions were $359/1144 = 0.31$ and $334/1144 = 0.29$.

These marginal proportions are correlated. We see that $227 + 678$ subjects had the same opinion on both questions. They compose most of the sample, since fewer people answered "yes" on one and "no" on the other. A strong association exists between the opinions on the two questions, the sample odds ratio being $(227 \times 678)/(132 \times 107) = 10.9$.

Let π_{ij} denote the probability of outcome i for question 1 and j for question 2. The probabilities of a "yes" outcome are π_{1+} for question 1 and π_{+1} for question 2. When these are identical, the probabilities of a "no" outcome are also identical. There is then said to be *marginal homogeneity*. Since

$$\pi_{1+} - \pi_{+1} = (\pi_{11} + \pi_{12}) - (\pi_{11} + \pi_{21}) = \pi_{12} - \pi_{21}$$

marginal homogeneity in 2×2 tables is equivalent to $\pi_{12} = \pi_{21}$.

8.1.1 McNemar Test Comparing Marginal Proportions

For matched-pairs data with a binary response, a test of marginal homogeneity has null hypothesis

$$H_0\colon \pi_{1+} = \pi_{+1}, \quad \text{or equivalently } H_0\colon \pi_{12} = \pi_{21}$$

When H_0 is true, we expect similar values for n_{12} and n_{21}. Let $n^* = n_{12} + n_{21}$ denote the total count in these two cells. Their allocations to those cells are outcomes of a binomial variate. Under H_0, each of these n^* observations has a $1/2$ chance of contributing to n_{12} and a $1/2$ chance of contributing to n_{21}. Therefore, n_{12} and n_{21}

are numbers of "successes" and "failures" for a binomial distribution having n^* trials and success probability $\frac{1}{2}$.

When $n^* > 10$, the binomial distribution has a similar shape to the normal distribution with the same mean, which is $\frac{1}{2}n^*$, and standard deviation, which is $\sqrt{[n^*(\frac{1}{2})(\frac{1}{2})]}$. The standardized normal test statistic equals

$$z = \frac{n_{12} - (\frac{1}{2})n^*}{\sqrt{n^*(\frac{1}{2})(\frac{1}{2})}} = \frac{n_{12} - n_{21}}{\sqrt{n_{12} + n_{21}}} \tag{8.1}$$

The square of this statistic has an approximate chi-squared distribution with $df = 1$. The chi-squared test for a comparison of two dependent proportions is called the *McNemar test.*

For Table 8.1, the z test statistic equals $z = (132 - 107)/\sqrt{(132 + 107)} = 1.62$. The one-sided P-value is 0.053 and the two-sided P-value is 0.106. There is slight evidence that the probability of approval was greater for higher taxes than for a lower standard of living.

If the samples of subjects for the two questions were *separate* rather than the same, the samples would be *independent* rather than dependent. A different 2×2 contingency table would then summarize the data: The two rows would represent the two questions and the two columns would represent the (yes, no) response categories. We would then compare the rows of the table rather than the marginal distributions. For example, the tests of independence from Section 2.4 would analyze whether the probability of a "yes" outcome was the same for each question. Those chi-squared tests are not relevant for dependent-samples data of form Table 8.1. We naturally expect an association between the row and column classifications, because of the matched-pairs connection. The more relevant question concerns whether the marginal distributions differ.

8.1.2 Estimating Differences of Proportions

A confidence interval for the true difference of proportions is more informative than a significance test. Let $\{p_{ij} = n_{ij}/n\}$ denote the sample cell proportions. The difference $p_{1+} - p_{+1}$ between the sample marginal proportions estimates the true difference $\pi_{1+} - \pi_{+1}$. The estimated variance of the sample difference equals

$$[p_{1+}(1 - p_{1+}) + p_{+1}(1 - p_{+1}) - 2(p_{11}p_{22} - p_{12}p_{21})]/n \tag{8.2}$$

Its square root is the SE for a confidence interval. In terms of the cell counts McNemar's test uses, this standard error is

$$SE = \sqrt{(n_{12} + n_{21}) - (n_{12} - n_{21})^2/n}/n$$

For Table 8.1, the difference of sample proportions equals $0.314 - 0.292 = 0.022$. For $n = 1144$ observations with $n_{12} = 132$ and $n_{21} = 107$,

$$SE = \sqrt{(132 + 107) - (132 - 107)^2/1144}/1144 = 0.0135$$

A 95% confidence interval equals $0.022 \pm 1.96(0.0135)$, or $(-0.004, 0.048)$. We infer that the probability of a "yes" response was between 0.004 less and 0.048 higher for paying higher taxes than for accepting a cut in living standards. If the probabilities differ, the difference is apparently small.

This is a Wald confidence interval. For small samples the actual coverage probability is closer to the nominal confidence level if you add 0.50 to every cell before finding the standard error.

The parenthetical part of the last term in the estimated variance (8.2) represents the effect of the dependence of the marginal proportions through their *covariance*. Matched-pairs data usually exhibit a strong positive association. Responses for most subjects are the same for the column and the row classification. A sample odds ratio exceeding 1.0 corresponds to $p_{11}p_{22} > p_{12}p_{21}$ and hence a negative contribution from the third term in this variance expression. Thus, an advantage of using dependent samples, rather than independent samples, is a smaller variance for the estimated difference in proportions.

8.2 LOGISTIC REGRESSION FOR MATCHED PAIRS

Logistic regression extends to handle matched-pairs responses. This section presents such models for binary data.

8.2.1 Marginal Models for Marginal Proportions

Let us first see how the analyses of the previous section relate to models. Let (Y_1, Y_2) denote the two responses, where a "1" denotes category 1 (success) and "0" denotes category 2. In Table 8.1, suppose Y_1 is opinion about raising taxes and Y_2 is opinion about accepting cuts in living standards. Then, $P(Y_1 = 1) = \pi_{1+}$ and $P(Y_2 = 1) = \pi_{+1}$ are marginal probabilities estimated by $359/1144 = 0.31$ and $334/1144 = 0.29$.

The difference between the marginal probabilities occurs as a parameter in a model using identity link function. For the model

$$P(Y_1 = 1) = \alpha + \delta, \quad P(Y_2 = 1) = \alpha$$

$\delta = P(Y_1 = 1) - P(Y_2 = 1)$. The ML estimate of δ is the difference between the sample marginal proportions. For Table 8.1, $\hat{\delta} = 0.31 - 0.29 = 0.02$. The hypothesis of equal marginal probabilities for the McNemar test is, in terms of this model, H_0: $\delta = 0$.

An alternative model applies the logit link,

$$\text{logit}[P(Y_1 = 1)] = \alpha + \beta, \quad \text{logit}[P(Y_2 = 1)] = \alpha \tag{8.3}$$

This is equivalent to

$$\text{logit}[P(Y_t = 1)] = \alpha + \beta x_t$$

where x_t is an indicator variable that equals 1 when $t = 1$ and 0 when $t = 2$. The parameter e^{β} is the odds ratio comparing the marginal distributions. Its ML estimate is the odds ratio for the sample marginal distributions. For Table 8.1, $\exp(\hat{\beta}) = [(359/785)/(334/810)] = 1.11$. The population odds of willingness to pay higher taxes are estimated to be 11% higher than the population odds of willingness to accept cuts in living standards.

These models are called *marginal models*. They focus on the marginal distributions of responses for the two observations.

8.2.2 Subject-Specific and Population-Averaged Tables

Next, we show a three-way representation of binary matched-pairs data that motivates a different type of model. This display presents the data as n separate 2×2 partial tables, one for each matched pair. The kth partial table shows the responses (Y_1, Y_2) for the kth matched pair. It has columns that are the two possible outcomes (e.g., "yes" and "no") for each observation. It shows the outcome of Y_1 (e.g., response to question 1) in row 1 and the outcome of Y_2 (e.g., response to question 2) in row 2.

Table 8.1 cross-classified results on two environmental questions for 1144 subjects. Table 8.2 shows a partial table for a subject who answers "yes" on both questions. The full three-way table corresponding to Table 8.1 has 1144 partial tables. Of them, 227 look like Table 8.2, 132 have first row (1, 0) and second row (0, 1), representing "yes" on question 1 and "no" on question 2, 107 have first row (0, 1) and second row (1, 0), and 678 have (0, 1) in each row.

Each subject has a partial table, displaying the two matched observations. The 1144 subjects provide 2288 observations in a $2 \times 2 \times 1144$ contingency table. Collapsing this table over the 1144 partial tables yields a 2×2 table with first row equal to

Table 8.2. Representation of Matched Pair Contributing to Count n_{11} in Table 8.1

	Response	
Issue	Yes	No
Pay higher taxes	1	0
Cut living standards	1	0

(359, 785) and second row equal to (334, 810). These are the total number of "yes" and "no" outcomes for the two questions. They were the marginal counts in Table 8.1.

We refer to the $2 \times 2 \times n$ table with a separate partial table for each of n matched pairs as the *subject-specific table*. Models that refer to it, such as the one in the next subsection, are called *conditional models*; because the data are stratified by subject, the effect comparing the responses is *conditional* on the subject. By contrast, the 2×2 table that cross-classifies in a single table the two responses for all subjects is called a *population-averaged table*. Table 8.1 is an example. Its margins provide estimates of population marginal probabilities. Models for the margins of such a table, such as model (8.3), are *marginal models*. Chapter 9 discusses marginal models in detail, and Chapter 10 presents conditional models, in each case also permitting explanatory variables.

8.2.3 Conditional Logistic Regression for Matched-Pairs

A conditional model for matched-pairs data refers to the subject-specific partial tables of form Table 8.2. It differs from other models studied so far by permitting each subject to have their own probability distribution. Refer to the matched observations as observation 1 and observation 2. Let Y_{it} denote observation t for subject i, where $y_{it} = 1$ denotes a success. The model has the form

$$\text{logit}[P(Y_{i1} = 1)] = \alpha_i + \beta, \quad \text{logit}[P(Y_{i2} = 1)] = \alpha_i \tag{8.4}$$

Equivalently, $\text{logit}[P(Y_{it} = 1)] = \alpha_i + \beta x_{it}$ with $x_{i1} = 1$ and $x_{i2} = 0$.

The probabilities of success for subject i for the two observations equal

$$P(Y_{i1} = 1) = \frac{\exp(\alpha_i + \beta)}{1 + \exp(\alpha_i + \beta)}, \quad P(Y_{i2} = 1) = \frac{\exp(\alpha_i)}{1 + \exp(\alpha_i)}$$

The $\{\alpha_i\}$ parameters permit the probabilities to vary among subjects. A subject with a relatively large positive α_i (compared to the magnitude of β) has a high probability of success for each observation. Such a subject is likely to have a success for each observation. A subject with a relatively large negative α_i has a low probability of success for each observation and is likely to have a failure for each observation. The greater the variability in these parameters, the greater the overall positive association between the two observations, successes (failures) for observation 1 tending to occur with successes (failures) for observation 2.

Model (8.4) assumes that, for each subject, the odds of success for observation 1 are $\exp(\beta)$ times the odds of success for observation 2. Since each partial table refers to a single subject, this conditional association is a *subject-specific effect*. The value $\beta = 0$ implies marginal homogeneity. In that case, for each subject, the probability of success is the same for both observations.

Inference for this model focuses on β for comparing the distributions for $t = 1$ and $t = 2$. However, the model has as many subject parameters $\{\alpha_i\}$ as subjects.

This causes difficulties with the fitting process and with the properties of ordinary ML estimators. One remedy, *conditional maximum likelihood*, maximizes the likelihood function and finds $\hat{\beta}$ for a conditional distribution that eliminates the subject parameters. (Section 5.4 discussed this method for conducting small-sample inference for logistic regression.) Consider tables with counts $\{n_{ij}\}$ summarizing the cross-classification of the two observations, such as Table 8.1, which was

Pay Higher Taxes	Cut Living Standards	
	Yes	No
Yes	227	132
No	107	678

The conditional ML estimate of the odds ratio $\exp(\beta)$ for model (8.4) equals n_{12}/n_{21}. For Table 8.1, $\exp(\hat{\beta}) = 132/107 = 1.23$. Assuming the model holds, a subject's estimated odds of a "yes" response are 23% higher for raising taxes (question 1) than for accepting a lower standard of living.

By contrast, the odds ratio of 1.11 found above in Section 8.2.1 refers to the margins of this table, which equivalently are the rows of the marginal table obtained by collapsing the $2 \times 2 \times 1144$ contingency table with subject-specific strata. That these odds ratios take different values merely reflects how conditional odds ratios can differ from marginal odds ratios. Sections 10.1.2–10.1.4 discuss further the distinction between conditional and marginal models and their odds ratios.

As in the McNemar test, n_{12} and n_{21} provide all the information needed for inference about β for logit model (8.4). An alternative way of fitting model (8.4), which Chapter 10 presents, treats $\{\alpha_i\}$ as *random effects*. This approach treats $\{\alpha_i\}$ as an unobserved sample having a particular probability distribution. In most cases the estimate of β is the same as with the conditional ML approach.

When the matched-pairs responses have k predictors, model (8.4) generalizes to

$$\text{logit}[P(Y_{it} = 1)] = \alpha_i + \beta_1 x_{1it} + \beta_2 x_{2it} + \cdots + \beta_k x_{kit}, \quad t = 1, 2 \qquad (8.5)$$

Typically, one explanatory variable is of special interest, such as observation time or treatment. The others are covariates being controlled. Software exists for finding conditional ML estimates of $\{\beta_j\}$ (e.g., *LogXact* and PROC LOGISTIC in SAS). For matched pairs, Problem 8.28 mentions a simple way to obtain conditional ML $\{\hat{\beta}_j\}$ using software for ordinary logistic regression.

8.2.4 Logistic Regression for Matched Case–Control Studies*

Case–control studies that match a single control with each case are an important application having matched-pairs data. For a binary response Y, each case $(Y = 1)$ is matched with a control $(Y = 0)$ according to certain criteria that could affect the response. The study observes cases and controls on the predictor variable X and analyzes the XY association.

Table 8.3 illustrates results of a matched case–control study. A study of acute myocardial infarction (MI) among Navajo Indians matched 144 victims of MI according to age and gender with 144 individuals free of heart disease. Subjects were then asked whether they had ever been diagnosed as having diabetes ($x = 0$, no; $x = 1$, yes). Table 8.3 has the same form as Table 8.1 (shown also in the previous subsection), except that the levels of X rather than the levels of Y form the two rows and the two columns.

Table 8.3. Previous Diagnoses of Diabetes for Myocardial Infarction Case–Control Pairs

	MI Cases		
MI Controls	Diabetes	No diabetes	Total
Diabetes	9	16	25
No diabetes	37	82	119
Total	46	98	144

Source: Coulehan et al., *Am. J. Public Health*, **76**: 412–414, 1986. Reprinted with permission by the American Public Health Association.

A display of the data using a partial table (similar to Table 8.2) for each matched case–control pair reverses the roles of X and Y. In each matched pair, one subject has $Y = 1$ (the case) and one subject has $Y = 0$ (the control). Table 8.4 shows the four possible patterns of X values. There are nine partial tables of type 8.4a, since for nine pairs both the case and the control had diabetes, 16 partial tables of type 8.4b, 37 of type 8.4c, and 82 of type 8.4d.

Now, for a subject i, consider the model

$$\text{logit}[P(Y_i = 1)] = \alpha_i + \beta x$$

The odds that a subject with diabetes ($x = 1$) is an MI case equal $\exp(\beta)$ times the odds that a subject without diabetes ($x = 0$) is an MI case. The probabilities $\{P(Y_i = 1)\}$ refer to the distribution of Y given X, but these retrospective data provide information

Table 8.4. Possible Case–Control Pairs for Table 8.3

	a		b		c		d	
Diabetes	Case	Control	Case	Control	Case	Control	Case	Control
Yes	1	1	0	1	1	0	0	0
No	0	0	1	0	0	1	1	1

only about the distribution of X given Y. One can estimate $\exp(\beta)$, however, since it refers to the XY odds ratio, which relates to both types of conditional distribution (Section 2.3.5). Even though a case–control study reverses the roles of X and Y in terms of which is fixed and which is random, the conditional ML estimate of the odds ratio $\exp(\beta)$ for Table 8.3 is $n_{12}/n_{21} = 37/16 = 2.3$.

For details about conditional logistic regression for case–control studies, see Breslow and Day (1980), Fleiss, Levin, and Paik (2003, Chapter 14), and Hosmer and Lemeshow (2000, Chapter 7).

8.2.5 Connection between McNemar and Cochran–Mantel–Haenszel Tests*

We have seen that the Cochran–Mantel–Haenszel (CMH) chi-squared statistic (4.9) tests conditional independence in three-way tables. Suppose we apply this statistic to the $2 \times 2 \times n$ subject-specific table that relates the response outcome and the observation, for each matched pair. In fact, that CMH statistic is algebraically identical to the McNemar statistic, namely $(n_{12} - n_{21})^2/(n_{12} + n_{21})$ for tables of the form of Table 8.1. That is, the McNemar test is a special case of the CMH test applied to the binary responses of n matched pairs displayed in n partial tables.

This connection is not needed for computations, because the McNemar statistic is so simple, but it does suggest ways of constructing statistics to handle more complex types of matched data. For a matched set of T observations, a generalized CMH test of conditional independence (Section 6.4.2) can be applied to a $T \times 2 \times n$ table. The test statistic for that case is sometimes called *Cochran's Q*.

The CMH representation also suggests ways to test marginal homogeneity for tables having possibly several response categories as well as several observations. Consider a matched set of T observations and a response scale having I categories. One can display the data in a $T \times I \times n$ table. A partial table displays the T observations for a given matched set, one observation in each row. A generalized CMH test of conditional independence provides a test of marginal homogeneity of the T marginal distributions.

The CMH representation is also helpful for more complex forms of case–control studies. For instance, suppose a study matches each case with several controls. With n matched sets, one displays each matched set as a stratum of a $2 \times 2 \times n$ table. Each stratum has one observation in column 1 (the case) and several observations in column 2 (the controls). The McNemar test no longer applies, but the ordinary CMH test can perform the analysis.

Methods for binary matched pairs extend to multiple outcome categories. The next section shows some ways to do this.

8.3 COMPARING MARGINS OF SQUARE CONTINGENCY TABLES

Matched pairs analyses generalize to $I > 2$ outcome categories. Let (Y_1, Y_2) denote the observations for a randomly selected subject. A square $I \times I$ table $\{n_{ij}\}$ shows counts of possible outcomes (i, j) for (Y_1, Y_2).

Let $\pi_{ij} = P(Y_1 = i, Y_2 = j)$. Marginal homogeneity is

$$P(Y_1 = i) = P(Y_2 = i) \quad \text{for } i = 1, \ldots, I$$

that is, each row marginal probability equals the corresponding column marginal probability.

8.3.1 Marginal Homogeneity and Nominal Classifications

One way to test H_0: marginal homogeneity compares ML fitted values $\{\hat{\mu}_{ij}\}$ that satisfy marginal homogeneity to $\{n_{ij}\}$ using G^2 or X^2 statistics. The $df = I - 1$. The ML fit of marginal homogeneity is obtained iteratively.

Another way generalizes the McNemar test. It tests H_0: marginal homogeneity by exploiting the large-sample normality of marginal proportions. Let $d_i = p_{i+} - p_{+i}$ compare the marginal proportions in column i and row i. Let d be a vector of the first $I - 1$ differences. It is redundant to include d_I, since $\sum d_i = 0$. Under H_0, $E(\mathrm{d}) = 0$ and the estimated covariance matrix of d is $\hat{\mathrm{V}}_0/n$, where $\hat{\mathrm{V}}_0$ has elements

$$\hat{v}_{ij0} = -(p_{ij} + p_{ji}) \quad \text{for } i \neq j$$
$$\hat{v}_{ii0} = p_{i+} + p_{+i} - 2p_{ii}$$

Now, d has a large-sample multivariate normal distribution. The quadratic form

$$W_0 = n\mathrm{d}'\hat{\mathrm{V}}_0^{-1}\mathrm{d} \qquad (8.6)$$

is a score test statistic. It is asymptotically chi-squared with $df = I - 1$. For $I = 2$, W_0 simplifies to the McNemar statistic, the square of equation (8.1).

8.3.2 Example: Coffee Brand Market Share

A survey recorded the brand choice for a sample of buyers of instant coffee. At a later coffee purchase by these subjects, the brand choice was again recorded. Table 8.5 shows results for five brands of decaffinated coffee. The cell counts on the "main diagonal" (the cells for which the row variable outcome is the same as the column variable outcome) are relatively large. This indicates that most buyers did not change their brand choice.

The table also shows the ML fitted values that satisfy marginal homogeneity. Comparing these to the observed cell counts gives $G^2 = 12.6$ and $X^2 = 12.4$ ($df = 4$). The P-values are less than 0.015 for testing H_0: marginal homogeneity. (Table A.11 in the Appendix shows how software can obtain the ML fit and test statistics.) The statistic (8.6) using differences in sample marginal proportions gives similar results, equaling 12.3 with $df = 4$.

Table 8.5. Choice of Decaffeinated Coffee at Two Purchase Dates, with ML Fit Satisfying Marginal Homogeneity in Parentheses

First Purchase	Second Purchase				
	High Point	Taster's	Sanka	Nescafe	Brim
High Point	93 (93)	17 (13.2)	44 (32.5)	7 (6.1)	10 (7.8)
Taster's choice	9 (12.7)	46 (46)	11 (10.5)	0 (0.0)	9 (9.1)
Sanka	17 (26.0)	11 (11.6)	155 (155)	9 (11.3)	12 (12.8)
Nescafe	6 (7.0)	4 (3.5)	9 (7.5)	15 (15)	2 (1.8)
Brim	10 (14.0)	4 (4.0)	12 (11.3)	2 (2.3)	27 (27)

Source: Based on data from R. Grover and V. Srinivasan, *J. Marketing Res.*, **24**: 139–153, 1987. Reprinted with permission by the American Marketing Association.

The sample marginal proportions for brands (High Point, Taster's Choice, Sanka, Nescafe, Brim) were (0.32, 0.14, 0.38, 0.07, 0.10) for the first purchase and (0.25, 0.15. 0.43, 0.06, 0.11) for the second purchase. To estimate the change for a given brand, we can combine the other categories and use the methods of Section 8.1.

We illustrate by comparing the proportions selecting High Point at the two times. We construct the table with row and column categories (High Point, Others). This table has counts, by row, of (93, 78/42, 328). The McNemar z statistic (8.1) equals $(78 - 42)/\sqrt{(78 + 42)} = 3.3$. There is strong evidence of a change in the population proportion choosing this brand ($P = 0.001$). The estimated difference is $0.32 - 0.25 = 0.07$, and the 95% confidence interval is 0.07 ± 0.04. The small P-value for the overall test of marginal homogeneity mainly reflects a decrease in the proportion choosing High Point and an increase in the proportion choosing Sanka, with no evidence of change for the other coffees.

8.3.3 Marginal Homogeneity and Ordered Categories

The tests of H_0: marginal homogeneity above, having $df = I - 1$, are designed to detect *any* difference between the margins. They treat the categories as unordered, using all $I - 1$ degrees of freedom available for comparisons of I pairs of marginal proportions. Test statistics designed to detect a *particular* difference can be more powerful. For example, when response categories are ordered, tests can analyze whether responses tend to be higher on the ordinal scale in one margin than the other. Ordinal tests, which have $df = 1$, are usually much more powerful. This is especially true when I is large and the association between classifications is strong.

An ordinal model comparison of the margins can use ordinal logits, such as logits of cumulative probabilities (Section 6.2). The model

$$\text{logit}[P(Y_{i1} \leq j)] = \alpha_{ij} + \beta, \quad \text{logit}[P(Y_{i2} \leq j)] = \alpha_{ij}$$

is a generalization of binary model (8.4) that expresses each cumulative logit in terms of subject effects and a margin effect. Like the cumulative logit models of Section 6.2, it makes the *proportional odds* assumption by which the effect β is assumed to be the same for each cumulative probability. The model states that, for each matched pair, the odds that observation 1 falls in category j or below (instead of above category j) are $\exp(\beta)$ times the odds for observation 2.

An estimate of β in this model is

$$\hat{\beta} = \log\left(\frac{\sum\sum_{i<j}(j-i)n_{ij}}{\sum\sum_{i>j}(i-j)n_{ij}}\right) \tag{8.7}$$

The numerator sum weights each cell count above the main diagonal by its distance $(j-i)$ from that diagonal. The denominator sum refers to cells below the main diagonal. An ordinal test of marginal homogeneity ($\beta = 0$) uses this effect. Estimator (8.7) of β has

$$SE = \sqrt{\frac{\sum\sum_{i<j}(j-i)^2 n_{ij}}{\left[\sum\sum_{i<j}(j-i)n_{ij}\right]^2} + \frac{\sum\sum_{i>j}(i-j)^2 n_{ij}}{\left[\sum\sum_{i>j}(i-j)n_{ij}\right]^2}}$$

The ratio $\hat{\beta}/SE$ is an approximate standard normal test statistic.

A simple alternative test compares the sample means for the two margins, for ordered category scores $\{u_i\}$. Denote the sample means for the rows (X) and columns (Y) by $\bar{x} = \sum_i u_i p_{i+}$ and $\bar{y} = \sum_i u_i p_{+i}$. The difference $(\bar{x} - \bar{y})$ divided by its estimated standard error under marginal homogeneity, which is the square root of

$$(1/n)\left[\sum_i \sum_j (u_i - u_j)^2 p_{ij}\right]$$

has an approximate null standard normal distribution. This test is designed to detect differences between true marginal means.

8.3.4 Example: Recycle or Drive Less to Help Environment?

Table 8.6 is from a General Social Survey. Subjects were asked "How often do you cut back on driving a car for environmental reasons?" and "How often do you make a special effort to sort glass or cans or plastic or papers and so on for recycling?"

For Table 8.6, the numerator of (8.7) equals

$$1(43 + 99 + 230) + 2(163 + 185) + 3(233) = 1767$$

Table 8.6. Behaviors on Recycling and Driving Less to Help Environment, with Fit of Ordinal Quasi-Symmetry Model

	Drive Less			
Recycle	Always	Often	Sometimes	Never
Always	12 (12)	43 (43.1)	163 (165.6)	233 (232.8)
Often	4 (3.9)	21 (21)	99 (98.0)	185 (184.5)
Sometimes	4 (1.4)	8 (9.0)	77 (77)	230 (227.3)
Never	0 (0.2)	1 (1.5)	18 (20.7)	132 (132)

and the denominator equals

$$1(4 + 8 + 18) + 2(4 + 1) + 3(0) = 40$$

Thus, $\hat{\beta} = \log(1767/40) = 3.79$. The estimated odds ratio is $\exp(\hat{\beta}) = 1767/40 = 44.2$. For each subject the estimated odds of response "always" (instead of the other three categories) on recycling are 44.2 times the estimated odds of that response for driving less. This very large estimated odds ratio indicates a substantial effect.

For Table 8.6, $\hat{\beta} = 3.79$ has $SE = 0.180$. For H_0: $\beta = 0$, $z = 3.79/0.180 = 21.0$ provides extremely strong evidence against the null hypothesis of marginal homogeneity. Strong evidence also results from the comparison of mean scores. For the scores $\{1, 2, 3, 4\}$, the mean for driving less is $[20 + 2(73) + 3(357) + 4(780)]/1230 = 3.54$, and the mean for recycling is $[451 + 2(309) + 3(319) + 4(151)]/1230 = 2.14$. The test statistic is $z = (\bar{x} - \bar{y})/SE = (2.14 - 3.54)/0.0508 = -27.6$. The sample marginal means also indicate that responses tended to be considerably more toward the low end of the response scale (i.e., more frequent) on recycling than on driving less.

8.4 SYMMETRY AND QUASI-SYMMETRY MODELS FOR SQUARE TABLES*

The probabilities in a square table satisfy *symmetry* if

$$\pi_{ij} = \pi_{ji} \tag{8.8}$$

for all pairs of cells. Cell probabilities on one side of the main diagonal are a mirror image of those on the other side. When symmetry holds, necessarily marginal homogeneity also holds. When $I > 2$, though, marginal homogeneity can occur without symmetry. This section shows how to compare marginal distributions using logistic models for pairs of cells in square tables.

8.4.1 Symmetry as a Logistic Model

The symmetry condition has the simple logistic form

$$\log(\pi_{ij}/\pi_{ji}) = 0 \quad \text{for all } i \text{ and } j$$

The ML fit of the symmetry model has expected frequency estimates

$$\hat{\mu}_{ij} = (n_{ij} + n_{ji})/2$$

The fit satisfies $\hat{\mu}_{ij} = \hat{\mu}_{ji}$. It has $\hat{\mu}_{ii} = n_{ii}$, a perfect fit on the main diagonal. The residual df for chi-squared goodness-of-fit tests equal $I(I-1)/2$.

The standardized residuals for the symmetry model equal

$$r_{ij} = (n_{ij} - n_{ji})/\sqrt{n_{ij} + n_{ji}} \tag{8.9}$$

The two residuals for each pair of categories are redundant, since $r_{ij} = -r_{ji}$. The sum of squared standardized residuals, one for each pair of categories, equals X^2 for testing the model fit.

8.4.2 Quasi-Symmetry

The symmetry model is so simple that it rarely fits well. For instance, when the marginal distributions differ substantially, the model fits poorly. One can accommodate marginal heterogeneity by the *quasi-symmetry* model,

$$\log(\pi_{ij}/\pi_{ji}) = \beta_i - \beta_j \quad \text{for all } i \text{ and } j \tag{8.10}$$

One parameter is redundant, and we set $\beta_I = 0$. The symmetry model is the special case of equation (8.10) in which all $\beta_i = 0$. Roughly speaking, the higher the value of $\hat{\beta}_i$, relatively more observations fall in row i compared to column i.

Fitting the quasi-symmetry model requires iterative methods. To use software, treat each separate pair of cell counts (n_{ij}, n_{ji}) as an independent binomial variate, ignoring the main-diagonal counts. Set up I artificial explanatory variables, corresponding to the coefficients of the $\{\beta_i\}$ parameters. For the logit $\log(\pi_{ij}/\pi_{ji})$ for a given pair of categories, the variable for β_i is 1, the variable for β_j is -1, and the variables for the other parameters equal 0. (Table A.13 in the Appendix illustrates this with SAS code.) One explanatory variable is redundant, corresponding to the redundant parameter. The fitted marginal totals equal the observed totals. Its residual $df = (I-1)(I-2)/2$.

8.4.3 Example: Coffee Brand Market Share Revisited

Table 8.5 in Section 8.3.2 summarized coffee purchases at two times. The symmetry model has $G^2 = 22.5$ and $X^2 = 20.4$, with $df = 10$. The lack of fit results primarily from the discrepancy between n_{13} and n_{31}. For that pair, the standardized residual equals $(44 - 17)/\sqrt{(44 + 17)} = 3.5$. Consumers of High Point changed to Sanka more often than the reverse. Otherwise, the symmetry model fits most of the table fairly well.

The quasi-symmetry model has $G^2 = 10.0$ and $X^2 = 8.5$, with $df = 6$. Permitting the marginal distributions to differ yields a better fit than the symmetry model provides. We will next see how to use this information to construct a test of marginal homogeneity.

8.4.4 Testing Marginal Homogeneity Using Symmetry and Quasi-Symmetry

For the quasi-symmetry model, $\log(\pi_{ij}/\pi_{ji}) = \beta_i - \beta_j$ for all i and j, marginal homogeneity is the special case in which all $\beta_i = 0$. This special case is the symmetry model. In other words, for the quasi-symmetry model, marginal homogeneity is equivalent to symmetry.

To test marginal homogeneity, we can test the null hypothesis that the symmetry (S) model holds against the alternative hypothesis of quasi symmetry (QS). The likelihood-ratio test compares the G^2 goodness-of-fit statistics,

$$G^2(S \mid QS) = G^2(S) - G^2(QS)$$

For $I \times I$ tables, the test has $df = I - 1$.

For Table 8.5, the 5×5 table on choice of coffee brand at two purchases, $G^2(S) = 22.5$ and $G^2(QS) = 10.0$. The difference $G^2(S \mid QS) = 12.5$, based on $df = 4$, provides evidence of differing marginal distributions ($P = 0.014$).

Section 8.3.1 described other tests of H_0: marginal homogeneity, based on the ML fit under H_0 and using $p_{i+} - p_{+i}$, $i = 1, \ldots, I$. For the coffee data, these gave similar results as using $G^2(S \mid QS)$. Those other tests do not assume that the quasi-symmetry model holds. In practice, however, for nominal classifications the statistic $G^2(S \mid QS)$ usually captures most of the information about marginal heterogeneity even if the quasi-symmetry model shows lack of fit.

8.4.5 An Ordinal Quasi-Symmetry Model

The symmetry and quasi-symmetry models treat the classifications as nominal. A special case of quasi-symmetry often is useful when the categories are ordinal. Let $u_1 \leq u_2 \leq \cdots \leq u_I$ denote ordered scores for both the row and column categories. The *ordinal quasi-symmetry model* is

$$\log(\pi_{ij}/\pi_{ji}) = \beta(u_j - u_i) \tag{8.11}$$

This is a special case of the quasi-symmetry model (8.10) in which $\{\beta_i\}$ have a linear trend. The symmetry model is the special case $\beta = 0$.

Model (8.11) has the form of the usual logistic model, $\text{logit}(\pi) = \alpha + \beta x$, with $\alpha = 0$, $x = u_j - u_i$ and π equal to the conditional probability for cell (i, j), given response in cell (i, j) or cell (j, i). The greater the value of $|\beta|$, the greater the difference between π_{ij} and π_{ji} and between the marginal distributions. With scores $\{u_i = i\}$, the probability that the second observation is x categories higher than the first observation equals $\exp(x\beta)$ times the probability that the first observation is x categories higher than the second observation.

For this model, the fitted marginal counts have the same means as the observed marginal counts. For the chosen category scores $\{u_i\}$, the sample mean for the row variable is $\sum_i u_i p_{i+}$. This equals the row mean $\sum_i u_i \hat{\pi}_{i+}$ for the fitted values. A similar equality holds for the column means. When responses in one margin tend to be higher on the ordinal scale than those in the other margin, the fit of model (8.11) exhibits this same ordering. When $\hat{\beta} > 0$, the mean response is lower for the row variable. When $\hat{\beta} < 0$, the mean response is higher for the row variable.

To estimate β in the ordinal quasi-symmetry model, fit model (8.11) using logit model software. Identify (n_{ij}, n_{ji}) as binomial numbers of successes and failures in $n_{ij} + n_{ji}$ trials, and fit a logit model with intercept forced to equal 0 (which most software can do with a "no intercept" option) and with value of the predictor x equal to $u_j - u_i$. (Table A.12 in the Appendix illustrates this with SAS code.)

8.4.6 Example: Recycle or Drive Less?

We illustrate with Table 8.6 from Section 8.3.4 about behaviors on recycling or driving less to help the environment. A cursory glance at the data reveals that the symmetry model is doomed. Indeed, $G^2 = 1106.1$ and $X^2 = 857.4$ for testing its fit, with $df = 6$. By comparison, the quasi-symmetry model fits well, having $G^2 = 2.7$ and $X^2 = 2.7$ with $df = 3$. The simpler ordinal quasi-symmetry model also fits well. For the scores $\{1, 2, 3, 4\}$, $G^2 = 4.4$ and $X^2 = 5.9$, with $df = 5$. Table 8.6 displays its fitted values.

For the ordinal quasi-symmetry model, $\hat{\beta} = 2.39$. From equation (8.11), the estimated probability that response on driving less is x categories higher than the response on recycling equals $\exp(2.39x)$ times the reverse probability. Responses on recycling tend to be lower on the ordinal scale (i.e., more frequent) than those on driving less. The mean for recycling is 2.1, close to the "often" score, whereas the mean for driving less is 3.5, midway between the "sometimes" and "never" scores.

8.4.7 Testing Marginal Homogeneity Using Symmetry and Ordinal Quasi-Symmetry

For the ordinal quasi-symmetry model (8.11), symmetry and thus marginal homogeneity is the special case $\beta = 0$. A likelihood-ratio test of marginal homogeneity uses the difference between the G^2 values for the symmetry and ordinal quasi-symmetry models, with $df = 1$.

For Table 8.6 on recycling and driving less, the symmetry model has $G^2 = 1106.1$ $(df = 6)$, and the ordinal quasi-symmetry model has $G^2 = 4.4$ $(df = 5)$. The likelihood-ratio statistic for testing marginal homogeneity is $1106.1 - 4.4 = 1101.7$, with $df = 1$. There is extremely strong evidence of heterogeneity ($P < 0.0001$).

Alternatively, a Wald statistic for an ordinal test of marginal homogeneity treats $(\hat{\beta}/SE)^2$ as chi-squared with $df = 1$. For these data, $(\hat{\beta}/SE)^2 = (2.39/0.151)^2 = 252.0$, also giving extremely strong evidence. A third ordinal chi-squared test does not require fitting this model, but is related to it, being its score test of marginal homogeneity. The test statistic is the square of the statistic described in Section 8.3.3 that compares sample means for the margins, for category scores $\{u_i\}$. For these data, $z = (\bar{x} - \bar{y})/SE = (2.14 - 3.54)/0.0508 = -27.6$, and $z^2 = 762.6$ with $df = 1$.

8.5 ANALYZING RATER AGREEMENT*

Table 8.7 shows ratings by two pathologists, labeled X and Y, who separately classified 118 slides on the presence and extent of carcinoma of the uterine cervix. The rating scale has the ordered categories (1) negative, (2) atypical squamous hyperplasia, (3) carcinoma *in situ* and (4) squamous or invasive carcinoma. This table illustrates another type of matched-pairs data, referring to separate ratings of a sample by two observers using the same categorical scale. Each matched pair consists of the ratings by the two observers for a particular slide.

Let $\pi_{ij} = P(X = i, Y = j)$ denote the probability that observer X classifies a slide in category i and observer Y classifies it in category j. Their ratings of a particular subject *agree* if their classifications are in the same category. In the square table, the main diagonal $\{i = j\}$ represents observer agreement. The term π_{ii} is the probability

Table 8.7. Diagnoses of Carcinoma, with Standardized Residuals for Independence Model

	Pathologist Y				
Pathologist X	1	2	3	4	Total
1	22 (8.5)	2 (−0.5)	2 (−5.9)	0 (−1.8)	26
2	5 (−0.5)	7 (3.2)	14 (−0.5)	0 (−1.8)	26
3	0 (−4.1)	2 (−1.2)	36 (5.5)	0 (−2.3)	38
4	0 (−3.3)	1 (−1.3)	17 (0.3)	10 (5.9)	28
Total	27	12	69	10	118

Source: N. S. Holmquist, C. A. McMahon, and O. D. Williams, *Arch. Pathol.*, **84**: 334–345, 1967. Reprinted with permission by the American Medical Association.

that they both classify a subject in category i. The sum $\sum_i \pi_{ii}$ is the total probability of agreement. Perfect agreement occurs when $\sum_i \pi_{ii} = 1$.

Many categorical scales are quite subjective, and perfect agreement is rare. This section presents ways to measure strength of agreement and detect patterns of disagreement. *Agreement* is distinct from *association.* Strong agreement requires strong association, but strong association can exist without strong agreement. If observer X consistently classifies subjects one level higher than observer Y, the strength of agreement is poor even though the association is strong.

8.5.1 Cell Residuals for Independence Model

One way of evaluating agreement compares the cell counts $\{n_{ij}\}$ to the values $\{n_{i+}n_{+j}/n\}$ predicted by the loglinear model of independence (7.1). That model provides a baseline, showing the degree of agreement expected if no association existed between the ratings. Normally it would fit poorly if there is even only mild agreement, but its cell standardized residuals (Section 2.4.5) provide information about patterns of agreement and disagreement.

Cells with positive standardized residuals have higher frequencies than expected under independence. Ideally, large positive standardized residuals occur on the main diagonal and large negative standardized residuals occur off that diagonal. The sizes are influenced, however, by the sample size n, larger values tending to occur as n increases.

In fact, the independence model fits Table 8.7 poorly ($G^2 = 118.0, df = 9$). Table 8.7 reports the standardized residuals in parentheses. The large positive standardized residuals on the main diagonal indicate that agreement for each category is greater than expected by chance, especially for the first category. The off-main-diagonal residuals are primarily negative. Disagreements occurred less than expected under independence, although the evidence of this is weaker for categories closer together. Inspection of cell counts reveals that the most common disagreements refer to observer Y choosing category 3 and observer X instead choosing category 2 or 4.

8.5.2 Quasi-Independence Model

A more useful loglinear model adds a term that describes agreement beyond that expected under independence. This *quasi-independence model* is

$$\log \mu_{ij} = \lambda + \lambda_i^X + \lambda_j^Y + \delta_i I(i = j) \tag{8.12}$$

where the indicator $I(i = j)$ equals 1 when $i = j$ and equals 0 when $i \neq j$. This model adds to the independence model a parameter δ_1 for cell (1, 1) (in row 1 and column 1), a parameter δ_2 for cell (2, 2), and so forth. When $\delta_i > 0$, more agreements regarding outcome i occur than would be expected under independence. Because of the addition of this term, the quasi-independence model treats the main diagonal differently from the rest of the table. The *ML* fit in those cells is perfect, with $\hat{\mu}_{ii} = n_{ii}$ for all i. For the remaining cells, the independence model still applies. In other words, conditional on observer disagreement, the rating by X is independent of the rating by Y.

The quasi-independence model has I more parameters than the independence model, so its residual $df = (I-1)^2 - I$. One can fit it using iterative methods in GLM software. Besides the row and column explanatory factors, you set up I indicator variables for the main-diagonal cells. The indicator variable i is 1 for cell (i, i) and 0 otherwise. The estimate of its coefficient is $\hat{\delta}_i$. (Table A.12 in the Appendix illustrates this representation in SAS code.)

For Table 8.7, the quasi-independence model has $G^2 = 13.2$ and $X^2 = 11.5$, with $df = 5$. It fits much better than the independence model, but some lack of fit remains. Table 8.8 displays the fit. The fitted counts have the same main-diagonal values and the same row and column totals as the observed data, but satisfy independence for cells not on the main diagonal.

Table 8.8. Fitted Values for Quasi-Independence and Quasi-Symmetry Models with Table 8.7

	Pathologist Y			
Pathologist X	1	2	3	4
1	22	2	2	0
	(22)[a]	(0.7)	(3.3)	(0.0)
	(22)[b]	(2.4)	(1.6)	(0.0)
2	5	7	14	0
	(2.4)	(7)	(16.6)	(0.0)
	(4.6)	(7)	(14.4)	(0.0)
3	0	2	36	0
	(0.8)	(1.2)	(36)	(0.0)
	(0.4)	(1.6)	(36)	(0.0)
4	0	1	17	10
	(1.9)	(3.0)	(13.1)	(10)
	(0.0)	(1.0)	(17.0)	(10)

[a]Quasi-independence model.
[b]Quasi-symmetry model.

For Table 8.5 in Section 8.3.2 on choice of coffee brand at two occasions, the quasi-independence model has $G^2 = 13.8$ with $df = 11$. This is a dramatic improvement over independence, which has $G^2 = 346.4$ with $df = 16$. Given a change in brands, the new choice of coffee brand is plausibly independent of the original choice.

8.5.3 Odds Ratios Summarizing Agreement

For a pair of subjects, consider the event that each observer classifies one subject in category a and one subject in category b. The odds that the two observers agree rather than disagree on which subject is in category a and which is in category b equal

$$\tau_{ab} = \frac{\pi_{aa}\pi_{bb}}{\pi_{ab}\pi_{ba}} = \frac{\mu_{aa}\mu_{bb}}{\mu_{ab}\mu_{ba}} \tag{8.13}$$

As τ_{ab} increases, the observers are more likely to agree on which subject receives each designation.

For the quasi-independence model, the odds (8.13) summarizing agreement for categories a and b equal

$$\tau_{ab} = \exp(\delta_a + \delta_b)$$

These increase as the diagonal parameters increase, so larger $\{\delta_i\}$ represent stronger agreement. For instance, categories 2 and 3 in Table 8.7 have $\hat{\delta}_2 = 0.6$ and $\hat{\delta}_3 = 1.9$. The estimated odds that one observer's rating is category 2 rather than 3 are $\hat{\tau}_{23} = \exp(0.6 + 1.9) = 12.3$ times as high when the other observer's rating is 2 than when it is 3. The degree of agreement seems fairly strong, which also happens for the other pairs of categories.

8.5.4 Quasi-Symmetry and Agreement Modeling

For Table 8.7, the quasi-independence model shows some lack of fit. This model is often inadequate for ordinal scales, which almost always exhibit a positive association between ratings. Conditional on observer disagreement, a tendency usually remains for high (low) ratings by one observer to occur with relatively high (low) ratings by the other observer.

The quasi-symmetry model (8.10) is more complex than the quasi-independence model. It also fits the main diagonal perfectly, but it permits association off the main diagonal. It often fits much better. For Table 8.7, it has $G^2 = 1.0$ and $X^2 = 0.6$, based on $df = 2$. Table 8.8 displays the fit. To estimate the agreement odds (8.13), we substitute the fitted values $\{\hat{\mu}_{ij}\}$ into equation (8.13). For categories 2 and 3 of Table 8.7, for example, $\hat{\tau}_{23} = 10.7$. It is not unusual for observer agreement tables to have many empty cells. When $n_{ij} + n_{ji} = 0$ for any pair (such as categories 1 and 4 in Table 8.7), the ML fitted values in those cells must also be zero.

The symmetry model fits Table 8.7 poorly, with $G^2 = 39.2$ and $X^2 = 30.3$ having $df = 5$. The statistic $G^2(S|QS) = 39.2 - 1.0 = 38.2$, with $df = 3$, provides strong evidence of marginal heterogeneity. The lack of perfect agreement reflects differences in marginal distributions. Table 8.7 reveals these to be substantial in each category but the first.

The ordinal quasi-symmetry model uses the category orderings. This model fits Table 8.7 poorly, partly because ratings do not tend to be consistently higher by one observer than the other.

8.5.5 Kappa Measure of Agreement

An alternative approach describes strength of agreement using a single summary index, rather than a model. The most popular index is *Cohen's kappa*. It compares the agreement with that expected if the ratings were independent. The probability of agreement equals $\sum_i \pi_{ii}$. If the observers' ratings were independent, then $\pi_{ii} = \pi_{i+}\pi_{+i}$ and the probability of agreement equals $\sum_i \pi_{i+}\pi_{+i}$.

Cohen's kappa is defined by

$$\kappa = \frac{\sum \pi_{ii} - \sum \pi_{i+}\pi_{+i}}{1 - \sum \pi_{i+}\pi_{+i}}$$

The numerator compares the probability of agreement with that expected under independence. The denominator replaces $\sum \pi_{ii}$ with its maximum possible value of 1, corresponding to perfect agreement. Kappa equals 0 when the agreement merely equals that expected under independence, and it equals 1.0 when perfect agreement occurs. The stronger the agreement is, for a given pair of marginal distributions, the higher the value of kappa.

For Table 8.7, $\sum \hat{\pi}_{ii} = (22 + 7 + 36 + 10)/118 = 0.636$, whereas $\sum \hat{\pi}_{i+}\hat{\pi}_{+i} = [(26)(27) + (26)(12) + (38)(69) + (28)(10)]/(118)^2 = 0.281$. Sample kappa equals

$$\hat{\kappa} = (0.636 - 0.281)/(1 - 0.281) = 0.49$$

The difference between the observed agreement and that expected under independence is about 50% of the maximum possible difference.

Kappa treats the variables as nominal, in the sense that, when categories are ordered, it treats a disagreement for categories that are close the same as for categories that are far apart. For ordinal scales, a *weighted kappa* extension gives more weight to disagreements for categories that are farther apart.

Controversy surrounds the usefulness of kappa, primarily because its value depends strongly on the marginal distributions. The same diagnostic rating process can yield quite different values of kappa, depending on the proportions of cases of the various types. We prefer to construct models describing the structure of agreement and disagreement, rather than to depend solely on this summary index.

8.6 BRADLEY–TERRY MODEL FOR PAIRED PREFERENCES*

Table 8.9 summarizes results of matches among five professional tennis players during 2004 and 2005. For instance, Roger Federer won three of the four matches that he and Tim Henman played. This section presents a model that applies to data of this sort, in which observations consist of pairwise comparisons that result in a preference for one category over another. The fitted model provides a ranking of the players. It also estimates the probabilities of win and of loss for matches between each pair of players.

Table 8.9. Results of 2004–2005 Tennis Matches for Men Players

	Loser				
Winner	Agassi	Federer	Henman	Hewitt	Roddick
Agassi	–	0	0	1	1
Federer	6	–	3	9	5
Henman	0	1	–	0	1
Hewitt	0	0	2	–	3
Roddick	0	0	1	2	–

Source: www.atptennis.com.

The model is often applied in product comparisons. For instance, a wine-tasting session comparing several brands of Brunello di Montalcino wine from Tuscany, Italy might consist of a series of pairwise competitions. For each pair of wines, raters taste each wine and indicate a preference for one of them. Based on results of several pairwise evaluations, the model fit establishes a ranking of the wines.

8.6.1 The Bradley–Terry Model

The Bradley–Terry model is a logistic model for paired preference data. For Table 8.9, let Π_{ij} denote the probability that player i is the victor when i and j play. The probability that player j wins is $\Pi_{ji} = 1 - \Pi_{ij}$ (ties cannot occur). For instance, when Agassi (player 1) and Federer (player 2) played, Π_{12} is the probability that Agassi won and $\Pi_{21} = 1 - \Pi_{12}$ is the probability that Federer won.

The Bradley–Terry model has player parameters $\{\beta_i\}$ such that

$$\text{logit}(\Pi_{ij}) = \log(\Pi_{ij}/\Pi_{ji}) = \beta_i - \beta_j \tag{8.14}$$

The probability that player i wins equals $1/2$ when $\beta_i = \beta_j$ and exceeds $1/2$ when $\beta_i > \beta_j$. One parameter is redundant, and software imposes constraints such as setting the last one equal to 0 and deleting the usual intercept term.

Logit model (8.14) is equivalent to the quasi-symmetry model (8.10). To fit it, we treat each separate pair of cell counts (n_{ij}, n_{ji}) as an independent binomial variate, as Section 8.4.3 described. For instance, from Federer's perspective, the (Federer, Henman) results correspond to three successes and one failure in four trials. From the model fit, the estimate of $\hat{\Pi}_{ij}$ is

$$\hat{\Pi}_{ij} = \exp(\hat{\beta}_i - \hat{\beta}_j)/[1 + \exp(\hat{\beta}_i - \hat{\beta}_j)]$$

8.6.2 Example: Ranking Men Tennis Players

For Table 8.9, if software sets $\beta_5 = 0$ (for Roddick), the estimates of the other parameters are $\hat{\beta}_1 = 1.45$ (Agassi), $\hat{\beta}_2 = 3.88$ (Federer), $\hat{\beta}_3 = 0.19$ (Henman), and

$\hat{\beta}_4 = 0.57$ (Hewitt). Federer is ranked highest of the players, by far, and Roddick the lowest.

As well as providing a player ranking, the model fit yields estimated probabilities of victory. To illustrate, when Federer played Agassi in 2004–2005, model (8.14) estimates the probability of a Federer win to be

$$\hat{\Pi}_{21} = \frac{\exp(\hat{\beta}_2 - \hat{\beta}_1)}{1 + \exp(\hat{\beta}_2 - \hat{\beta}_1)} = \frac{\exp(3.88 - 1.45)}{1 + \exp(3.88 - 1.45)} = 0.92$$

For such small data sets, the model smoothing provides estimates that are more pleasing and realistic than the sample proportions. For instance, Federer beat Agassi in all six of their matches, but the model estimates the probability of a Federer victory to be 0.92 rather than 1.00. Also, the model can provide estimates for pairs of players who did not play. Agassi did not play Henman in 2004–2005, but for such a match the estimated probability of a Agassi victory was 0.78.

To check whether the difference between two players is statistically significant, we compare $(\hat{\beta}_i - \hat{\beta}_j)$ to its SE. From the covariance matrix of parameter estimates, the SE equals the square root of $[\text{Var}(\hat{\beta}_i) + \text{Var}(\hat{\beta}_j) - 2\,\text{Cov}(\hat{\beta}_i, \hat{\beta}_j)]$. For instance, for comparing Agassi and Roddick, $\hat{\beta}_1 - \hat{\beta}_5 = 1.449$ has $SE = 1.390$, indicating an insignificant difference. Estimates are imprecise for this small data set. The only comparisons showing strong evidence of a true difference are those between Federer and the other players.

A confidence interval for $\beta_i - \beta_j$ translates directly to one for Π_{ij}. For Federer and Roddick, a 95% confidence interval for $\beta_2 - \beta_5$ is $3.88 \pm 1.96(1.317)$, or (1.30, 6.46). This translates to (0.79, 0.998) for the probability Π_{25} of a Federer win $\{\text{e.g., } \exp(1.30)/[1 + \exp(1.30)] = 0.79\}$.

The assumption of independent, identical trials that leads to the binomial distribution and the usual fit of the logistic model is overly simplistic for this application. For instance, the probability Π_{ij} that player i beats player j may vary according to whether the court is clay, grass, or hard, and it would vary somewhat over time. In fact, the model does show some lack of fit. The goodness-of-fit statistics are $G^2 = 8.2$ and $X^2 = 11.6$, with $df = 5$.

PROBLEMS

8.1 Apply the McNemar test to Table 8.3. Interpret.

8.2 A recent General Social Survey asked subjects whether they believed in heaven and whether they believed in hell. Table 8.10 shows the results.

- **a.** Test the hypothesis that the population proportions answering "yes" were identical for heaven and hell. Use a two-sided alternative.
- **b.** Find a 90% confidence interval for the difference between the population proportions. Interpret.

Table 8.10. Data from General Social Survey for Problem 8.2

Believe in Heaven	Believe in Hell	
	Yes	No
Yes	833	125
No	2	160

8.3 Refer to the previous exercise. Estimate and interpret the odds ratio for a logistic model for the probability of a "yes" response as a function of the item (heaven or hell), using (**a**) the marginal model (8.3) and (**b**) the conditional model (8.4).

8.4 Explain the following analogy: The McNemar test is to binary data as the paired difference t test is to normally distributed data.

8.5 Section 8.1.1 gave the large-sample z or chi-squared McNemar test for comparing dependent proportions. The exact P-value, needed for small samples, uses the binomial distribution. For Table 8.1, consider H_a: $\pi_{1+} > \pi_{+1}$, or equivalently, H_a: $\pi_{12} > \pi_{21}$.

a. The exact P-value is the binomial probability of at least 132 successes out of 239 trials, when the parameter is 0.50. Explain why. (Software reports P-value = 0.060.)

b. For these data, how is the mid P-value (Section 1.4.5) defined in terms of binomial probabilities? (This P-value = 0.053.)

c. Explain why H_a: $\pi_{1+} \neq \pi_{+1}$ has ordinary P-value $= 0.120$ and mid P-value $= 0.106$. (The large-sample McNemar test has P-value that is an approximation for the binomial mid P-value. It is also 0.106 for these data.)

8.6 For Table 7.19 on opinions about measures to deal with AIDS, treat the data as matched pairs on opinion, stratified by gender.

a. For females, test the equality of the true proportions supporting government action for the two items.

b. Refer to (**a**). Construct a 90% confidence interval for the difference between the true proportions of support. Interpret.

c. For females, estimate the odds ratio $\exp(\beta)$ for (i) marginal model (8.3), (ii) conditional model (8.4). Interpret.

d. Explain how you could construct a 90% confidence interval for the difference between males and females in their differences of proportions of support for a particular item. (Hint: The gender samples are independent.)

8.7 Refer to Table 8.1 on ways to help the environment. Suppose sample proportions of approval of 0.314 and 0.292 were based on *independent* samples of size 1144 each. Construct a 95% confidence interval for the true difference of proportions. Compare with the result in Section 8.1.2, and comment on how the use of dependent samples can improve precision.

8.8 A crossover experiment with 100 subjects compares two treatments for migraine headaches. The response scale is success (+) or failure (−). Half the study subjects, randomly selected, used drug A the first time they got a migraine headache and drug B the next time. For them, six had responses $(A+, B+)$, 25 had responses $(A+, B-)$, 10 had responses $(A-, B+)$, and nine had responses $(A-, B-)$. The other 50 subjects took the drugs in the reverse order. For them, 10 were $(A+, B+)$, 20 were $(A+, B-)$, 12 were $(A-, B+)$, and eight were $(A-, B-)$.

a. Ignoring treatment order, use the McNemar test to compare the success probabilities for the two treatments. Interpret.

b. The McNemar test uses only the pairs of responses that differ. For this study, Table 8.11 shows such data from both treatment orders. Explain why a test of independence for this table tests the hypothesis that success rates are identical for the two treatments. Analyze these data, and interpret.

Table 8.11. Data for Problem 8.8

Treatment Order	Treatment That is Better	
	First	Second
A then B	25	10
B then A	12	20

8.9 Estimate β in model (8.4) applied to Table 8.1 on helping the environment. Interpret.

8.10 A case–control study has eight pairs of subjects. The cases have colon cancer, and the controls are matched with the cases on gender and age. A possible explanatory variable is the extent of red meat in a subject's diet, measured as "low" or "high." For three pairs, both the case and the control were high; for one pair, both the case and the control were low; for three pairs, the case was high and the control was low; for one pair, the case was low and the control was high.

a. Display the data in a 2×2 cross-classification of diet for the case against diet for the control. Display the $2 \times 2 \times 8$ table with partial tables relating diet to response (case, control) for the matched pairs. Successive parts refer to these as Table A and Table B.

b. Find the McNemar z^2 statistic for Table A and the CMH statistic (4.9) for Table B. Compare.

c. For Table B, show that the *CMH* statistic does not change if you delete pairs from the data set in which both the case and the control had the same diet.

d. This sample size is too small for these large-sample tests. Find the exact P-value for testing marginal homogeneity against the alternative hypothesis of a higher incidence of colon cancer for the "high" red meat diet. (See Problem 8.5.)

8.11 For the subject-specific model (8.4) for matched pairs,

$$\text{logit}[P(Y_{i1} = 1)] = \alpha_i + \beta, \quad \text{logit}[P(Y_{i2} = 1)] = \alpha_i$$

the estimated variance for the conditional ML estimate $\hat{\beta} = \log(n_{12}/n_{21})$ of β is $(1/n_{12} + 1/n_{21})$. Find a 95% confidence interval for the odds ratio $\exp(\beta)$ for Table 8.1 on helping the environment. Interpret.

8.12 For Table 7.3 on the student survey, viewing the table as matched triplets, you can compare the proportion of "yes" responses among alcohol, cigarettes, and marijuana.

a. Construct the marginal distribution for each substance, and find the three sample proportions of "yes" responses.

b. Explain how you could represent the data with a three-way contingency table in order to use a generalized *CMH* procedure (see Section 6.4.2) to test marginal homogeneity.

8.13 Table 8.12, from the 2004 General Social Survey, reports subjects' religious affiliation in 2004 and at age 16, for categories (1) Protestant, (2) Catholic, (3) Jewish, (4) None or Other.

Table 8.12. Data for Problem 8.13

Affiliation	Religious Affiliation Now			
at Age 16	1	2	3	4
1	1228	39	2	158
2	100	649	1	107
3	1	0	54	9
4	73	12	4	137

Source: 2004 General Social Survey.

a. The symmetry model has deviance $G^2 = 150.6$ with $df = 6$. Use residuals for the model [see equation (8.9)] to analyze transition patterns between pairs of religions.

b. The quasi-symmetry model has deviance $G^2 = 2.3$ with $df = 3$. Interpret.

c. Test marginal homogeneity by comparing fits in (**a**) and (**b**). (The small P-value mainly reflects the large sample size and is due to a small decrease in the proportion classified Catholic and increase in the proportion classified None or Other, with little evidence of change for other categories.)

8.14 Table 8.13, from the 2004 General Social Survey, reports respondents' region of residence in 2004 and at age 16.

a. Fit the symmetry and quasi-symmetry models. Interpret results.

b. Test marginal homogeneity by comparing the fits of these models.

Table 8.13. Data for Problem 8.14

Residence at Age 16	Residence in 2004			
	Northeast	Midwest	South	West
Northeast	425	17	80	36
Midwest	10	555	74	47
South	7	34	771	33
West	5	14	29	452

Source: 2004 General Social Survey.

8.15 Table 8.14 is from a General Social Survey. Subjects were asked their opinion about a man and a woman having sexual relations before marriage and a married person having sexual relations with someone other than the marriage partner. The response categories are 1 = always wrong, 2 = almost always wrong, 3 = wrong only sometimes, 4 = not wrong at all.

a. The estimate (8.7) for the subject-specific cumulative logit model is $\hat{\beta} = -4.91$. Interpret.

b. The estimate $\hat{\beta} = -4.91$ for the model in (**a**) has $SE = 0.45$. Conduct a Wald test for H_0: $\beta = 0$. Interpret.

Table 8.14. Data for Problem 8.15

Premarital Sex	Extramarital Sex			
	1	2	3	4
1	144	2	0	0
2	33	4	2	0
3	84	14	6	1
4	126	29	25	5

Source: General Social Survey.

c. For the symmetry model, $G^2 = 402.2$, with $df = 6$. For the quasi-symmetry model, $G^2 = 1.4$, with $df = 3$. Interpret, and compare the fits to test marginal homogeneity.

d. The ordinal quasi-symmetry model with scores $\{1, 2, 3, 4\}$ has $G^2 = 2.1$, with $df = 5$. Interpret, and show how to compare to the symmetry model to test marginal homogeneity.

e. Based on $\hat{\beta} = -2.86$ for the model in (**d**), explain why responses on extramarital sex tend to be lower on the ordinal scale (i.e., more negative) than those on premarital sex. (The mean scores are 1.28 for extramarital sex and 2.69 for premarital sex.)

8.16 Table 8.15 is from a General Social Survey. Subjects were asked "How often do you make a special effort to buy fruits and vegetables grown without pesticides or chemicals?" and "How often do you make a special effort to sort glass or cans or plastic or papers and so on for recycling?" The categories are 1 = always, 2 = often or sometimes, 3 = never. Analyze these data using the (**a**) symmetry, (**b**) quasi-symmetry, (**c**) ordinal quasi-symmetry models. Prepare a two-page report summarizing your analyses.

Table 8.15. Data for Problem 8.16

Chemical	Recycle		
Free	1	2	3
1	66	39	3
2	227	359	48
3	150	216	108

8.17 Table 8.16 is from the 2000 General Social Survey. Subjects were asked whether danger to the environment was caused by car pollution and/or by a rise in the world's temperature caused by the "greenhouse effect." The response categories are 1 = extremely dangerous, 2 = very dangerous, 3 = somewhat dangerous, 4 = not or not very dangerous. Analyze these data by fitting a model, interpreting parameter estimates, and conducting inference. Prepare a one-page report summarizing your analyses.

8.18 Refer to Problem 6.16 with Table 6.19 on a study about whether cereal containing psyllium had a desirable effect in lowering LDL cholesterol. For both the control and treatment groups, use methods of this chapter to compare the beginning and ending cholesterol levels. Compare the changes in cholesterol levels for the two groups. Interpret.

Table 8.16. Data for Problem 8.17

Car Pollution	Greenhouse Effect 1	2	3	4
1	95	72	32	8
2	66	129	116	13
3	31	101	233	82
4	5	4	24	26

Source: 2000 General Social Survey.

8.19 Refer to Table 8.13 on regional mobility. Fit the independence model and the quasi-independence (QI) model. Explain why there is a dramatic improvement in fit with the QI model. (Hint: For the independence model, the standardized residuals are about 40 for the cells on the main diagonal; what happens with these cells for the QI model?)

8.20 Table 8.17 displays diagnoses of multiple sclerosis for two neurologists. The categories are (1) Certain multiple sclerosis, (2) Probable multiple sclerosis, (3) Possible multiple sclerosis, and (4) Doubtful, unlikely, or definitely not multiple sclerosis.

a. Use the independence model and residuals to study the pattern of agreement. Interpret.

b. Use a more complex model to study the pattern and strength of agreement between the neurologists. Interpret results.

c. Use kappa to describe agreement. Interpret.

Table 8.17. Data for Problem 8.20

Neurologist A	Neurologist B 1	2	3	4
1	38	5	0	1
2	33	11	3	0
3	10	14	5	6
4	3	7	3	10

Source: based on data in J. R. Landis and G. Koch, *Biometrics*, **33**: 159–174, 1977. Reprinted with permission from the Biometric Society.

8.21 Refer to Table 8.5. Fit the quasi-independence model. Calculate the fitted odds ratio for the four cells in the first two rows and the last two columns. Interpret. Analyze the data from the perspective of describing agreement between choice of coffee at the two times.

8.22 In 1990, a sample of psychology graduate students at the University of Florida made blind, pairwise preference tests of three cola drinks. For 49 comparisons of Coke and Pepsi, Coke was preferred 29 times. For 47 comparisons of Classic Coke and Pepsi, Classic Coke was preferred 19 times. For 50 comparisons of Coke and Classic Coke, Coke was preferred 31 times. Comparisons resulting in ties are not reported.

a. Fit the Bradley–Terry model and establish a ranking of the drinks.

b. Estimate the probability that Coke is preferred to Pepsi, using the model fit, and compare with the sample proportion.

8.23 Table 8.18 refers to journal citations among four statistical theory and methods journals (*Biometrika, Communications in Statistics, Journal of the American Statistical Association, Journal of the Royal Statistical Society Series B*) during 1987–1989. The more often that articles in a particular journal are cited, the more prestige that journal accrues. For citations involving a pair of journals X and Y, view it as a victory for X if it is cited by Y and a defeat for X if it cites Y.

a. Fit the Bradley–Terry model. Interpret the fit, and give a prestige ranking of the journals.

b. For citations involving *Commun. Statist.* and *JRSS-B*, estimate the probability that the *Commun. Statist.* article cites the *JRSS-B* article.

Table 8.18. Data for Problem 8.23

	Cited Journal			
Citing Journal	*Biometrika*	*Commun. Statist.*	*JASA*	*JRSS-B*
Biometrika	714	33	320	284
Commun. Statist.	730	425	813	276
JASA	498	68	1072	325
JRSS-B	221	17	142	188

Source: S. M. Stigler, *Statist. Sci.*, **9**: 94–108, 1994. Reprinted with permission from the Institute of Mathematical Statistics.

8.24 Table 8.19 summarizes results of tennis matches for several women professional players between 2003 and 2005.

a. Fit the Bradley–Terry model. Report the parameter estimates, and rank the players.

b. Estimate the probability that Serena Williams beats Venus Williams. Compare the model estimate to the sample proportion.

c. Construct a 90% confidence interval for the probability that Serena Williams beats Venus Williams. Interpret.

Table 8.19. Women's Tennis Data for Problem 8.24

	Loser				
Winner	Clijsters	Davenport	Pierce	S. Williams	V. Williams
Clijsters	—	6	3	0	2
Davenport	2	—	0	2	4
Pierce	1	2	—	0	1
S. Williams	2	2	2	—	2
V. Williams	3	2	2	2	—

Source: www.wtatour.com.

d. Show that the likelihood-ratio test for testing that all $\beta_i = 0$ has test statistic 2.6, based on $df = 4$. Hence, it is plausible, based on this small sample, that no differences exist among the players in the chance of victory.

8.25 Refer to the fit of the Bradley–Terry model to Table 8.9.

a. Agassi did not play Henman in 2004–2005, but if they did play, show that the estimated probability of a Agassi victory is 0.78.

b. The likelihood-ratio statistic for testing H_0: $\beta_1 = \cdots = \beta_5$ equals 26.3 with $df = 4$. Interpret.

8.26 When the Bradley–Terry model holds, explain why it is not possible that A could be preferred to B (i.e., $\Pi_{AB} > \frac{1}{2}$) and B could be preferred to C, yet C could be preferred to A.

8.27 In loglinear model form, the quasi-symmetry (QS) model is

$$\log \mu_{ij} = \lambda + \lambda_i^X + \lambda_j^Y + \lambda_{ij}$$

where $\lambda_{ij} = \lambda_{ji}$ for all i and j.

a. For this model, by finding $\log(\mu_{ij}/\mu_{ji})$ show that the model implies a logit model of form (8.10), which is

$$\log(\pi_{ij}/\pi_{ji}) = \beta_i - \beta_j \quad \text{for all } i \text{ and } j$$

b. Show that the special case of QS with $\lambda_i^X = \lambda_i^Y$ for all i is the symmetry model in loglinear form.

c. Show that the quasi-independence model is the special case in which $\lambda_{ij} = 0$ for $i \neq j$.

8.28 For matched pairs, to obtain conditional ML $\{\hat{\beta}_j\}$ for model (8.5) using software for ordinary logistic regression, let

$$y_i^* = 1 \text{ when } (y_{i1} = 1, y_{i2} = 0), \quad y_i^* = 0 \text{ when } (y_{i1} = 0, y_{i2} = 1)$$

Let $x_{1i}^* = x_{1i1}^* - x_{1i2}^*, \ldots, x_{ki}^* = x_{ki1}^* - x_{ki2}^*$. Fit the ordinary logistic model to y^* with predictors $\{x_1^*, \ldots, x_k^*\}$, forcing the intercept parameter α to equal zero. This works because the likelihood is the same as the conditional likelihood for model (8.5) after eliminating $\{\alpha_i\}$.

a. Apply this approach to model (8.4) with Table 8.1 and report $\hat{\beta}$ and its SE.

b. The pairs $(y_{i1} = 1, y_{i2} = 1)$ and $(y_{i1} = 0, y_{i2} = 0)$ do not contribute to the likelihood or to estimating $\{\beta_j\}$. Identify the counts for such pairs in Table 8.1. Do these counts contribute to McNemar's test?

CHAPTER 9

Modeling Correlated, Clustered Responses

Many studies observe the response variable for each subject repeatedly, at several times (such as in *longitudinal* studies) or under various conditions. Repeated measurement occurs commonly in health-related applications. For example, a physician might evaluate patients at weekly intervals regarding whether a new drug treatment is successful. Repeated observations on a subject are typically correlated.

Correlated observations can also occur when the response variable is observed for *matched sets* of subjects. For example, a study of factors that affect childhood obesity might sample families and then observe the children in each family. A matched set consists of children within a particular family. Children from the same family may tend to respond more similarly than children from different families. Another example is a (survival, nonsurvival) response for each fetus in a litter of a pregnant mouse, for a sample of pregnant mice exposed to various dosages of a toxin. Fetuses from the same litter are likely to be more similar than fetuses from different litters.

We will refer to a matched set of observations as a *cluster*. For repeated measurement on subjects, the set of observations for a given subject forms a cluster. Observations within a cluster are usually positively correlated. Analyses should take the correlation into account. Analyses that ignore the correlation can estimate model parameters well, but the standard error estimators can be badly biased.

The next two chapters generalize methods of the previous chapter for matched pairs to matched sets and to include explanatory variables. Section 9.1 describes a class of *marginal models* and contrasts them with *conditional models*, which Chapter 10 presents. To fit marginal models, Section 9.2 discusses a method using *generalized estimating equations* (*GEE*). This is a multivariate method that, for discrete data, is computationally simpler than ML. Section 9.2 models a data set with a binary response, and Section 9.3 considers multinomial responses. The final section introduces a *transitional* approach that models observations in a longitudinal study using explanatory variables that include previous response outcomes.

9.1 MARGINAL MODELS VERSUS CONDITIONAL MODELS

As with independent observations, with clustered observations models focus on how the probability of a particular outcome (e.g., "success") depends on explanatory variables. For longitudinal studies, one explanatory variable is the time of each observation. For instance, in treating a chronic condition (such as a phobia) with one of two treatments, the model might describe how the probability of success depends on the treatment and on the length of time for which that treatment has been used.

9.1.1 Marginal Models for a Clustered Binary Response

Let T denote the number of observations in each cluster. (In practice, the number of observations often varies by cluster, but it is simpler to use notation that ignores that.) Denote the T observations by $(Y_1, Y_2, \ldots, Y_T)$.

For binary responses, the T success probabilities $\{P(Y_1 = 1), P(Y_2 = 1), \ldots, P(Y_T = 1)\}$ are marginal probabilities of a T-dimensional contingency table that cross classifies the T observations. Marginal models describe how the logits of the marginal probabilities, $\{\text{logit}[P(Y_t = 1)]\}$, depend on explanatory variables. To illustrate the models and questions of interest, let us consider an example to be analyzed in Section 9.2.

9.1.2 Example: Longitudinal Study of Treatments for Depression

Table 9.1 refers to a longitudinal study comparing a new drug with a standard drug for treating subjects suffering mental depression. Subjects were classified into two groups according to whether the initial severity of depression was mild or severe.

Table 9.1. Cross-classification of Responses on Depression at Three Times (N = Normal, A = Abnormal) by Treatment and Diagnosis Severity

Diagnosis		Response at Three Times							
Severity	Treatment	NNN	NNA	NAN	NAA	ANN	ANA	AAN	AAA
Mild	Standard	16	13	9	3	14	4	15	6
Mild	New drug	31	0	6	0	22	2	9	0
Severe	Standard	2	2	8	9	9	15	27	28
Severe	New drug	7	2	5	2	31	5	32	6

Source: Reprinted with permission from the Biometric Society (G. G. Koch et al., *Biometrics*, **33**: 133–158, 1977).

In each group, subjects were randomly assigned to one of the two drugs. Following 1 week, 2 weeks, and 4 weeks of treatment, each subject's extent of suffering from mental depression was classified as normal or abnormal.

Table 9.1 shows four groups, the combinations of categories of two explanatory variables – treatment type and severity of depression. Since the study observed the binary response (depression assessment) at $T = 3$ occasions, Table 9.1 shows a $2 \times 2 \times 2$ table for each group. The three depression assessments form a multivariate response with three components, with $Y_t = 1$ for normal and 0 for abnormal at time t. The 12 marginal distributions result from three repeated observations for each of the four groups.

Table 9.2 shows sample proportions of normal responses for the 12 marginal distributions. For instance, from Table 9.1, the sample proportion of normal responses after week 1 for subjects with mild depression using the standard drug was

$$(16 + 13 + 9 + 3)/(16 + 13 + 9 + 3 + 14 + 4 + 15 + 6) = 0.51$$

We see that the sample proportion of normal responses (1) increased over time for each group, (2) increased at a faster rate for the new drug than the standard, for each initial severity of depression, and (3) was higher for the mild than the severe cases of depression, for each treatment at each occasion. In such a study, the company that developed the new drug would hope to show that patients have a significantly higher rate of improvement with it.

Let s denote the initial severity of depression, with $s = 1$ for severe and $s = 0$ for mild. Let d denote the drug, with $d = 1$ for new and $d = 0$ for standard. Let t denote the time of measurement. When the time metric reflects cumulative drug dosage, a logit scale often has an approximate linear effect for the logarithm of time. We use scores (0, 1, 2), the logs to base 2 of the week numbers (1, 2, and 4). Similar substantive results occur using the week numbers themselves.

Let $P(Y_t = 1)$ denote the probability of a normal response at time t for a randomly selected subject. One possible model for how Y_t depends on the severity s, drug d, and the time t is the main effects model,

$$\text{logit}[P(Y_t = 1)] = \alpha + \beta_1 s + \beta_2 d + \beta_3 t$$

Table 9.2. Sample Marginal Proportions of Normal Response for Depression Data of Table 9.1

Diagnosis Severity	Treatment	Sample Proportion Week 1	Week 2	Week 4
Mild	Standard	0.51	0.59	0.68
	New drug	0.53	0.79	0.97
Severe	Standard	0.21	0.28	0.46
	New drug	0.18	0.50	0.83

This model assumes that the linear time effect β_3 is the same for each group. The sample proportions in Table 9.2, however, show a higher rate of improvement for the new drug. A more realistic model permits the time effect to differ by drug. We do this by including a drug-by-time interaction term,

$$\text{logit}[P(Y_t = 1)] = \alpha + \beta_1 s + \beta_2 d + \beta_3 t + \beta_4 (d \times t)$$

Here, β_3 describes the time effect for the standard drug ($d = 0$) and $\beta_3 + \beta_4$ describes the time effect for the new drug ($d = 1$).

We will fit this model, interpret the estimates, and make inferences in Section 9.2. We will see that an estimated slope (on the logit scale) for the standard drug is $\hat{\beta}_3 = 0.48$. For the new drug the estimated slope increases by $\hat{\beta}_4 = 1.02$, yielding an estimated slope of $\hat{\beta}_3 + \hat{\beta}_4 = 1.50$.

9.1.3 Conditional Models for a Repeated Response

The models just considered describe how $P(Y_t = 1)$, the probability of a normal response at time t, depends on severity, drug, and time, for a randomly selected subject. By contrast, for matched pairs Section 8.2.3 presented a different type of model that describes probabilities at the subject level. That model permits heterogeneity among subjects, even at fixed levels of the explanatory variables.

Let Y_{it} denote the response for subject i at time t. For the depression data, a subject-specific analog of the model just considered is

$$\text{logit}[P(Y_{it} = 1)] = \alpha_i + \beta_1 s + \beta_2 d + \beta_3 t + \beta_4 (d \times t)$$

Each subject has their own intercept (α_i), reflecting variability in the probability among subjects at a particular setting (s, d, t) for the explanatory variables.

This is called a *conditional model*, because the effects are defined conditional on the subject. For example, the model identifies β_3 as the time effect for a given subject using the standard drug. The effect is *subject-specific*, because it is defined at the subject level. By contrast, the effects in the marginal models specified in the previous subsection are *population-averaged*, because they refer to averaging over the entire population rather than to individual subjects.

The remainder of this chapter focuses only on marginal models. The following chapter presents conditional models and also discusses issues relating to the choice of model.

9.2 MARGINAL MODELING: THE GENERALIZED ESTIMATING EQUATIONS (GEE) APPROACH

ML fitting of marginal logit models is difficult. We will not explore the technical reasons here, but basically, it is because the models refer to *marginal* probabilities whereas the likelihood function refers to the *joint* distribution of the clustered responses. With a few explanatory variables (such as in Table 9.1), ML model fitting is available with some specialized software (see www.stat.ufl.edu/~aa/cda/software.html).

9.2.1 Quasi-Likelihood Methods

A GLM specifies a probability distribution for Y and provides a formula for how its mean $E(Y) = \mu$ depends on the explanatory variables by using a link function to connect the mean to a linear predictor. The choice of distribution for Y determines the relationship between μ and Var(Y). For binary data with success probability π, for example, an observation Y has $E(Y) = \pi$ and Var(Y) $= \pi(1 - \pi)$, which is $\mu(1 - \mu)$. For count data with the Poisson distribution, Var(Y) $= \mu$.

For a given formula for how μ depends on the explanatory variables, the ML method must assume a particular type of probability distribution for Y, in order to determine the likelihood function. By contrast, the *quasi-likelihood* approach assumes only a relationship between μ and Var(Y) rather than a specific probability distribution for Y. It allows for departures from the usual assumptions, such as overdispersion caused by correlated observations or unobserved explanatory variables. To do this, the quasi-likelihood approach takes the usual variance formula but multiplies it by a constant that is itself estimated using the data.

Consider clustered binary data, with n observations in a cluster. The observations within a cluster are likely to be correlated. So, the variance of the number of successes in a cluster may differ from the variance $n\pi(1 - \pi)$ for a binomial distribution, which assumes independent trials. The quasi-likelihood approach permits the variance of the number of successes to be some multiple ϕ of the usual variance, that is, $\phi n\pi(1 - \pi)$, where ϕ is estimated based on the variability observed in the sample data. Overdispersion occurs when $\phi > 1$. The quasi-likelihood estimates are not maximum likelihood (ML), however, because the method does not completely specify a distribution for Y, and thus, there is not a likelihood function.

9.2.2 Generalized Estimating Equation Methodology: Basic Ideas

A computationally simple alternative to ML for clustered categorical data is a multivariate generalization of quasi likelihood. Rather than assuming a particular type of distribution for $(Y_1, \ldots, Y_T)$, this method only links each marginal mean to a linear predictor and provides a guess for the variance–covariance structure of $(Y_1, \ldots, Y_T)$. The method uses the observed variability to help generate appropriate standard errors. The method is called the *GEE method* because the estimates are solutions of *generalized estimating equations*. These equations are multivariate generalizations of the equations solved to find ML estimates for GLMs.

Once we have specified a marginal model for each Y_t, for the GEE method we must:

- Assume a particular distribution (e.g., binomial) for each Y_t. This determines how Var(Y_t) depends on $E(Y_t)$.

- Make an educated guess for the correlation structure among $\{Y_t\}$. This is called the *working correlation* matrix.

One possible working correlation has *exchangeable* structure. This treats $\rho = \text{Corr}(Y_s, Y_t)$ as identical (but unknown) for all pairs s and t. Another possibility, often used for time series data, has *autoregressive* structure. This has the form $\text{Corr}(Y_s, Y_t) = \rho^{t-s}$. For example, $\text{Corr}(Y_1, Y_2) = \rho$, $\text{Corr}(Y_1, Y_3) = \rho^2$, $\text{Corr}(Y_1, Y_4) = \rho^3, \ldots$, with observations farther apart in time being more weakly correlated. The *independence* working correlation structure assumes $\text{Corr}(Y_s, Y_t) = 0$ for each pair. This treats the observations in a cluster as uncorrelated, that is, the same as if they were from separate clusters. At the other extreme, the *unstructured* working correlation matrix permits $\text{Corr}(Y_s, Y_t)$ to differ for each pair.

For the assumed working correlation structure, the GEE method uses the data to estimate the correlations. Those correlation estimates also impact the estimates of model parameters and their standard errors. In practice, usually little if any *a priori* information is available about the correlation structure. The lack of assumption needed for the unstructured case seems desirable, but this has the disadvantage of several extra parameters to estimate, especially when T is large. When the correlations are small, all working correlation structures yield similar GEE estimates and standard errors. Unless one expects dramatic differences among the correlations, we recommend using the exchangeable working correlation structure.

Even if your guess about the correlation structure is poor, valid standard errors result from an adjustment the GEE method makes using the empirical dependence the actual data exhibit. That is, the naive standard errors based on the assumed correlation structure are updated using the information the sample data provide about the dependence. The result is *robust* standard errors that are usually more appropriate than ones based solely on the assumed correlation structure. For example, the GEE method provides reasonable estimates and standard errors even if we use the independence working correlation structure, which is usually implausible. A sensible choice of working correlation, however, can result in slightly more efficient estimates.

The GEE method assumes a probability distribution for each *marginal* distribution, but it makes no assumption about the *joint* distribution of $(Y_1, \ldots, Y_T)$ other than to select a working correlation structure. This is helpful. For continuous multivariate responses it is common to assume a multivariate normal distribution. However, for discrete data, such as a categorical response or a count response, there is no multivariate generalization of standard univariate distributions such as the binomial and Poisson that provides simple specification of correlation structure.

9.2.3 GEE for Binary Data: Depression Study

Let us again consider Table 9.1, from a study with 340 patients that compared two treatments for mental depression. With exchangeable correlation structure, the estimated common correlation between pairs of the three responses is -0.003. The successive observations apparently have pairwise appearance like independent observations. This is unusual for repeated measurement data.

Table 9.3 reports the GEE estimates based on the independence working correlations. For that case, the GEE estimates equal those obtained from ordinary logistic regression, that is, using ML with $3 \times 340 = 1020$ independent observations rather than treating the data as three dependent observations for each of 340 subjects. The empirical standard errors incorporate the sample dependence to adjust the independence-based standard errors.

Table 9.3. Output from Using GEE to Fit Logistic Model to Table 9.1

Initial Parameter Estimates			GEE Parameter Estimates Empirical Std Error Estimates		
Parameter	Estimate	Std error	Parameter	Estimate	Std Error
Intercept	−0.0280	0.1639	Intercept	-0.0280	0.1742
severity	−1.3139	0.1464	severity	-1.3139	0.1460
drug	−0.0596	0.2222	drug	-0.0596	0.2285
time	0.4824	0.1148	time	0.4824	0.1199
drug*time	1.0174	0.1888	drug*time	1.0174	0.1877

Working Correlation Matrix

	Col1	Col2	Col3
Row1	1.0000	0.0000	0.0000
Row2	0.0000	1.0000	0.0000
Row3	0.0000	0.0000	1.0000

The estimated time effect is $\hat{\beta}_3 = 0.482$ for the standard drug ($d = 0$) and $\hat{\beta}_3 + \hat{\beta}_4 = 1.500$ for the new one ($d = 1$). For the new drug, the slope is $\hat{\beta}_4 = 1.017$ ($SE = 0.188$) higher than for the standard drug. The Wald test of no interaction, H_0: $\beta_4 = 0$, tests a common time effect for each drug. It has z test statistic equal to $1.017/0.188 = 5.4$ (P-value < 0.0001). Therefore, there is strong evidence of faster improvement for the new drug. It would be inadequate to use the simpler model lacking the drug-by-time interaction term.

The severity of depression estimate is $\hat{\beta}_1 = -1.314$ ($SE = 0.146$). For each drug–time combination, the estimated odds of a normal response when the initial diagnosis was severe equal $\exp(-1.314) = 0.27$ times the estimated odds when the initial diagnosis was mild. The estimate $\hat{\beta}_2 = -0.060$ ($SE = 0.228$) for the drug effect applies only when $t = 0$ (i.e., after one week), for which the interaction term does not contribute to the drug effect. It indicates an insignificant difference between the drugs after 1 week. At time t, the estimated odds of normal response with the new drug are $\exp(-0.060 + 1.017t)$ times the estimated odds for the standard drug, for each initial diagnosis level. By the final week ($t = 2$), this estimated odds ratio has increased to 7.2.

In summary, severity, drug treatment, and time all have substantial effects on the probability of a normal response. The chance of a normal response is similar for the two drugs initially and increases with time, but it increases more quickly for those taking the new drug than the standard drug.

9.2.4 Example: Teratology Overdispersion

Table 9.4 shows results of a teratology experiment. Female rats on iron-deficient diets were assigned to four groups. Group 1 received only placebo injections. The other groups received injections of an iron supplement according to various schedules. The rats were made pregnant and then sacrificed after 3 weeks. For each fetus in each rat's litter, the response was whether the fetus was dead.

We treat the fetuses in a given litter as a cluster. Let y_i denote the number of dead fetuses for the T_i fetuses in litter i. Let π_{it} denote the probability of death for fetus t in litter i. Let $z_{ig} = 1$ if litter i is in group g and 0 otherwise.

First, we ignore the clustering and suppose that y_i is a $bin(T_i, \pi_{it})$ variate. The model

$$\text{logit}(\pi_{it}) = \alpha + \beta_2 z_{i2} + \beta_3 z_{i3} + \beta_4 z_{i4}$$

treats all litters in a group g as having the same probability of death, $\exp(\alpha + \beta_g)/[1 + \exp(\alpha + \beta_g)]$, where $\beta_1 = 0$. Here, β_i is a log odds ratio comparing group i with the placebo group (group number 1). Table 9.5 shows ML estimates and standard errors. There is strong evidence that the probability of death is substantially lower for each treatment group than the placebo group.

Because of unmeasured covariates that affect the response, it is natural to expect that the actual probability of death varies from litter to litter within a particular treatment group. In fact, the data show evidence of overdispersion, with goodness-of-fit statistics $X^2 = 154.7$ and $G^2 = 173.5$ ($df = 54$). For comparison, Table 9.5 also shows results with the GEE approach to fitting the logit model, assuming an exchangeable working correlation structure for observations within a litter. The estimated within-litter correlation between the binary responses is 0.19.

Table 9.4. Response Counts of (Litter Size, Number Dead) for 58 Litters of Rats in a Low-Iron Teratology Study

Group 1: untreated (low iron)
(10, 1) (11, 4) (12, 9) (4, 4) (10, 10) (11, 9) (9, 9) (11, 11) (10, 10) (10, 7) (12, 12)
(10, 9) (8, 8) (11, 9) (6, 4) (9, 7) (14, 14) (12, 7) (11, 9) (13, 8) (14, 5) (10, 10)
(12, 10) (13, 8) (10, 10) (14, 3) (13, 13) (4, 3) (8, 8) (13, 5) (12, 12)

Group 2: injections days 7 and 10
(10, 1) (3, 1) (13, 1) (12, 0) (14, 4) (9, 2) (13, 2) (16, 1) (11, 0) (4, 0) (1, 0) (12, 0)

Group 3: injections days 0 and 7
(8, 0) (11, 1) (14, 0) (14, 1) (11, 0)

Group 4: injections weekly
(3, 0) (13, 0) (9, 2) (17, 2) (15, 0) (2, 0) (14, 1) (8, 0) (6, 0) (17, 0)

Source: D. F. Moore and A. Tsiatis, *Biometrics*, **47**: 383–401, 1991.

Table 9.5. Estimates and Standard Errors (in Parentheses) for Logistic Models Fitted to Teratology Data of Table 9.4

	Type of Logistic Model Fitting	
Parameter	Binomial ML	GEE
Intercept	1.14 (0.13)	1.21 (0.27)
Group 2	−3.32 (0.33)	−3.37 (0.43)
Group 3	−4.48 (0.73)	−4.58 (0.62)
Group 4	−4.13 (0.48)	−4.25 (0.60)
Overdispersion	None	$\hat{\rho} = 0.19$

Note: Binomial ML assumes no overdispersion; GEE has exchangeable working correlation. The intercept term gives result for group 1 (placebo) alone.

Suppose an application has positive within-cluster correlation, as often happens in practice and as seems to be the case here. Then, standard errors for *between-cluster* effects (such as comparisons of separate treatment groups) and standard errors of estimated means within clusters tend to be larger than when the observations are independent. We see this in Table 9.5 except for group 3 and its comparison with the placebo group. With positive within-cluster correlation, standard errors for *within-cluster* effects, such as a slope for a trend in the repeated measurements on a subject, tend to be smaller than when the observations are independent.

9.2.5 Limitations of GEE Compared with ML

Because the GEE method specifies the marginal distributions and the correlation structure but not the complete multivariate distribution, there is no likelihood function. In this sense, the GEE method is a multivariate type of quasi-likelihood method. So, its estimates are not ML estimates.

For clustered data, the GEE method is much simpler computationally than ML and much more readily available in software. However, it has limitations. Because it does not have a likelihood function, likelihood-ratio methods are not available for checking fit, comparing models, and conducting inference about parameters. Instead inference uses statistics, such as Wald statistics, based on the approximate normality of the estimators together with their estimated covariance matrix. Such inference is reliable mainly for very large samples. Otherwise, the empirically based standard errors tend to underestimate the true ones.

Some software attempts to improve on Wald-type inference by making tests available that mimic the way score tests are constructed, if one had a likelihood function. These generalized score tests also incorporate empirical information in forming standard error estimates, and they are preferable to Wald tests.

9.3 EXTENDING GEE: MULTINOMIAL RESPONSES

Since its introduction about 20 years ago, the GEE method has been extended in various ways. Extensions include model-fitting for clustered multinomial data, modelling association as well as marginal distributions, and accounting for missing data. This section discusses these extensions.

9.3.1 Marginal Modeling of a Clustered Multinomial Response

Models for marginal distributions of a clustered binary response generalize to multicategory responses. With nominal responses, baseline-category logit models describe the odds of each outcome relative to a baseline. For ordinal responses, cumulative logit models describe odds for the cumulative probabilities.

The GEE methodology was originally specified for modeling univariate marginal distributions, such as the binomial and Poisson. It has since been extended to marginal modeling of multinomial responses. With this approach, for each pair of outcome categories one selects a working correlation matrix for the pairs of repeated observations. Currently, for multinomial data, most software supports only the independence working correlation structure. As in the univariate case, the GEE method uses the empirical dependence to find standard errors that are appropriate even if this working correlation guess is poor. Standard errors based on assuming independent observations would usually be invalid.

9.3.2 Example: Insomnia Study

For a sample of patients with insomnia problems, Table 9.6 shows results of a randomized, double-blind clinical trial comparing an active hypnotic drug with a placebo.

Table 9.6. Time to Falling Asleep, by Treatment and Occasion

	Time to Falling Asleep				
		Follow-up			
Treatment	Initial	<20	20–30	30–60	>60
Active	<20	7	4	1	0
	20–30	11	5	2	2
	30–60	13	23	3	1
	>60	9	17	13	8
Placebo	<20	7	4	2	1
	20–30	14	5	1	0
	30–60	6	9	18	2
	>60	4	11	14	22

Source: From S. F. Francom, C. Chuang-Stein, and J. R. Landis, *Statist. Med.*, **8**: 571–582, 1989. Reprinted with permission from John Wiley & Sons, Ltd.

The response is the patient's reported time (in minutes) to fall asleep after going to bed. Patients responded before and following a 2 week treatment period. The two treatments, active drug and placebo, form a binary explanatory variable. The subjects were randomly allocated to the treatment groups. Here, each subject forms a cluster, with the observations in a cluster being the ordinal response at the two occasions of observation.

Table 9.7 displays sample marginal distributions for the four treatment – occasion combinations. From the initial to follow-up occasion, time to falling asleep seems to shift downwards for both treatments. The degree of shift seems greater for the active drug, indicating possible interaction. Let t denote the occasion ($0 =$ initial, $1 =$ follow-up) and let x denote the treatment ($0 =$ placebo, $1 =$ active drug). The cumulative logit model

$$\text{logit}[P(Y_t \leq j)] = \alpha_j + \beta_1 t + \beta_2 x + \beta_3 (t \times x) \tag{9.1}$$

permits interaction between occasion and treatment. Like the cumulative logit models of Section 6.2, it makes the *proportional odds* assumption of the same effects for each response cutpoint.

Table 9.7. Sample Marginal Distributions of Table 9.6

		Response			
Treatment	Occasion	<20	20–30	30–60	>60
Active	Initial	0.101	0.168	0.336	0.395
	Follow-up	0.336	0.412	0.160	0.092
Placebo	Initial	0.117	0.167	0.292	0.425
	Follow-up	0.258	0.242	0.292	0.208

For independence working correlation, the GEE estimates (with *SE* values in parentheses) are:

$$\hat{\beta}_1 = 1.038(0.168), \quad \hat{\beta}_2 = 0.034(0.238), \quad \hat{\beta}_3 = 0.708(0.244)$$

The SE values are not the naive ones assuming independence, but the ones adjusted for the actual empirical dependence. At the initial observation, the estimated odds that time to falling asleep for the active treatment is below any fixed level equal $\exp(0.034) = 1.03$ times the estimated odds for the placebo treatment. In other words, initially the two groups had similar distributions, as expected by the randomization of subjects to treatments. At the follow-up observation, the effect is $\exp(0.034 + 0.708) = 2.1$. Those taking the active drug tended to fall asleep more quickly.

The $\hat{\beta}_3$ and SE values indicate considerable evidence of interaction. The test statistic $z = 0.708/0.244 = 2.9$ provides strong evidence that the distribution of time to fall asleep decreased more for the treatment group than for the placebo group (two-sided P-value $= 0.004$).

For simpler interpretation, it can be helpful to assign scores to the ordered categories and report the sample marginal means and their differences. With response scores $\{10, 25, 45, 75\}$ for time to fall asleep, the initial means were 50.0 for the active group and 50.3 for the placebo. The difference in means between the initial and follow-up responses was 22.2 for the active group and 13.0 for the placebo.

If we had naively treated repeated responses as independent for the entire analysis, we would have obtained the same estimates as in the GEE analysis but the SE values for within-subject time effects would have been misleadingly large. For example, the interaction effect estimate of 0.708 would have had a SE of 0.334 rather than 0.244. With positive within-cluster correlation, standard errors for within-cluster effects tend to be smaller than when the observations are independent.

9.3.3 Another Way of Modeling Association with GEE

For categorical data, one aspect of GEE that some statisticians find unsatisfactory is specifying a correlation matrix for the clustered responses. For binary responses, unlike continuous responses, correlations cannot take value over the entire $[-1, +1]$ range. The actual range depends on the marginal probabilities. The odds ratio is a more suitable measure of the association.

An alternative version of GEE specifies a working association matrix using the odds ratio. For example, the exchangeable structure states that the odds ratio is the same for each pair of observations. Some software gives the option of an iterative *alternating logistic regression* algorithm. It alternates between a GEE step for finding the regression parameter estimates and a step for an association model for the log odds ratio. This is particularly useful in those cases in which the study of the association is also a major focus.

9.3.4 Dealing with Missing Data

Studies with repeated measurement often have cases for which at least one observation in a cluster is missing. In a longitudinal study, for instance, some subjects may drop out before the study's end. With the GEE method, the clusters can have different numbers of observations. The data input file has a separate line for each observation, and for longitudinal studies computations use those times for which a subject has an observation.

When data are missing, analyzing the observed data alone as if no data are missing can result in biased estimates. Bias does not occur in the rather stringent case in which the data are *missing completely at random*. This means that whether an observation is missing is statistically independent of the value of that observation.

Often, missingness depends on the missing values. For instance, in a longitudinal study measuring pain, perhaps a subject dropped out when the pain exceeded some threshhold. Then, more complex analyses are needed that model the joint distribution of the responses and the binary outcome for each potential observation on whether the observation was actually made or it is missing. Ways to do this are beyond the scope of this book. For details, see Molenberghs and Verbeke (2005).

Analyses when many data are missing should be made with caution. At a minimum, one should compare results of the analysis using all available cases for all clusters to the analysis using only clusters having no missing observations. If results differ substantially, conclusions should be tentative until the reasons for missingness can be studied.

9.4 TRANSITIONAL MODELING, GIVEN THE PAST

Let Y_t denote the response at time t, $t = 1, 2, \ldots$, in a longitudinal study. Some studies focus on the dependence of Y_t on the previously observed responses $\{y_1, y_2, \ldots, y_{t-1}\}$ as well as any explanatory variables. Models that include past observations as predictors are called *transitional models*.

A *Markov chain* is a transitional model for which, for all t, the conditional distribution of Y_t, given $Y_1, \ldots, Y_{t-1}$, is assumed identical to the conditional distribution of Y_t given Y_{t-1} alone. That is, given Y_{t-1}, Y_t is conditionally independent of $Y_1, \ldots, Y_{t-2}$. Knowing the most recent observation, information about previous observations before that one does not help with predicting the next observation. A Markov chain model is adequate for modeling Y_t if the model with y_{t-1} as the only past observation used as a predictor fits as well, for practical purposes, as a model with $\{y_1, y_2, \ldots, y_{t-1}\}$ as predictors.

9.4.1 Transitional Models with Explanatory Variables

Transitional models usually also include explanatory variables other than past observations. With binary y and k such explanatory variables, one might specify a logistic regression model for each t,

$$\text{logit}[P(Y_t = 1)] = \alpha + \beta y_{t-1} + \beta_1 x_1 + \cdots + \beta_k x_k$$

This Markov chain model is called a *regressive logistic model*. Given the predictor values, the model treats repeated observations by a subject as independent. Thus, one can fit the model with ordinary GLM software, treating each observation separately.

This model generalizes so a predictor x_j can take a different value for each t. For example, in a longitudinal medical study, a subject's values for predictors such as blood pressure could change over time. A higher-order Markov model could also include in the predictor set y_{t-2} and possibly other previous observations.

9.4.2 Example: Respiratory Illness and Maternal Smoking

Table 9.8 is from the Harvard study of air pollution and health. At ages 7–10 children were evaluated annually on whether they had a respiratory illness. Explanatory variables are the age of the child t ($t = 7, 8, 9, 10$) and maternal smoking at the start of the study ($s = 1$ for smoking regularly, $s = 0$ otherwise).

Table 9.8. Child's Respiratory Illness by Age and Maternal Smoking

Child's Respiratory Illness			No Maternal Smoking Age 10		Maternal Smoking Age 10	
Age 7	Age 8	Age 9	No	Yes	No	Yes
No	No	No	237	10	118	6
		Yes	15	4	8	2
	Yes	No	16	2	11	1
		Yes	7	3	6	4
Yes	No	No	24	3	7	3
		Yes	3	2	3	1
	Yes	No	6	2	4	2
		Yes	5	11	4	7

Source: Thanks to Dr. James Ware for these data.

Let y_t denote the response on respiratory illness at age t. For the regressive logistic model

$$\text{logit}[P(Y_t = 1)] = \alpha + \beta y_{t-1} + \beta_1 s + \beta_2 t, \quad t = 8, 9, 10$$

each subject contributes three observations to the model fitting. The data set consists of 12 binomials, for the $2 \times 3 \times 2$ combinations of (s, t, y_{t-1}). For instance, for the combination (0, 8, 0), from Table 9.8 we see that $y_8 = 0$ for $237 + 10 + 15 + 4 = 266$ subjects and $y_8 = 1$ for $16 + 2 + 7 + 3 = 28$ subjects.

The ML fit of this regressive logistic model is

$$\text{logit}[\hat{P}(Y_t = 1)] = -0.293 + 2.210 y_{t-1} + 0.296 s - 0.243 t$$

The SE values are 0.158 for the y_{t-1} effect, 0.156 for the s effect, and 0.095 for the t effect. Not surprisingly, y_{t-1} has a strong effect – a multiplicative impact of $e^{2.21} = 9.1$ on the odds. Given that and the child's age, there is slight evidence of a positive effect of maternal smoking: The likelihood-ratio statistic for H_0: $\beta_1 = 0$ is 3.55 ($df = 1$, $P = 0.06$). The maternal smoking effect weakens further if we add y_{t-2} to the model (Problem 9.13).

9.4.3 Comparisons that Control for Initial Response

The transitional type of model can be especially useful for matched-pairs data. The marginal models that are the main focus of this chapter would evaluate how the marginal distributions of Y_1 and Y_2 depend on explanatory variables. It is often more relevant to treat Y_2 as a univariate response, evaluating effects of explanatory variables while controlling for the initial response y_1. That is the focus of a transitional model.

Consider the insomnia study of Problem 9.3.2 from the previous section. Let y_1 be the initial time to fall asleep, let Y_2 be the follow-up time, with explanatory variable x defining the two treatment groups ($1 =$ active drug, $0 =$ placebo). We will now treat Y_2 as an ordinal response and y_1 as an explanatory variable, using scores $\{10, 25, 45, 75\}$. In the model

$$\text{logit}[P(Y_2 \leq j)] = \alpha_j + \beta_1 x + \beta_2 y_1 \tag{9.2}$$

β_1 compares the follow-up distributions for the treatments, controlling for the initial observation. This models the follow-up response (Y_2), conditional on y_1, rather than marginal distributions of (Y_1, Y_2). It's the type of model for an ordinal response that Section 6.2 discussed. Here, the initial response y_1 plays the role of an explanatory variable.

From software for ordinary cumulative logit models, the ML treatment effect estimate is $\hat{\beta}_1 = 0.885$ ($SE = 0.246$). This provides strong evidence that follow-up time to fall asleep is lower for the active drug group. For any given value for the initial response, the estimated odds of falling asleep by a particular time for the active treatment are $\exp(0.885) = 2.4$ times those for the placebo group. Exercise 9.12 considers alternative analyses for these data.

9.4.4 Transitional Models Relate to Loglinear Models

Effects in transitional models differ from effects in marginal models, both in magnitude and in their interpretation. The effect of a predictor x_j on Y_t is conditional on y_{t-1} in a transitional model, but it ignores y_{t-1} in a marginal model. Effects in transitional models are often considerably weaker than effects in marginal models, because conditioning on a previous response attenuates the effect of a predictor.

Transitional models have connections with the loglinear models of Chapter 7, which described joint distributions. Associations in loglinear models *are* conditional on the other response variables. In addition, a joint distribution of $(Y_1, Y_2, \ldots, Y_T)$ can be factored into the distribution of Y_1, the distribution of Y_2 given Y_1, the distribution of Y_3 given Y_1 and Y_2, and so forth.

PROBLEMS

9.1 Refer to Table 7.3 on high school students' use of alcohol, cigarettes, and marijuana. View the table as matched triplets.

- **a.** Construct the marginal distribution for each substance. Find the sample proportions of students who used (i) alcohol, (ii) cigarettes, (iii) marijuana.
- **b.** Specify a marginal model that could be fitted as a way of comparing the margins. Explain how to interpret the parameters in the model. State the hypothesis, in terms of the model parameters, that corresponds to marginal homogeneity.

9.2 Refer to Table 7.13. Fit a marginal model to describe main effects of race, gender, and substance type (alcohol, cigarettes, marijuana) on whether a subject had used that substance. Summarize effects.

9.3 Refer to the previous exercise. Further study shows evidence of an interaction between gender and substance type. Using GEE with exchangeable working correlation, the estimated probability $\hat{\pi}$ of using a particular substance satisfies

$$\text{logit}(\hat{\pi}) = -0.57 + 1.93S_1 + 0.86S_2 + 0.38R - 0.20G + 0.37G \times S_1 + 0.22G \times S_2$$

where R, G, S_1, S_2 are dummy variables for race (1 = white, 0 = nonwhite), gender (1 = female, 0 = male), and substance type ($S_1 = 1$, $S_2 = 0$ for alcohol; $S_1 = 0$, $S_2 = 1$ for cigarettes; $S_1 = S_2 = 0$ for marijuana). Show that:

a. The group with highest estimated probability of use of marijuana is white males. What group is it for alcohol?

b. Given gender, the estimated odds a white subject used a given substance are 1.46 times the estimated odds for a nonwhite subject.

c. Given race, the estimated odds a female has used alcohol are 1.19 times the estimated odds for males; for cigarettes and for marijuana, the odds ratios are 1.02 and 0.82.

d. Given race, the estimated odds a female has used alcohol (cigarettes) are 9.97 (2.94) times the estimated odds she has used marijuana.

e. Given race, the estimated odds a male has used alcohol (cigarettes) are 6.89 (2.36) times the estimated odds he has used marijuana. Interpret the interaction.

9.4 Refer to Table 9.1. Analyze the depression data (available at the text web site) using GEE assuming exchangeable correlation and with the time scores $(1, 2, 4)$. Interpret model parameter estimates and compare substantive results to those in the text with scores $(0, 1, 2)$.

9.5 Analyze Table 9.8 using a marginal logit model with age and maternal smoking as predictors. Report the prediction equation, and compare interpretations to the regressive logistic Markov model of Section 9.4.2.

9.6 Table 9.9 refers to a three-period crossover trial to compare placebo (treatment A) with a low-dose analgesic (treatment B) and high-dose analgesic (treatment C) for relief of primary dysmenorrhea. Subjects in the study were divided randomly into six groups, the possible sequences for administering the treatments. At the end of each period, each subject rated the treatment as giving no relief (0) or some relief (1). Let $y_{i(k)t} = 1$ denote relief for subject i using treatment t ($t = A, B, C$), where subject i is nested in treatment sequence k ($k = 1, \ldots, 6$).

Table 9.9. Crossover Data for Problem 9.6

Treatment	Response Pattern for Treatments (A, B, C)							
Sequence	000	001	010	011	100	101	110	111
A B C	0	2	2	9	0	0	1	1
A C B	2	0	0	9	1	0	0	4
B A C	0	1	1	8	1	3	0	1
B C A	0	1	1	8	1	0	0	1
C A B	3	0	0	7	0	1	2	1
C B A	1	5	0	4	0	3	1	0

Source: B. Jones and M. G. Kenward, *Statist. Med.*, **6**: 171–181, 1987.

a. Assuming common treatment effects for each sequence and setting $\beta_A = 0$, use GEE to obtain and interpret $\{\hat{\beta}_t\}$ for the model

$$\text{logit}[P(Y_{i(k)t} = 1)] = \alpha_k + \beta_t$$

b. How would you order the drugs, taking significance into account?

9.7 Table 9.10 is from a Kansas State University survey of 262 pig farmers. For the question "What are your primary sources of veterinary information"?, the categories were (A) Professional Consultant, (B) Veterinarian, (C) State or Local Extension Service, (D) Magazines, and (E) Feed Companies and Reps. Farmers sampled were asked to select all relevant categories. The $2^5 \times 2 \times 4$ table shows the (yes, no) counts for each of these five sources cross-classified with the farmers' education (whether they had at least some college education) and size of farm (number of pigs marketed annually, in thousands).

a. Explain why it is not proper to analyze the data by fitting a multinomial model to the counts in the $2 \times 4 \times 5$ contingency table cross-classifying education by size of farm by the source of veterinary information, treating source as the response variable. (This table contains 453 positive responses of sources from the 262 farmers.)

b. For a farmer with education i and size of farm s, let $y_{ist} = 1$ for response "yes" on source t and 0 for response "no." Table 9.11 shows output for using GEE with exchangeable working correlation to estimate parameters in the model lacking an education effect,

$$\text{logit}[P(Y_{ist} = 1)] = \alpha_t + \beta_t s, \quad s = 1, 2, 3, 4$$

Explain why the results suggest a strong positive size of farm effect for source A and perhaps a weak negative size effect of similar magnitude for sources C, D, and E.

Table 9.10. Pig Farmer Data for Problem 9.7

			A = yes								A = no							
			B = yes				B = no				B = yes				B = no			
			C = yes		C = no		C = yes		C = no		C = yes		C = no		C = yes		C = no	
			Response on D															
Education	Pigs	E	Y	N	Y	N	Y	N	Y	N	Y	N	Y	N	Y	N	Y	N
No	<1	Y	1	0	0	0	0	0	0	0	2	1	1	2	1	1	5	3
		N	0	0	0	0	0	0	0	1	1	0	0	5	4	7	7	0
	1–2	Y	2	0	0	0	0	0	0	0	4	0	0	4	1	0	0	4
		N	0	0	0	0	0	0	0	0	0	0	0	5	0	3	4	0
	2–5	Y	3	0	0	0	0	0	0	0	3	0	0	1	2	0	1	1
		N	1	0	0	0	0	0	0	3	0	0	0	2	0	1	4	0
	>5	Y	2	0	0	0	0	0	0	0	1	0	1	0	0	1	0	2
		N	1	0	0	2	1	0	1	6	0	1	1	1	0	0	6	0
Some	<1	Y	3	0	0	0	0	0	0	0	4	0	1	1	0	0	2	11
		N	0	0	0	0	0	0	0	0	4	0	1	2	4	6	14	0
	1–2	Y	0	0	0	0	0	0	0	0	2	0	0	1	0	0	1	6
		N	0	0	0	0	1	0	0	1	2	1	0	4	2	7	14	0
	2–5	Y	0	0	0	0	0	0	0	0	1	0	0	0	0	1	1	3
		N	1	0	0	0	0	0	0	0	0	0	0	5	0	4	4	0
	>5	Y	1	0	0	0	0	0	0	0	0	0	1	1	0	0	0	2
		N	1	1	0	0	0	1	0	10	0	0	0	4	1	2	4	0

Source: Thanks to Professor Tom Loughin, Kansas State University, for showing me these data.

Table 9.11. Computer Output for Problem 9.7

Working Correlation Matrix

	Col1	Col2	Col3	Col4	Col5
Row1	1.0000	0.0997	0.0997	0.0997	0.0997
Row2	0.0997	1.0000	0.0997	0.0997	0.0997
Row3	0.0997	0.0997	1.0000	0.0997	0.0997
Row4	0.0997	0.0997	0.0997	1.0000	0.0997
Row5	0.0997	0.0997	0.0997	0.0997	1.0000

Analysis Of GEE Parameter Estimates
Empirical Standard Error Estimates

Parameter		Estimate	Std Error	Z	Pr > |Z|
source	1	−4.4994	0.6457	−6.97	<0.0001
source	2	−0.8279	0.2809	−2.95	0.0032
source	3	−0.1526	0.2744	−0.56	0.5780
source	4	0.4875	0.2698	1.81	0.0708
source	5	−0.0808	0.2738	−0.30	0.7680
size*source	1	1.0812	0.1979	5.46	<0.0001
size*source	2	0.0792	0.1105	0.72	0.4738
size*source	3	−0.1894	0.1121	−1.69	0.0912
size*source	4	−0.2206	0.1081	−2.04	0.0412
size*source	5	−0.2387	0.1126	−2.12	0.0341

9.8 Table 10.4 in Chapter 10 shows General Social Survey responses on attitudes toward legalized abortion. For the response Y_t about legalization (1 = support, 0 = oppose) for question t ($t = 1, 2, 3$) and for gender g (1 = female, 0 = male), consider the model $\text{logit}[P(Y_t = 1)] = \alpha + \gamma g + \beta_t$ with $\beta_3 = 0$.

a. A GEE analysis using unstructured working correlation gives correlation estimates 0.826 for questions 1 and 2, 0.797 for 1 and 3, and 0.832 for 2 and 3. What does this suggest about a reasonable working correlation structure?

b. Table 9.12 shows a GEE analysis with exchangeable working correlation. Interpret effects.

9.9 Refer to the clinical trials data in Table 10.8 available at the text web site, which are analyzed with random effects models in Section 10.3.2. Use GEE methods to analyze the data from the 41 centers, treating each center as a cluster.

a. Specify a working correlation and fit a model.

b. Explain how to compare the two surgeries with a confidence interval. Interpret.

c. Show how results compare to those from using ML with a model that treats all observations as independent and has additive center and treatment effects.

Table 9.12. Computer Output for Abortion Survey Data of Problem 9.8

```
                    Working Correlation Matrix

                        Col1            Col2            Col3

         Row1         1.0000          0.8173          0.8173
         Row2         0.8173          1.0000          0.8173
         Row3         0.8173          0.8173          1.0000

               Analysis Of GEE Parameter Estimates
                Empirical Standard Error Estimates

Parameter             Estimate      Std Error        Z        Pr > |Z|

Intercept              -0.1253        0.0676      -1.85        0.0637
question        1       0.1493        0.0297       5.02       <0.0001
question        2       0.0520        0.0270       1.92        0.0544
question        3       0.0000        0.0000        .             .
female                  0.0034        0.0878       0.04        0.9688
```

9.10 Refer to the GSS data on sex in Table 8.14 in Exercise 8.15. Using GEE methods with cumulative logits, compare the two marginal distributions. Compare the results with those in Problem 8.15.

9.11 Analyze the data in the $3 \times 3 \times 3 \times 3$ table on government spending in Table 7.25 with a marginal cumulative logit model. Interpret the effects.

9.12 For the insomnia study summarized in Table 9.6, model (9.2) compared treatments while controlling for initial response of time to fall asleep.

a. Add an interaction term to model (9.2). Summarize how the estimated treatment effect varies according to the initial responses by showing that the estimated treatment log odds ratio changes from 0.00 to 1.41 as the initial response score goes from 10 to 75.

b. Now fit the model without interaction by treating initial response as qualitative, using dummy variables. Show that the estimated treatment log odds ratio is 0.911 ($SE = 0.249$), and interpret.

c. Now fit the model with interaction terms by treating initial response as qualitative. Explain why the results suggest that the active treatment seems relatively more successful at the two highest initial response levels.

9.13 Analyze Table 9.8 from Section 9.4.2 using a transitional model with *two* previous responses.

a. Given that y_{t-1} is in the model, does y_{t-2} provide additional predictive power?

b. How does the maternal smoking effect compare with the model using only y_{t-1} of the past responses?

9.14 Analyze the depression data in Table 9.1 using a Markov transitional model. Compare results and interpretations to those in this chapter using marginal models.

9.15 Table 9.13 is from a longitudinal study of coronary risk factors in school children. A sample of children aged 10–13 in 1977 were classified by gender and by relative weight (obese, not obese) in 1977, 1979, and 1981. Analyze these data, summarizing results in a one-page report.

Table 9.13. Children Classified by Gender and Relative Weight

	Responses[a]							
Gender	NNN	NNO	NON	NOO	ONN	ONO	OON	OOO
Male	119	7	8	3	13	4	11	16
Female	129	8	7	9	6	2	7	14

Source: From R. F. Woolson and W. R. Clarke, *J. R. Statist. Soc.*, **A147**: 87–99, 1984. Reproduced with permission from the Royal Statistical Society, London.
[a]NNN indicates not obese in 1977, 1979, and 1981, NNO indicates not obese in 1977 and 1979, but obese in 1981, and so forth.

9.16 Refer to the cereal diet and cholesterol study of Problem 6.16 (Table 6.19). Analyze these data with marginal models, summarizing results in a one-page report.

9.17 What is wrong with this statement: "For a first-order Markov chain, Y_t is independent of Y_{t-2}"?

9.18 True, or false? With repeated measures data having multiple observations per subject, one can treat the observations as independent and still get valid estimates, but the standard errors based on the independence assumption may be badly biased.

CHAPTER 10

Random Effects: Generalized Linear Mixed Models

Chapter 9 focused on modeling the *marginal* distributions of clustered responses. This chapter presents an alternative model type that has a term in the model for each cluster. The cluster-specific term takes the same value for each observation in a cluster. This term is treated as varying randomly among clusters. It is called a *random effect.*

Section 8.2.3 introduced cluster-specific terms in a model for matched pairs. Such models have *conditional* interpretations, the effects being conditional on the cluster. The effects are called *cluster-specific*, or *subject-specific* when each cluster is a subject. This contrasts with marginal models, which have *population-averaged* interpretations in which effects are averaged over the clusters.

The *generalized linear mixed model*, introduced in Section 10.1, extends generalized linear models to include random effects. Section 10.2 shows examples of logistic regression models with random effects and discusses connections and comparisons with marginal models. Section 10.3 shows examples of multinomial models and models with multiple random effect terms. Section 10.4 introduces *multilevel models* having random effects at different levels of a hierarchy. For example, an educational study could include a random effect for each student as well as for each school that the students attend. Section 10.5 summarizes model fitting methods. The appendix shows how to use software to do the analyses in this chapter.

10.1 RANDOM EFFECTS MODELING OF CLUSTERED CATEGORICAL DATA

Parameters that describe a factor's effects in ordinary linear models are called *fixed effects*. They apply to *all* categories of interest, such as genders, age groupings, or

An Introduction to Categorical Data Analysis, Second Edition. By Alan Agresti

treatments. By contrast, *random effects* apply to a *sample*. For a study with repeated measurement of subjects, for example, a cluster is a set of observations for a particular subject, and the model contains a random effect term for each subject. The random effects refer to a sample of clusters from all the possible clusters.

10.1.1 The Generalized Linear Mixed Model

Generalized linear models (GLMs) extend ordinary regression by allowing nonnormal responses and a link function of the mean (recall Chapter 3). The *generalized linear mixed model*, denoted by *GLMM*, is a further extension that permits random effects as well as fixed effects in the linear predictor. Denote the random effect for cluster i by u_i. We begin with the most common case, in which u_i is an intercept term in the model.

Let y_{it} denote observation t in cluster i. Let x_{it} be the value of the explanatory variable for that observation. (The model extends in an obvious way for multiple predictors.) Conditional on u_i, a GLMM resembles an ordinary GLM. Let $\mu_{it} = E(Y_{it}|u_i)$, the mean of the response variable for a given value of the random effect. With the link function $g(\cdot)$, the GLMM has the form

$$g(\mu_{it}) = u_i + \beta x_{it}, \quad i = 1, \ldots, n, \quad t = 1, \ldots, T$$

A GLMM with random effect as an intercept term is called a *random intercept* model. In practice, the random effect u_i is unknown, like the usual intercept parameter. It is treated as a random variable and is assumed to have a normal $N(\alpha, \sigma)$ distribution, with unknown parameters. The variance σ^2 is referred to as a *variance component*.

When $x_{it} = 0$ in this model, the expected value of the linear predictor is α, the mean of the probability distribution of u_i. An equivalent model enters α explicitly in the linear predictor,

$$g(\mu_{it}) = u_i + \alpha + \beta x_{it} \tag{10.1}$$

compensating by taking u_i to have an expected value of 0. The ML estimate of α is identical either way. However, it would be redundant to have both the α term in the model and permit an unspecified mean for the distribution of u_i. We will specify models this second way, taking u_i to have a $N(0, \sigma)$ distribution. The separate α parameter in the model then is the value of the linear predictor when $x_{it} = 0$ and u_i takes value at its mean of 0.

Why not treat the cluster-specific $\{u_i\}$ terms as fixed effects (parameters)? Usually a study has a large number of clusters, and so the model would then contain a large number of parameters. Treating $\{u_i\}$ as random effects, we have only a single additional parameter (σ) in the model, describing their dispersion.

Section 10.5 outlines the model-fitting process for GLMMs. As in ordinary models for a univariate response, for given predictor values ML fitting treats the observations as independent. For the GLMM, this independence is assumed conditional on the

$\{u_i\}$ as well as the ordinary predictor values. In practice, $\{u_i\}$ are unknown. Averaged with respect to the distribution of the $\{u_i\}$, the model implies nonnegative correlation among observations within a cluster, as discussed in the next subsection.

The model-fitting process estimates the fixed effects, the parameter σ of the normal distribution for the random effects, and provides predictions $\{\hat{u}_i\}$ of the random effects. We can substitute the fixed effect estimates and $\{\hat{u}_i\}$ in the linear predictor to estimate the means for particular clusters. The estimate $\hat{\sigma}$ describes the variability among clusters. In some studies, this variability might represent heterogeneity caused by not including certain explanatory variables that are associated with the response. The random effect then reflects terms that would be in the fixed effects part of the model if those explanatory variables had been included.

10.1.2 A Logistic GLMM for Binary Matched Pairs

We illustrate the GLMM expression (10.1) using a simple case – binary matched pairs, which Sections 8.1 and 8.2 considered. Cluster i consists of the observations (y_{i1}, y_{i2}) for matched pair i. Observation t in cluster i has $y_{it} = 1$ (a success) or 0 (a failure), for $t = 1, 2$.

Section 8.2.3 introduced a logistic model that permits heterogeneity among the clusters,

$$\text{logit}[P(Y_{i1} = 1)] = \alpha_i + \beta, \quad \text{logit}[P(Y_{i2} = 1)] = \alpha_i \tag{10.2}$$

The fixed effect β represents a log odds ratio, given the cluster. That is, e^{β} is the odds ratio comparing the response distribution at $t = 1$ with the response distribution at $t = 2$. Section 8.2.3 treated α_i as a fixed effect. With n clusters, there are $n + 1$ parameters. The conditional ML method estimated β after eliminating $\{\alpha_i\}$ from the likelihood function.

The random effects approach instead replaces α_i in model (10.2) by $u_i + \alpha$, where u_i is a random variable with mean 0. The model then is

$$\text{logit}[P(Y_{i1} = 1)] = u_i + \alpha + \beta, \quad \text{logit}[P(Y_{i2} = 1)] = u_i + \alpha \tag{10.3}$$

The $\{u_i\}$ are treated as independent from a $N(0, \sigma)$ distribution, with unknown σ. This is the special case of the GLMM (10.1) with logit link function g, $T = 2$, and x_{it} an indicator variable that is 1 for $t = 1$ and 0 for $t = 2$. Logistic regression models that contain a random effect assumed to have a normal distribution are an important class of models for binary data called *logistic-normal models*.

This model implies a nonnegative correlation between observations within a cluster. This reflects that observations from the same cluster usually tend to be more alike than observations from different clusters. Clusters with a relatively large positive u_i have a relatively large $P(Y_{it} = 1)$ for each t, whereas clusters with a relatively large negative u_i have a relatively small $P(Y_{it} = 1)$ for each t. For example, suppose $\alpha + \beta$ and α were close to 0. Then, with a large positive u_i it is common to see outcomes

$(y_{i1} = 1, y_{i2} = 1)$, whereas with a large negative u_i it is common to see outcomes $(y_{i1} = 0, y_{i2} = 0)$. When a high proportion of cases have these outcomes, the association between the repeated responses is positive. Greater association results from greater heterogeneity (i.e., larger σ).

10.1.3 Example: Sacrifices for the Environment Revisited

Table 10.1 shows a 2×2 table from the General Social Survey, analyzed originally in Chapter 8. Subjects were asked whether, to help the environment, they were willing to (1) raise taxes, (2) accept a cut in living standards. The ML fit of model (10.3), treating $\{u_i\}$ as normal, yields $\hat{\beta} = 0.210$ ($SE = 0.130$), with $\hat{\sigma} = 2.85$. For a given subject, the estimated odds of a "yes" response on accepting higher taxes equal $\exp(0.210) = 1.23$ times the odds of a "yes" response on accepting a lower standard of living.

Table 10.1. Opinions Relating to Environment

	Cut Living Standards		
Pay Higher Taxes	Yes	No	Total
Yes	227	132	359
No	107	678	785
Total	334	810	1144

The relatively large $\hat{\sigma}$ value of 2.85 reflects a strong association between the two responses. In fact, Table 10.1 has a sample odds ratio of 10.9. Whenever the sample log odds ratio in such a table is nonnegative, as it usually is, the ML estimate of β with this random effects approach is identical to the conditional ML estimate from treating $\{\alpha_i\}$ in model (10.2) as fixed effects. Section 8.2.3 presented this conditional ML approach. For these data the conditional ML estimate is $\hat{\beta} = \log(132/107) = 0.210$, with $SE = [(1/107) + (1/132)]^{1/2} = 0.130$.

10.1.4 Differing Effects in Conditional Models and Marginal Models

As Sections 9.1.3 and 8.2.2 discussed, parameters in GLMMs and marginal models have different interpretations. The parameters in GLMMs have conditional (cluster-specific) intepretations, given the random effect. By contrast, effects in marginal models are averaged over all clusters (i.e., population-averaged), and so those effects do not refer to a comparison at a fixed value of a random effect.

Section 8.2.2 noted that the cluster-specific model (10.2) applies naturally to the data as displayed in a separate partial table for each cluster, displaying the two matched responses. For the survey data on the environmental issues, each subject is a cluster and has their own table. The first row shows the response on taxes (a 1 in the first column for "yes" or in the second column for "no"), and the second row shows the response on lowering living standards. (Recall the form of Table 8.2.) The 1144 subjects provide 2288 observations in a $2 \times 2 \times 1144$ contingency table. Collapsing this table over the 1144 partial tables yields a 2×2 table with first row equal to (359, 785) and second row equal to (334, 810). These are the total number of "yes" and "no" responses for the two items. They are the marginal counts in Table 10.1.

Marginal models apply to the collapsed table, summarized over the subjects. The marginal model that corresponds to the subject-specific model (10.2) is

$$\text{logit}[P(Y_1 = 1)] = \alpha + \beta, \quad \text{logit}[P(Y_2 = 1)] = \alpha$$

where Y_1 is the response about higher taxes for a randomly selected subject and Y_2 is the response about lower standard of living for a different randomly selected subject. From Table 10.1, the estimated log odds of a "yes" response was $\hat{\alpha} + \hat{\beta} = \log(359/785) = -0.782$ for higher taxes and $\hat{\alpha} = \log(334/810) = -0.886$ for a lower standard of living. The estimate of β is the difference between these log odds. This is the log odds ratio, $\hat{\beta} = \log[(359 \times 810)/(785 \times 334)] = 0.104$, using the marginal counts of Table 10.1.

This estimated effect $\hat{\beta} = 0.104$ for the marginal model has the same sign but is weaker in magnitude than the estimated effect $\hat{\beta} = 0.210$ for the conditional model (10.3). The estimated effect for the conditional model says that for any given subject, the estimated odds of a "yes" response on higher taxes are $\exp(0.210) = 1.23$ times the estimated odds of a "yes" response on lower standard of living. The estimated effect for the marginal model says that the estimated odds of a "yes" response on higher taxes for a randomly selected subject are $\exp(0.104) = 1.11$ times the estimated odds of a "yes" response on lower standard of living for a different randomly selected subject.

When the link function is nonlinear, such as the logit, the population-averaged effects of marginal models are typically smaller in magnitude than the cluster-specific effects of GLMMs. Figure 10.1 illustrates why this happens. For a single quantitative

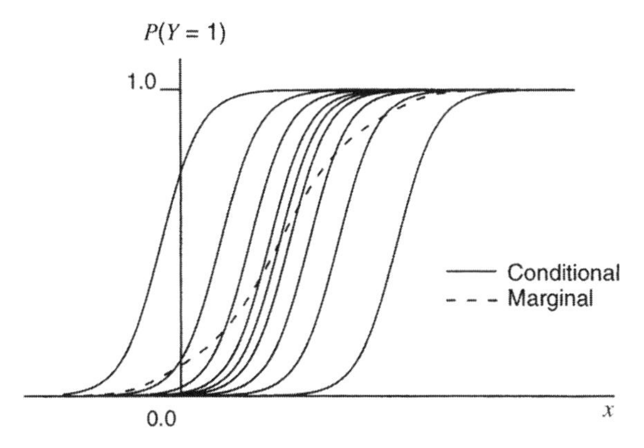

Figure 10.1. Logistic random-intercept model, showing the conditional (subject-specific) curves and the marginal (population-averaged) curve averaging over these.

explanatory variable x, the figure shows cluster-specific logistic regression curves for $P(Y_{it} = 1 \mid u_i)$ for several clusters when considerable heterogeneity exists. This corresponds to a relatively large σ for the random effects. At any fixed value of x, variability occurs in the conditional means, $E(Y_{it} \mid u_i) = P(Y_{it} = 1 \mid u_i)$. The average of these is the marginal mean, $E(Y_{it})$. These averages for various x values yield the superimposed dashed curve. That curve shows a weaker effect than each separate curve has. The difference between the two effects is greater as the cluster-specific curves are more spread out, that is, as the spread σ of the random effects is greater.

10.2 EXAMPLES OF RANDOM EFFECTS MODELS FOR BINARY DATA

This section presents examples of random effects models for binary responses. These are special cases of the logistic-normal model.

10.2.1 Small-Area Estimation of Binomial Probabilities

Small-area estimation refers to estimation of parameters for many geographical areas when each may have relatively few observations. For example, a study might find county-specific estimates of characteristics such as the unemployment rate or the proportion of families having health insurance coverage. With a national or statewide survey, counties with small populations may have few observations.

Let π_i denote the true probability of "success" in area i, $i = 1, \ldots, n$. These areas may be all the ones of interest, or only a sample. The fixed effects model

$$\text{logit}(\pi_i) = \beta_i, \quad i = 1, \ldots, n,$$

treats the areas as levels of a single factor. The model is saturated, having n parameters for the n binomial observations. Let T_i denote the number of observations from area i, of which y_i are successes. When we treat $\{y_i\}$ as independent binomial variates, the sample proportions $\{p_i = y_i/T_i\}$ are ML estimates of $\{\pi_i\}$.

When some areas have few observations, sample proportions in those areas may poorly estimate $\{\pi_i\}$. For small $\{T_i\}$, the sample proportions have large standard errors. They may display much more variability than $\{\pi_i\}$, especially when $\{\pi_i\}$ are similar (see Problem 10.5).

Random effects models that treat each area as a cluster can provide improved estimates. With random effects for the areas, the model is

$$\text{logit}(\pi_i) = u_i + \alpha \tag{10.4}$$

where $\{u_i\}$ are independent $N(0, \sigma)$ variates. The model now has two parameters (α and σ) instead of n parameters. When $\sigma = 0$, all π_i are identical. In assuming that the logits of the probabilities vary according to a normal distribution, the fitting process "borrows from the whole," using data from all the areas to estimate the probability in any given one. The estimate for a given area is then a weighted average of the sample proportion for that area alone and the overall proportion for all the areas.

Software provides ML estimates $\hat{\alpha}$ and $\hat{\sigma}$ and predicted values $\{\hat{u}_i\}$ for the random effects. The predicted value $\hat{u}_i$ depends on *all* the data, not only the data for area i. The estimate of the probability π_i in area i is then

$$\hat{\pi}_i = \exp(\hat{u}_i + \hat{\alpha})/[1 + \exp(\hat{u}_i + \hat{\alpha})]$$

A benefit of using data from all the areas instead of only area i to estimate π_i is that the estimator $\hat{\pi}_i$ tends to be closer than the sample proportion p_i to π_i. The $\{\hat{\pi}_i\}$ result from shrinking the sample proportions toward the overall sample proportion. The amount of shrinkage increases as $\hat{\sigma}$ decreases. If $\hat{\sigma} = 0$, then $\{\hat{\pi}_i\}$ are identical. In fact, they then equal the overall sample proportion after pooling all n samples. When truly all π_i are equal, $\hat{\pi}_i$ is a much better estimator of that common value than the sample proportion from sample i alone.

For a given $\hat{\sigma} > 0$, the $\{\hat{\pi}_i\}$ give more weight to the sample proportions as $\{T_i\}$ grows. As each sample has more data, we put more trust in the separate sample proportions.

The simple random effects model (10.4), which is natural for small-area estimation, can be useful for any application that estimates a large number of binomial parameters when the sample sizes are small. The following example illustrates this.

10.2.2 Example: Estimating Basketball Free Throw Success

In basketball, the person who plays center is usually the tallest player on the team. Often, centers shoot well from near the basket but not so well from greater distances. Table 10.2 shows results of free throws (a standardized shot taken from a distance of 15 feet from the basket) for the 15 top-scoring centers in the National Basketball Association after one week of the 2005–2006 season.

Let π_i denote the probability that player i makes a free throw ($i = 1, \ldots, 15$). For T_i observations of player i, we treat the number of successes y_i as binomial with index T_i and parameter π_i. Table 10.2 shows $\{T_i\}$ and $\{p_i = y_i/T_i\}$.

For the ML fit of model (10.4), $\hat{\alpha} = 0.908$ and $\hat{\sigma} = 0.422$. For a player with random effect $u_i = 0$, the estimated probability of making a free throw is $\exp(0.908)/[1 + \exp(0.908)] = 0.71$. We predict that 95% of the logits fall within $0.908 \pm 1.96(0.422)$, which is (0.08, 1.73). This interval corresponds to probabilities in the range (0.52, 0.85).

The predicted random effect values (obtained using PROC NLMIXED in SAS) yield probability estimates $\{\hat{\pi}_i\}$, also shown in Table 10.2. Since $\{T_i\}$ are small and since $\hat{\sigma}$ is relatively small, these estimates shrink the sample proportions substantially toward the overall sample proportion of free throws made, which was $101/143 = 0.706$. The $\{\hat{\pi}_i\}$ vary only between 0.61 and 0.76, whereas the sample proportions

Table 10.2. Estimates of Probability of Centers Making a Free Throw, Based on Data from First Week of 2005–2006 NBA Season

Player	n_i	p_i	$\hat{\pi}_i$	Player	n_i	p_i	$\hat{\pi}_i$
Yao	13	0.769	0.730	Curry	11	0.545	0.663
Frye	10	0.900	0.761	Miller	10	0.900	0.761
Camby	15	0.667	0.696	Haywood	8	0.500	0.663
Okur	14	0.643	0.689	Olowokandi	9	0.889	0.754
Blount	6	0.667	0.704	Mourning	9	0.778	0.728
Mihm	10	0.900	0.761	Wallace	8	0.625	0.692
Ilgauskas	10	0.600	0.682	Ostertag	6	0.167	0.608
Brown	4	1.000	0.748				

Note: p_i = sample, $\hat{\pi}_i$ = estimate using random effects model.
Source: nba.com.

vary between 0.17 and 1.0. Relatively extreme sample proportions based on few observations, such as the sample proportion of 0.17 for Ostertag, shrink more. If you are a basketball fan, which estimate would you think is more sensible for Ostertag's free throw shooting prowess, 0.17 or 0.61?

Are the data consistent with the simpler model, $\text{logit}(\pi_i) = \alpha$, in which π_i is identical for each player? To answer this, we could test H_0: $\sigma = 0$ for model (10.4). The usual tests do not apply to this hypothesis, however, because $\hat{\sigma}$ cannot be negative and so is not approximately normally distributed about σ under H_0. We will learn how to conduct the analysis in Section 10.5.2.

10.2.3 Example: Teratology Overdispersion Revisited

Section 9.2.4 showed results of a teratology experiment in which female rats on iron-deficient diets were assigned to four groups. Group 1 received only placebo injections. The other groups received injections of an iron supplement at various schedules. The rats were made pregnant and then sacrificed after 3 weeks. For each fetus in each rat's litter, the response was whether the fetus was dead. Because of unmeasured covariates that vary among rats in a given treatment, it is natural to permit the probability of death to vary from litter to litter within each treatment group.

Let y_i denote the number dead out of the T_i fetuses in litter i. Let π_{it} denote the probability of death for fetus t in litter i. Section 9.2.4 used the model

$$\text{logit}(\pi_{it}) = \alpha + \beta_2 z_{i2} + \beta_3 z_{i3} + \beta_4 z_{i4}$$

where $z_{ig} = 1$ if litter i is in group g and 0 otherwise. The estimates and standard errors treated the $\{y_i\}$ as binomial. This approach regards the outcomes for fetuses in a litter as independent and identical, with the same probability of death for each fetus in each litter within a given treatment group. This is probably unrealistic. Section 9.2.4 used the GEE approach to allow observations within a litter to be correlated.

Table 10.3. Estimates and Standard Errors (in Parentheses) for Logit Models Fitted to Table 9.4 from the Teratology Study

	Type of Logit Model		
Parameter	Binomial ML	GEE	GLMM
Intercept	1.144 (0.129)	1.144 (0.276)	1.802 (0.362)
Group 2	−3.322 (0.331)	−3.322 (0.440)	−4.515 (0.736)
Group 3	−4.476 (0.731)	−4.476 (0.610)	−5.855 (1.190)
Group 4	−4.130 (0.476)	−4.130 (0.576)	−5.594 (0.919)
Overdispersion	None	$\hat{\rho} = 0.185$	$\hat{\sigma} = 1.53$

Note: Binomial ML assumes no overdispersion; GEE (independence working equations) estimates are the same as binomial ML estimates.

Table 10.3 summarizes results for these approaches and for the GLMM that adds a normal random intercept u_i for litter i in the binomial logit model. This allows heterogeneity in the probability of death for different litters in a given treatment group. The estimated standard deviation of the random effect is $\hat{\sigma} = 1.53$. Results are similar in terms of significance of the treatment groups relative to placebo. Estimated effects are larger for this logistic-normal model than for the marginal model (estimated by GEE), because they are cluster-specific (i.e., litter-specific) rather than population-averaged. (Recall the discussion in Section 10.1.4, illustrated by Figure 10.1.)

10.2.4 Example: Repeated Responses on Similar Survey Items

An extension of the matched-pairs model (10.3) allows $T > 2$ observations in each cluster. We illustrate using Table 10.4, for which a cluster is a set of three observations for a subject. In a General Social Survey, the subjects indicated whether they supported legalizing abortion in each of three situations. The table cross classifies subjects by responses on the three abortion items and by their gender.

Let y_{it} denote the response for subject i on item t, with $y_{it} = 1$ representing support for legalized abortion. A random intercept model with main effects for the

Table 10.4. Support (1 = Yes, 2 = No) for Legalizing Abortion in Three Situations, by Gender

	Sequence of Responses on the Three Items							
Gender	(1,1,1)	(1,1,2)	(2,1,1)	(2,1,2)	(1,2,1)	(1,2,2)	(2,2,1)	(2,2,2)
Male	342	26	6	21	11	32	19	356
Female	440	25	14	18	14	47	22	457

Source: Data from 1994 General Social Survey. Items are (1) if the family has a very low income and cannot afford any more children, (2) when the woman is not married and does not want to marry the man, and (3) when the woman wants it for any reason.

abortion items and gender is

$$\text{logit}[P(Y_{it} = 1)] = u_i + \beta_t + \gamma x_i \tag{10.5}$$

where $x_i = 1$ for females and 0 for males, and where $\{u_i\}$ are independent $N(0, \sigma)$. The gender effect γ is assumed the same for each item, and the parameters $\{\beta_t\}$ for comparing the abortion items are assumed the same for each gender. Here, there is no constraint on $\{\beta_t\}$. If the model also contained a term α, it would need a constraint such as $\beta_3 = 0$.

Table 10.5 summarizes ML fitting results. The contrasts of $\{\hat{\beta}_t\}$ compare support for legalized abortion under different conditions. These indicate greater support with item 1 (when the family has a low income and cannot afford any more children) than the other two. There is slight evidence of greater support with item 2 (when the woman is not married and does not want to marry the man) than with item 3 (when the woman wants it for any reason).

The fixed effects estimates have subject-specific log odds ratio interpretations. For a given subject of either gender, for instance, the estimated odds of supporting legalized abortion for item 1 equal exp(0.83) = 2.3 times the estimated odds for item 3. This odds ratio also applies for sets of subjects who have the same random effect value. Since $\hat{\gamma} = 0.01$, for each item the estimated probability of supporting legalized abortion is similar for females and males with similar random effect values.

For these data, the random effects have estimated standard deviation $\hat{\sigma} = 8.6$. This is extremely high. It indicates that subjects are highly heterogeneous in their response probabilities for a given item. It also corresponds to strong associations among responses on the three items. This is reflected by 1595 of the 1850 subjects making the same response on all three items – that is, response patterns (0, 0, 0) and (1, 1, 1). In the United States, people tend to be either uniformly opposed to legalized abortion, regardless of the circumstances, or uniformly in favor of it.

Table 10.5. Summary of ML Estimates for Random Effects Model (10.5) and GEE Estimates for Corresponding Marginal Model with Exchangeable Working Correlation Matrix

		GLMM ML		Marginal Model GEE	
Effect	Parameter	Estimate	*SE*	Estimate	*SE*
Abortion	$\beta_1 - \beta_3$	0.83	0.16	0.149	0.030
	$\beta_1 - \beta_2$	0.54	0.16	0.097	0.028
	$\beta_2 - \beta_3$	0.29	0.16	0.052	0.027
Gender	γ	0.01	0.48	0.003	0.088
$\sqrt{\text{Var}(u_i)}$	σ	8.6	0.54		

To allow interaction between gender and item, a model uses different $\{\beta_t\}$ for men and women. This corresponds to having extra parameters that are the coefficients of cross products of the gender and the item indicator variables. Such a model does not fit better. The likelihood-ratio statistic comparing the two models (that is, double the difference in maximized log-likelihoods) equals 1.0 ($df = 2$) for testing that the extra parameters equal 0.

A marginal model analog of (10.5) is

$$\text{logit}[P(Y_t = 1)] = \beta_t + \gamma x$$

where Y_t is the response on item t for a randomly selected subject. Table 10.5 shows GEE estimates for the exchangeable working correlation structure. These population-averaged $\{\hat{\beta}_t\}$ are much smaller than the subject-specific $\{\hat{\beta}_t\}$ from the GLMM. This reflects the very large GLMM heterogeneity ($\hat{\sigma} = 8.6$) and the corresponding strong correlations among the three responses. For instance, the GEE analysis estimates a common correlation of 0.82 between pairs of responses. Although the GLMM $\{\hat{\beta}_t\}$ are about 5–6 times the marginal model $\{\hat{\beta}_t\}$, so are the standard errors. The two approaches provide similar substantive interpretations and conclusions.

10.2.5 Item Response Models: The Rasch Model

In the example just considered comparing three opinion items, we have seen that a GLMM without a gender effect,

$$\text{logit}[P(Y_{it} = 1)] = u_i + \beta_t \tag{10.6}$$

is adequate. Early applications of this form of GLMM were in psychometrics to describe responses to a battery of T questions on an exam. The probability $P(Y_{it} = 1 \mid u_i)$ that subject i makes the correct response on question t depends on the overall ability of subject i, characterized by u_i, and the easiness of question t, characterized by β_t. Such models are called *item-response models*.

The logit form (10.6) is often called the *Rasch model*, named after a Danish statistician who introduced it for such applications in 1961. Rasch treated the subject terms as fixed effects and used conditional ML methods. These days it is more common to treat subject terms as random effects.

10.2.6 Example: Depression Study Revisited

Table 9.1 showed data from a longitudinal study to compare a new drug with a standard drug for treating subjects suffering mental depression. Section 9.1.2 analyzed the data using marginal models. The response y_t for observation t on mental depression equals 1 for normal and 0 for abnormal, where $t = 1, 2, 3$ for three times of measurement. For severity of initial diagnosis s (1 = severe, 0 = mild), drug treatment d

(1 = new, 0 = standard), and time of observation t, we used the model

$$\text{logit}[P(Y_t = 1)] = \alpha + \beta_1 s + \beta_2 d + \beta_3 t + \beta_4 (d \times t)$$

to evaluate effects on the marginal distributions.

Now let y_{it} denote observation t for subject i. The model

$$\text{logit}[P(Y_{it} = 1)] = u_i + \alpha + \beta_1 s + \beta_2 d + \beta_3 t + \beta_4 (d \times t)$$

has subject-specific rather than population-averaged effects. Table 10.6 shows the ML estimates. The time trend estimates are $\hat{\beta}_3 = 0.48$ for the standard drug and $\hat{\beta}_3 + \hat{\beta}_4 = 1.50$ for the new one. These are nearly identical to the GEE estimates for the corresponding marginal model, which Table 10.6 also shows. (Sections 9.1.2 and 9.2.3 discussed these.) The reason is that the repeated observations are only weakly correlated, as the GEE analysis observed. Here, this is reflected by $\hat{\sigma} = 0.07$, which suggests little heterogeneity among subjects in their response probabilities.

Table 10.6. Model Parameter Estimates for Marginal and Conditional Models Fitted to Table 9.1 on Depression Longitudinal Study

Parameter	GEE Marginal Estimate	*SE*	Random Effects ML Estimate	*SE*
Diagnosis	−1.31	0.15	−1.32	0.15
Treatment	−0.06	0.23	−0.06	0.22
Time	0.48	0.12	0.48	0.12
Treat × time	1.02	0.19	1.02	0.19

When we assume $\sigma = 0$ in this model, the log-likelihood decreases by less than 0.001. For this special case of the model, the ML estimates and *SE* values are the same as if we used ordinary logistic regression without the random effect and ignored the clustering (e.g., acting as if each observation comes from a different subject).

10.2.7 Choosing Marginal or Conditional Models

Some statisticians prefer conditional models (usually with random effects) over marginal models, because they more fully describe the structure of the data. However, many statisticians believe that both model types are useful, depending on the application. We finish the section by considering issues in choosing one type over the other.

With the marginal model approach, ML is sometimes possible but the GEE approach is computationally simpler and more readily available with standard software. A drawback of the GEE approach is that likelihood-based inferences are not possible because the joint distribution of the responses is not specified. In addition, this approach does not explicitly include random effects and therefore does not allow these effects to be estimated.

The conditional modeling approach is preferable if one wants to fully model the joint distribution. The marginal modeling approach focuses only on the marginal distribution. The conditional modeling approach is also preferable if one wants to estimate cluster-specific effects or estimate their variability, or if one wants to specify a mechanism that could generate positive association among clustered observations. For example, some methodologists use conditional models whenever the main focus is on within-cluster effects. In the depression study (Section 10.2.6), the conditional model is appropriate if we want the estimate of the time effect to be "within-subject," describing the time trend at the subject level.

By contrast, if the main focus is on comparing groups that are independent samples, effects of interest are "between-cluster" rather than "within-cluster." It may then be adequate to estimate effects with a marginal model. For example, if after a period of time we mainly want to compare the rates of depression for those taking the new drug and for those taking the standard drug, a marginal model is adequate. In many surveys or epidemiological studies, a goal is to compare the relative frequency of occurrence of some outcome for different groups in a population. Then, quantities of primary interest include between-group odds ratios comparing marginal probabilities for the different groups.

When marginal effects are the main focus, it is simpler to model the margins directly. One can then parameterize the model so regression parameters have a direct marginal interpretation. Developing a more detailed model of the joint distribution that generates those margins, as a random effects model does, provides greater opportunity for misspecification. For instance, with longitudinal data the assumption that observations are independent, given the random effect, need not be realistic.

Latent variable constructions used to motivate model forms (such as the probit and cumulative logit) usually apply more naturally at the cluster level than the marginal level. Given a conditional model, one can in principle recover information about marginal distributions, although this may require extra work not readily done by standard software. That is, a conditional model implies a marginal model, but a marginal model does not itself imply a conditional model. In this sense, a conditional model has more information.

We have seen that parameters describing effects are usually larger in conditional models than marginal models, moreso as variance components increase. Usually, though, the significance of an effect (e.g., as determined by the ratio of estimate to *SE*) is similar for the two model types. If one effect seems more important than another in a conditional model, the same is usually true with a marginal model. The choice of the model is usually not crucial to inferential conclusions.

10.2.8 Conditional Models: Random Effects Versus Conditional ML

For the fixed effects approach with cluster-specific terms, a difficulty is that the model has a large number of parameters. To estimate the other effects in the model, the conditional ML approach removes the cluster-specific terms from the model. Section 8.2.3 introduced the conditional ML approach for binary matched pairs. Compared with the random effects approach, it has the advantage that it does not assume a parametric distribution for the cluster-specific terms.

However, the conditional ML approach has limitations and disadvantages. It is restricted to inference about within-cluster fixed effects. The conditioning removes the source of variability needed for estimating between-cluster effects. This approach does not provide information about cluster-specific terms, such as predictions of their values and estimates of their variability or of probabilities they determine. When the number of observations per cluster is large, it is computationally difficult to implement. Finally, conditional ML can be less efficient than the random effects approach for estimating the other fixed effects.

One application in which conditional ML with cluster-specific terms in logistic regression models has been popular is case–control studies. A case and the matching control or controls form a cluster. Section 8.2.4 discussed this for the matched-pairs case. For further details, see Breslow and Day (1980), Fleiss et al. (2003, Chapter 14), and Hosmer and Lemeshow (2000, Chapter 7).

10.3 EXTENSIONS TO MULTINOMIAL RESPONSES OR MULTIPLE RANDOM EFFECT TERMS

GLMMs extend directly from binary outcomes to multiple-category outcomes. Modeling is simpler with ordinal responses, because it is often adequate to use the same random effect term for each logit. With cumulative logits, this is the *proportional odds* structure that Section 6.2.1 used for fixed effects. However, GLMMs can have more than one random effect term in a model. Most commonly this is done to allow random slopes as well as random intercepts. We next show examples of these two cases.

10.3.1 Example: Insomnia Study Revisited

Table 9.6 in Section 9.3.2 showed results of a clinical trial at two occasions comparing a drug with placebo in treating insomnia patients. The response, time to fall asleep, fell in one of four ordered categories. We analyzed the data with marginal models in Section 9.3.2 and with transitional models in Section 9.4.3.

Let y_t = time to fall asleep at occasion t (0 = initial, 1 = follow-up), and let x = treatment (1 = active, 0 = placebo). The marginal model

$$\text{logit}[P(Y_t \le j)] = \alpha_j + \beta_1 t + \beta_2 x + \beta_3 (t \times x)$$

permits interaction. Table 10.7 shows GEE estimates.

Table 10.7. Results of Fitting Cumulative Logit Models (with Standard Errors in Parentheses) to Table 9.6

Effect	Marginal GEE	Random Effects (GLMM) ML
Treatment	0.034 (0.238)	0.058 (0.366)
Occasion	1.038 (0.168)	1.602 (0.283)
Treatment × occasion	0.708 (0.244)	1.081 (0.380)

Now, let y_{it} denote the response for subject i at occasion t. The random-intercept model

$$\text{logit}[P(Y_{it} \leq j)] = u_i + \alpha_j + \beta_1 t + \beta_2 x + \beta_3 (t \times x)$$

takes the linear predictor from the marginal model and adds a random effect u_i. The random effect is assumed to be the same for each cumulative probability. A subject with a relatively high u_i, for example, would have relatively high cumulative probabilities, and hence a relatively high chance of falling at the low end of the ordinal scale.

Table 10.7 also shows results of fitting this model. Results are substantively similar to the marginal model. The response distributions are similar initially for the two treatment groups, but the interaction suggests that at the follow-up response the active treatment group tends to fall asleep more quickly. We conclude that the time to fall asleep decreases more for the active treatment group than for the placebo group.

From Table 10.7, estimates and standard errors are about 50% larger for the GLMM than for the marginal model. This reflects the relatively large heterogeneity. The random effects have estimated standard deviation $\hat{\sigma} = 1.90$. This corresponds to a strong association between the responses at the two occasions.

10.3.2 Bivariate Random Effects and Association Heterogeneity

The examples so far have used univariate random effects, taking the form of random intercepts. Sometimes it is sensible to have a multivariate random effect, for example to allow a slope as well as an intercept to be random.

We illustrate using Table 10.8, from three of 41 studies that compared a new surgery with an older surgery for treating ulcers. The analyses below use data from all 41 studies, which you can see at the text web site. The response was whether the surgery resulted in the adverse event of recurrent bleeding (1 = yes, 0 = no).

As usual, to compare two groups on a binary response with data stratified on a third variable, we can analyze the strength of association in the 2 × 2 tables and investigate how that association varies (if at all) among the strata. When the strata are themselves a sample, such as different studies for a meta analysis, or schools, or medical clinics, a random effects approach is natural. We then use a separate random effect for each

Table 10.8. Tables Relating Treatment (New Surgery or Older Surgery) to Outcome on an Adverse Event, for Three Studies

		Adverse Event		Sample	Fitted
Study	Treatment	Yes	No	Odds Ratio	Odds Ratio
1	New surgery	7	8	0.159	0.147
	Old surgery	11	2		
5	New surgery	3	9	∞	2.59
	Old surgery	0	12		
6	New surgery	4	3	0.0	0.126
	Old surgery	4	0		

Note: From article by B. Efron, *J. Am. Statist. Assoc.*, **91**: 539, 1996. Complete data for 41 studies available at www.stat.ufl.edu/~aa/intro-cda/appendix.html.

stratum rather than for each subject. With a random sampling of strata, we can extend inferences to the population of strata.

Let y_{it} denote the response for a subject in study i using treatment t ($1 = $ new; $2 = $ old). One possible model is the logistic-normal random intercept model,

$$\begin{aligned} \text{logit}[P(Y_{i1} = 1)] &= u_i + \alpha + \beta \\ \text{logit}[P(Y_{i2} = 1)] &= u_i + \alpha \end{aligned} \tag{10.7}$$

where $\{u_i\}$ are $N(0, \sigma)$. This model assumes that the log odds ratio β between treatment and response is the same in each study. The parameter σ summarizes study-to-study heterogeneity in the logit-probabilities of adverse event. Note that the model treats each study as a cluster and gives it a random effect. The estimated treatment effect is $\hat{\beta} = -1.173$ ($SE = 0.118$). This is similar to the estimated treatment effect from treating the study terms as fixed rather than random ($\hat{\beta} = -1.220$, $SE = 0.119$).

It is more realistic to allow the treatment effect to vary across the 41 studies. A logistic-normal model permitting treatment-by-study interaction is

$$\begin{aligned} \text{logit}[P(Y_{i1} = 1)] &= u_i + \alpha + (\beta + v_i) \\ \text{logit}[P(Y_{i2} = 1)] &= u_i + \alpha \end{aligned} \tag{10.8}$$

Here, u_i is a random intercept, and v_i is a random slope in the sense that it is the coefficient of a treatment indicator variable ($1 = $ new treatment, $0 = $ old treatment). We assume that $\{(u_i, v_i)\}$ have a *bivariate normal* distribution. That distribution has means 0 and standard deviation σ_u for $\{u_i\}$ and σ_v for $\{v_i\}$, with a correlation ρ between u_i and v_i, $i = 1, \ldots, 41$.

For model (10.8), the log odds ratio between treatment and response equals $\beta + v_i$ in study i. So, β is the mean study-specific log odds ratio and σ_v describes variability in the log odds ratios. The fit of the model provides a simple summary of an estimated mean $\hat{\beta}$ and an estimated standard deviation $\hat{\sigma}_v$ of the log odds ratios for the population of strata.

In Table 10.8 the sample odds ratios vary considerably among studies, as is true also for all 41 studies. Some sample odds ratios even take the boundary values of 0 or ∞. For model (10.8), the summary treatment estimate is $\hat{\beta} = -1.299$ ($SE = 0.277$). This estimated mean log odds ratio corresponds to a summary odds ratio estimate of 0.27. There seems to be considerable heterogeneity in the true log odds ratios, suggested by $\hat{\sigma}_v = 1.52$ ($SE = 0.26$).

The hypothesis H_0: $\beta = 0$ states that there is a lack of association between treatment and response, in the sense of a mean log odds ratio of 0. The evidence against it is strong. For example, the Wald statistic is $z = \hat{\beta}/SE = -1.299/0.277 = -4.7$. However, the evidence is weaker than for model (10.7) without treatment-by-study interaction, for which $z = \hat{\beta}/SE = -1.173/0.118 = -10.0$. The extra variance component in the interaction model pertains to variability in the log odds ratios. As its estimate $\hat{\sigma}_v$ increases, so does the standard error of the estimated treatment effect $\hat{\beta}$ tend to increase. The more that the treatment effect varies among studies, the more difficult it is to estimate precisely the mean of that effect. When $\hat{\sigma}_v = 0$, the $\hat{\beta}$ and SE values are the same as for the simpler model (10.7).

The model fitting also provides predicted random effects. For stratum i with predicted random effect $\hat{v}_i$, the predicted log odds ratio is $\hat{\beta} + \hat{v}_i$. This shrinks the sample log odds ratio toward the mean of the sample log odds ratio for all the strata. This is especially useful when the sample size in a stratum is small, because the ordinary sample log odds ratio then has large standard error. Table 10.8 shows the sample odds ratios and the model predicted odds ratios for three studies. For all 41 studies, the sample odds ratios vary from 0.0 to ∞. Their random effects model counterparts (computed with PROC NLMIXED in SAS) vary only between 0.004 (for a study that reported 0 out of 34 adverse events for the new surgery and 34 out of 34 adverse events for the old surgery!) and 2.6 (for study 5). The smoothed estimates are much less variable and do not have the same ordering as the sample values, because the shrinkage tends to be greater for studies having smaller sample sizes.

10.4 MULTILEVEL (HIERARCHICAL) MODELS

Hierarchical models describe observations that have a nested nature: Units at one level are contained within units of another level. Hierarchical data are common in certain application areas, such as in educational studies.

A study of factors that affect student performance might measure, for each student and each exam in a battery of exams, whether the student passed. Students are nested within schools, and the model could study variability among students as well as variability among schools. The model could analyze the effect of the student's ability or past performance and of characteristics of the school the student attends. Just as two observations for the same student (on different exams) might tend to be more alike than observations for different students, so might two students in the same school tend to have more-alike observations than students from different schools. This could be because students within a school tend to be similar on various socioeconomic indices.

Hierarchical models contain terms for the different levels of units. For the example just mentioned, the model would contain terms for the student and for the school. Level 1 refers to measurements at the student level, and level 2 refers to measurements at the school level. GLMMs having a hierarchical structure of this sort are called *multilevel models*.

Multilevel models usually have a large number of terms. To limit the number of parameters, the model treats terms for the units as random effects rather than fixed effects. The random effects can enter the model at each level of the hierarchy. For example, random effects for students and random effects for schools refer to different levels of the model. Level 1 random effects can account for variability among students in student-specific characteristics not measured by the explanatory variables. These might include student ability and parents' socioeconomic status. The level 2 random effects account for variability among schools due to school-specific characteristics not measured by the explanatory variables. These might include the quality of the teaching staff, the teachers' average salary, the degree of drug-related problems in the school, and characteristics of the district for which the school enrolls students.

10.4.1 Example: Two-Level Model for Student Advancement

An educational study analyzes factors that affect student advancement in school from one grade to the next. For student t in school i, the response variable y_{it} measures whether the student passes to the next grade ($y_{it} = 1$) or fails. We will consider a model having two levels, one for students and one for schools. When there are many schools and we can regard them as approximately a random sample of schools that such a study could consider, we use random effects for the schools.

Let $\{x_{it1}, \ldots, x_{itk}\}$ denote the values of k explanatory variables that have values that vary at the student level. For example, for student t in school i, perhaps x_{it1} measures the student's performance on an achievement test, x_{it2} is gender, x_{it3} is race, and x_{it4} is whether he or she previously failed any grades. The level-one model is

$$\text{logit}[P(y_{it} = 1)] = \alpha_i + \beta_1 x_{it1} + \beta_2 x_{it2} + \cdots + \beta_k x_{itk}$$

The level-two model provides a linear predictor for the level-two (i.e., school-level) term in the level-one (i.e., student-level) model. That level-two term is the intercept, α_i. The level-two model has the form

$$\alpha_i = u_i + \alpha + \gamma_1 w_{i1} + \gamma_2 w_{i2} + \cdots + \gamma_t w_{i\ell}$$

Here, $\{w_{i1}, \ldots, w_{i\ell}\}$ are ℓ explanatory variables that have values that vary only at the school level, so they do not have a t subscript. For example, perhaps w_{i1} is per-student expenditure of school i. The term u_i is the random effect for school i.

Substituting the level-two model into the level-one model, we obtain

$$\text{logit}[P(y_{it} = 1)] = u_i + \alpha + \gamma_1 w_{i1} + \cdots + \gamma_t w_{i\ell} + \beta_1 x_{it1} + \cdots + \beta_k x_{itk}$$

This is a logistic-normal model with a random intercept (u_i). Here, a random effect enters only at level 2. More generally, the β parameters in the level-one model can depend on the school and themselves be modeled as part of the level-two model. This would be the case if we wanted to include random slopes in the model, for example, to allow the effect of race to vary by the school. More generally yet, the model can have more than two levels, or random effects can enter into two or more levels. When there are several observations per student, for example, the model can include random effects for students as well as for schools.

10.4.2 Example: Grade Retention

Raudenbush and Bryk (2002, pp. 296–304) analyzed data from a survey of 7516 sixth graders in 356 schools in Thailand. The response variable y_{it} measured whether student t in school i had to repeat at least one grade during the primary school years ($1 = \text{yes}, 0 = \text{no}$).

The level-one (student-level) variables were SES = socioeconomic status and whether the student was male (MALE = 1 if yes, 0 if female), spoke Central Thai dialect (DI = 1 if yes, 0 if no), had breakfast daily (BR = 1 if yes, 0 if no), and had some preprimary experience (PRE = 1, 0 if no). The level-one model was

$$\text{logit}[P(y_{it} = 1)] = \alpha_i + \beta_1 SES_{it} + \beta_2 MALE_{it} + \beta_3 DI_{it} + \beta_4 BR_{it} + \beta_5 PRE_{it}$$

The level-two (school-level) variables were MEANSES = the school mean SES, SIZE = size of school enrollment, and TEXTS = a measure of availability of textbooks in the school. The level-two model takes the school-specific term α_i from the level-one model and expresses it as

$$\alpha_i = u_i + \alpha + \gamma_1 MEANSES_i + \gamma_2 \, SIZE_i + \gamma_3 \, TEXTS_i$$

The random effect u_i reflects heterogeneity among the schools.

Raudenbush and Bryk first summarized results of a model that ignores the explanatory variables and analyzes only the variability in results among schools,

$$\text{logit}[P(y_{it} = 1)] = u_i + \alpha$$

They reported estimates $\hat{\alpha} = -2.22$ and $\hat{\sigma} = 1.30$ for a normal random effects distribution. For a student in a school having random effect at the mean ($u_i = 0$), the estimated probability of at least one retention is $\exp(-2.22)/[1 + \exp(-2.22)] = 0.10$. We predict that 95% of the logits fall within $-2.22 \pm 1.96(1.30)$, or $(-4.8, 0.34)$, which corresponds to probabilities between 0.01 and 0.58. This reflects substantial variability among schools.

For the multilevel model that includes both sets of explanatory variables, Raudenbush and Bryk reported the model fit,

$$\begin{aligned}\text{logit}[\hat{P}(y_{it} = 1)] &= \hat{u}_i - 2.18 - 0.69 MEANSES_i - 0.23 SIZE_i + 0.02 TEXTS_i \\ &\quad - 0.36 SES_{it} + 0.56 MALE_{it} + 0.31 DI_{it} - 0.35 BR_{it} - 0.49 PRE_{it}\end{aligned}$$

For example, for given values of the other explanatory variables and for a given value of the random effect, the estimated odds of retention for males were $\exp(0.56) = 1.75$ times those for females.

For between-groups comparisons with a GLMM, such as interpreting a gender effect, odds ratios relate to a fixed value for the random effect. For further discussion and several examples of multilevel models, both for discrete and continuous responses, see Snijders and Bosker (1999) and Raudenbush and Bryk (2002).

10.5 MODEL FITTING AND INFERENCE FOR GLMMS*

Now that we have seen several examples, we discuss a few issues involved in fitting GLMMs. Although technical details are beyond the scope of this text, it is not always simple to fit GLMMs and you should be aware of factors that affect the fit.

10.5.1 Fitting GLMMs

Conditional on $\{u_i\}$, model-fitting treats $\{y_{it}\}$ as independent over i and t. In practice, we do not know $\{u_i\}$. The model treats them as unobserved random variables. As Section 10.1.2 discussed, the variability among $\{u_i\}$ induces a positive correlation among the responses within a cluster.

The likelihood function for a GLMM refers to the fixed effects parameters $\{\alpha, \beta_1, \ldots, \beta_k\}$ and the parameter σ of the $N(0, \sigma)$ random effects distribution. To obtain the likelihood function, software eliminates $\{u_i\}$ by (1) forming the likelihood function as if the $\{u_i\}$ values were known, and then (2) averaging that function with respect to the $N(0, \sigma)$ distribution of $\{u_i\}$. Inference focuses primarily on the fixed effects, but the random effects or their distribution are often themselves of interest. Software predicts $\hat{u}_i$ by the estimated mean of the conditional distribution of u_i, given the data. This prediction depends on *all* the data, not just the data for cluster i.

The main difficulty in fitting GLMMs is step (2), the need to eliminate the random effects in order to obtain the likelihood function. The calculus-based integral used to average with respect to the normal distribution of the random effects does not have a closed form. Numerical methods for approximating it can be computationally intensive.

Gauss–Hermite quadrature is a method that uses a finite sum to do this. In essence, this method approximates the area under a curve by the area under a histogram with a particular number of bars. The approximation depends on the choice of the number of terms, which is the number q of *quadrature points* at which the function to be integrated is evaluated. As q increases, the approximation improves: The estimates and standard errors get closer to the actual ML estimates and their standard errors. Be careful not to use too few quadrature points. Most software will pick a default value for q, often small such as $q = 5$. As a check you can then increase q above that value to make sure the estimates and standard errors have stabilized to the desired degree of precision (e.g., that they do not change in the first few significant digits).

Gauss–Hermite quadrature is feasible when there is only a random intercept in the model or only a random intercept and a random slope. With complex random effect structure, other approaches are needed. *Monte Carlo* methods simulate in order to approximate the relevant integral. The Gauss–Hermite and Monte Carlo methods approximate the ML parameter estimates but converge to the ML estimates as they are applied more finely – for example, as the number of quadrature points increases for numerical integration. This is preferable to other approximate methods that are simpler but need not yield estimates near the ML estimates. Two such methods are *penalized quasi-likelihood* (PQL) and *Laplace approximation*. When true variance components are large, PQL can produce variance component estimates with substantial negative bias. Bias also occurs when the response distribution is far from normal (e.g., binary). The Laplace approximation replaces the function to be integrated by an approximation for which the integral has closed form.

Another approach to fitting of GLMMs is Bayesian. With it, the distinction between fixed and random effects no longer occurs. A probability distribution (the *prior distribution*) is assumed for each effect of either type. Prior distributions are usually chosen to be very spread out. Then, the data have the primary influence in determining the estimates.

10.5.2 Inference for Model Parameters

For GLMMs, inference about fixed effects proceeds in the usual way. For instance, likelihood-ratio tests can compare models when one model is the special case of the other in which one or more fixed effects equal 0.

Inference about random effects, such as whether a variance component equals zero, is more complex. For example, the basketball free-throw shooting example in Section 10.2.2 used a model

$$\text{logit}(\pi_i) = u_i + \alpha$$

that let the probability of success vary among players. The model

$$\text{logit}(\pi_i) = \alpha$$

in which the probability of success is the same for each player is the special case in which $\sigma = 0$ for the $N(0, \sigma)$ distribution of $\{u_i\}$. Since σ cannot be negative, the simpler model falls on the boundary of the parameter space for the more complex model. When this happens, the usual likelihood-ratio chi-squared test for comparing models is not valid. Likewise, a Wald statistic such as $\hat{\sigma}/SE$ does not have an approximate standard normal null distribution. (When $\sigma = 0$, because $\hat{\sigma} < 0$ is impossible, $\hat{\sigma}$ is not even approximately normally distributed around σ.)

The large-sample distribution of the likelihood-ratio statistic is known for the test of H_0: $\sigma = 0$ against H_a: $\sigma > 0$ for a model containing a single random effect term. The null distribution has probability 1/2 at 0 and 1/2 following the shape of a chi-squared distribution with $df = 1$. The test statistic value of 0 occurs when $\hat{\sigma} = 0$, in which case the maximum of the likelihood function is identical under H_0 and H_a. When $\hat{\sigma} > 0$ and the observed test statistic equals t, the P-value for this test is half the right-tail probability above t for a chi-squared distribution with $df = 1$.

For the basketball free-throw shooting example, the random effect has $\hat{\sigma} = 0.42$, with $SE = 0.39$. The likelihood-ratio test statistic comparing this model to the simpler model that assumes the same probability of success for each player equals 0.42. As usual, this equals double the difference between the maximized log likelihood values. The P-value is half the right-tail probability above 0.42 for a chi-squared distribution with $df = 1$, which is 0.26. It is plausible that all players have the same chance of success. However, the sample size was small, which is why an implausibly simplistic model seems adequate.

PROBLEMS

10.1 Refer back to Table 8.10 from a recent General Social Survey that asked subjects whether they believe in heaven and whether they believe in hell.

- **a.** Fit model (10.3). If your software uses numerical integration, report $\hat{\beta}$, $\hat{\sigma}$, and their standard errors for 2, 10, 100, 400, and 1000 quadrature points. Comment on convergence.
- **b.** Interpret $\hat{\beta}$.
- **c.** Compare $\hat{\beta}$ and its SE for this approach to their values for the conditional ML approach of Section 8.2.3.

10.2 You plan to apply the matched-pairs model (10.3) to a data set for which y_{i1} is whether the subject agrees that abortion should be legal if the woman cannot afford the child (1 = yes, 0 = no), and y_{i2} is whether the subject opposes abortion if a woman wants it because she is unmarried (1 = yes, 0 = no).

- **a.** Indicate a way in which this model would probably be inappropriate. (Hint: Do you think these variables would have a positive, or negative, log odds ratio?)
- **b.** How could you reword the second question so the model would be more appropriate?

10.3 A dataset on pregnancy rates among girls in 13 north central Florida counties has information on the total in 2005 for each county i on T_i = number of births and y_i = number of those for which mother's age was under 18. Let π_i be the probability that a pregnancy in county i is to a mother of age under 18. The logistic-normal model, $\text{logit}(\pi_i) = u_i + \alpha$, has $\hat{\alpha} = -3.24$ and $\hat{\sigma} = 0.33$.

a. Find $\hat{\pi}_i$ for a county estimated to be (i) at the mean, (ii) two standard deviations below the mean, (iii) two standard deviations above the mean on the random effects distribution.

b. For estimating $\{\pi_i\}$, what advantage does this model have over the fixed effects model, $\text{logit}(\pi_i) = \beta_i$?

10.4 Table 10.9 shows the free-throw shooting, by game, of Shaq O'Neal of the Los Angeles Lakers during the 2000 NBA (basketball) playoffs. In game i, let y_i = number made out of T_i attempts.

a. Fit the model, $\text{logit}(\pi_i) = u_i + \alpha$, where $\{u_i\}$ are independent $N(0, \sigma)$, and given $\{u_i\}$, $\{y_i\}$ are independent binomial variates for $\{T_i\}$ trials with success probabilities $\{\pi_i\}$. Report $\hat{\alpha}$ and $\hat{\sigma}$.

b. Use $\hat{\alpha}$ to summarize O'Neal's free-throw shooting in an average game (for which $u_i = 0$).

c. Use $\hat{\alpha}$ and $\hat{\sigma}$ to estimate how O'Neal's free-throw shooting varies among games.

Table 10.9. Shaq O'Neal Basketball Data for Problem 10.4

Game	No. Made	No. Attempts	Game	No. Made	No. Attempts	Game	No. Made	No. Attempts
1	4	5	9	4	12	17	8	12
2	5	11	10	1	4	18	1	6
3	5	14	11	13	27	19	18	39
4	5	12	12	5	17	20	3	13
5	2	7	13	6	12	21	10	17
6	7	10	14	9	9	22	1	6
7	6	14	15	7	12	23	3	12
8	9	15	16	3	10			

Source: www.nba.com.

10.5 For 10 coins, let π_i denote the probability of a head for coin i. You flip each coin five times. The sample numbers of heads are {2, 4, 1, 3, 3, 5, 4, 2, 3, 1}.

a. Report the sample proportion estimates of π_i. Formulate a model for which these are the ML estimates.

b. Formulate a random effects model for the data. Using software, find the ML estimates of the parameters. Interpret.

c. Using software, for the model in (**b**) obtain predicted values $\{\hat{\pi}_i\}$.

d. Which estimates would you prefer for $\{\pi_i\}$, those in (**a**) or those in (**c**)? Why?

e. Suppose all $\pi_i = 0.50$. Compare the estimates in (**a**) and in (**c**) by finding the average absolute distance of the estimates from 0.50 in each case. What does this suggest?

10.6 For Table 7.3 from the survey of high school students, let $y_{it} = 1$ when subject i used substance t ($t = 1$, cigarettes; $t = 2$, alcohol; $t = 3$, marijuana). Table 10.10 shows output for the logistic-normal model, $\text{logit}[P(Y_{it} = 1)] = u_i + \beta_t$.

a. Interpret $\{\hat{\beta}_t\}$. Illustrate odds ratio interpretations by comparing use of cigarettes and marijuana.

b. In practical terms, what does the large value for $\hat{\sigma}$ imply?

c. In practical terms, what does a large positive value for u_i for a particular student represent?

Table 10.10. Computer Output for GLMM for Student Survey in Problem 10.6

Description	Value	Parameter	Estimate	Std Error	t Value
Subjects	2276				
Max Obs Per Subject	3	beta1	4.2227	0.1824	23.15
Parameters	4	beta2	1.6209	0.1207	13.43
Quadrature Points	200	beta3	−0.7751	0.1061	−7.31
Log Likelihood	−3311	sigma	3.5496	0.1627	21.82

10.7 Refer to the previous exercise.

a. Compare $\{\hat{\beta}_t\}$ to the estimates for the marginal model in Problem 9.1. Why are they so different?

b. How is the focus different for the model in the previous exercise than for the loglinear model (*AC*, *AM*, *CM*) that Section 7.1.6 used?

c. If $\hat{\sigma} = 0$ in the GLMM in the previous exercise, the loglinear model (A, C, M) of mutual independence has the same fit as the GLMM. Why do you think this is so?

10.8 For the student survey summarized by Table 7.13, (**a**) analyze using GLMMs, (**b**) compare results and interpretations to those with marginal models in Problem 9.2.

10.9 For the crossover study summarized by Table 9.9 (Problem 9.6), fit the model

$$\text{logit}[P(Y_{i(k)t} = 1)] = u_{i(k)} + \alpha_k + \beta_t$$

where $\{u_{i(k)}\}$ are independent $N(0, \sigma)$. Interpret $\{\hat{\beta}_t\}$ and $\hat{\sigma}$.

10.10 For the previous exercise, compare estimates of $\beta_B - \beta_A$ and $\beta_C - \beta_A$ and their SE values to those using the corresponding marginal model of Problem 9.6.

10.11 Refer to Table 5.5 on admissions decisions for Florida graduate school applicants. For a subject in department i of gender g (1 = females, 0 = males), let $y_{ig} = 1$ denote being admitted.

a. For the fixed effects model, $\text{logit}[P(Y_{ig} = 1)] = \alpha + \beta g + \beta_i^D$, the estimated gender effect is $\hat{\beta} = 0.173$ ($SE = 0.112$). Interpret.

b. The corresponding model (10.7) that treats departments as a normal random effect has $\hat{\beta} = 0.163$ ($SE = 0.111$). Interpret.

c. The model of form (10.8) that allows the gender effect to vary randomly by department has $\hat{\beta} = 0.176$ ($SE = 0.132$), with $\hat{\sigma}_v = 0.20$. Interpret. Explain why the standard error of $\hat{\beta}$ is slightly larger than with the other analyses.

d. The marginal sample log odds ratio between gender and whether admitted equals -0.07. How could this take different sign from $\hat{\beta}$ in these models?

10.12 Consider Table 8.14 on premarital and extramarital sex. Table 10.11 shows the results of fitting a cumulative logit model with a random intercept.

a. Interpret $\hat{\beta}$.

b. What does the relatively large $\hat{\sigma}$ value suggest?

Table 10.11. Computer Output for Problem 10.12

Subjects	475	Parameter	Estimate	Std Error	t Value
Max Obs Per Subject	2	inter1	−1.5422	0.1826	−8.45
Parameters	5	inter2	−0.6682	0.1578	−4.24
Quadrature Points	100	inter3	0.9273	0.1673	5.54
Log Likelihood	−890.1	beta	4.1342	0.3296	12.54
		sigma	2.0757	0.2487	8.35

10.13 Refer to the previous exercise. Analyze these data with a corresponding cumulative logit marginal model.

a. Interpret $\hat{\beta}$.

b. Compare $\hat{\beta}$ to its value in the GLMM. Why are they so different?

10.14 Refer to Problem 9.11 for Table 7.25 on government spending.

a. Analyze these data using a cumulative logit model with random effects. Interpret.

b. Compare the results with those with a marginal model (Problem 9.11).

10.15 Refer to Table 4.16 and Problem 4.20, about an eight-center clinical trial comparing a drug with placebo for curing an infection. Model the data in a way that allows the odds ratio to vary by center. Summarize your analyses and interpretations in a one-page report.

10.16 See http://bmj.com/cgi/content/full/317/7153/235 for a meta analysis of studies about whether administering albumin to critically ill patients increases or decreases mortality. Analyze the data for the 13 studies with hypovolemia patients using logistic models with (i) fixed effects, (ii) random effects. Summarize your analyses in a two-page report.

10.17 Refer to the insomnia example in Section 10.3.1.

a. From results in Table 10.7 for the GLMM, explain how to get the interpretation quoted in the text that "The response distributions are similar initially for the two treatment groups, but the interaction suggests that at the follow-up response the active treatment group tends to fall asleep more quickly."

b. According to SAS, the maximized log likelihood equals -593.0, compared with -621.0 for the simpler model forcing $\sigma = 0$. Compare models, using a likelihood-ratio test. What do you conclude?

10.18 Analyze Table 9.8 with age and maternal smoking as predictors using a (**a**) logistic-normal model, (**b**) marginal model, (**c**) transitional model. Summarize your analyses in a two-page report, explaining how the interpretation of the maternal smoking effect differs for the three approaches.

10.19 Refer to the toxicity study with data summarized in Table 10.12. Collapsing the ordinal response to binary in terms of whether with data summarized in the outcome is normal, consider logistic models as a linear function of the dose level.

a. Does the ordinary binomial GLM show evidence of overdispersion?

b. Fit the logistic model using a GEE approach with exchangeable working correlation among fetuses in the same litter. Interpret and compare with results in (a).

c. Fit the logistic GLMM after adding a litter-specific normal random effect. Interpret and compare with previous results.

10.20 Refer to the previous exercise. Analyze the data with a marginal model and with a GLMM, both of cumulative logit form, for the ordinal response. Summarize analyses in a two-page report.

10.21 Table 10.13 reports results of a study of fish hatching under three environments. Eggs from seven clutches were randomly assigned to three treatments, and the response was whether an egg hatched by day 10. The three treatments were (1) carbon dioxide and oxygen removed, (2) carbon dioxide only removed, and (3) neither removed.

Table 10.12. Response Counts for 94 Litters of Mice on (Number Dead, Number Malformed, Number Normal), for Problem 10.19

Dose = 0.00 g/kg	Dose = 0.75 g/kg	Dose = 1.50 g/kg	Dose = 3.00 g/kg
(1, 0, 7), (0, 0, 14)	(0, 3, 7), (1, 3, 11)	(0, 8, 2), (0, 6, 5)	(0, 4, 3), (1, 9, 1)
(0, 0, 13), (0, 0, 10)	(0, 2, 9), (0, 0, 12)	(0, 5, 7), (0, 11, 2)	(0, 4, 8), (1, 11, 0)
(0, 1, 15), (1, 0, 14)	(0, 1, 11), (0, 3, 10)	(1, 6, 3), (0, 7, 6)	(0, 7, 3), (0, 9, 1)
(1, 0, 10), (0, 0, 12)	(0, 0, 15), (0, 0, 11)	(0, 0, 1), (0, 3, 8)	(0, 3, 1), (0, 7, 0)
(0, 0, 11), (0, 0, 8)	(2, 0, 8), (0, 1, 10)	(0, 8, 3), (0, 2, 12)	(0, 1, 3), (0, 12, 0)
(1, 0, 6), (0, 0, 15)	(0, 0, 10), (0, 1, 13)	(0, 1, 12), (0, 10, 5)	(2, 12, 0), (0, 11, 3)
(0, 0, 12), (0, 0, 12)	(0, 1, 9), (0, 0, 14)	(0, 5, 6), (0, 1, 11)	(0, 5, 6), (0, 4, 8)
(0, 0, 13), (0, 0, 10)	(1, 1, 11), (0, 1, 9)	(0, 3, 10), (0, 0, 13)	(0, 5, 7), (2, 3, 9)
(0, 0, 10), (1, 0, 11)	(0, 1, 10), (0, 0, 15)	(0, 6, 1), (0, 2, 6)	(0, 9, 1), (0, 0, 9)
(0, 0, 12), (0, 0, 13)	(0, 0, 15), (0, 3, 10)	(0, 1, 2), (0, 0, 7)	(0, 5, 4), (0, 2, 5)
(1, 0, 14), (0, 0, 13)	(0, 2, 5), (0, 1, 11)	(0, 4, 6), (0, 0, 12)	(1, 3, 9), (0, 2, 5)
(0, 0, 13), (1, 0, 14)	(0, 1, 6), (1, 1, 8)		(0, 1, 11)
(0, 0, 14)			

Source: Study described in article by C. J. Price, C. A Kimmel, R. W. Tyl, and M. C. Marr, *Toxicol. Appl. Pharmac.*, **81**: 113–127, 1985.

a. Let π_{it} denote the probability of hatching for an egg from clutch i in treatment t. Assuming independent binomial observations, fit the model

$$\text{logit}(\pi_{it}) = \alpha + \beta_1 z_1 + \beta_2 z_2$$

where $z_t = 1$ for treatment t and 0 otherwise. What does your software report for $\hat{\beta}_1$. (It should be $-\infty$, since treatment 1 has no successes.)

b. Model these data using random effects for the clutches. Compare the results with (**a**).

Table 10.13. Data on Fish Hatching for Problem 10.21

	Treatment 1		Treatment 2		Treatment 3	
Clutch	No. Hatched	Total	No. Hatched	Total	No. Hatched	Total
1	0	6	3	6	0	6
2	0	13	0	13	0	13
3	0	10	8	10	6	9
4	0	16	10	16	9	16
5	0	32	25	28	23	30
6	0	7	7	7	5	7
7	0	21	10	20	4	20

Source: Thanks to Becca Hale, Zoology Department, University of Florida for these data.

10.22 Problem 3.15 analyzed data from a General Social Survey on responses of 1308 subjects to the question, "Within the past 12 months, how many people have you known personally that were victims of homicide?" It used Poisson and negative binomial GLMs for count data. Here is a possible GLMM: For response y_i for subject i of race x_i (1 = black, 0 = white),

$$\log[E(Y_i)] = u_i + \alpha + \beta x_i$$

where conditional on u_i, y_i has a Poisson distribution, and where $\{u_i\}$ are independent $N(0, \sigma)$. Like the negative binomial GLM, unconditionally (when $\sigma > 0$) this model can allow more variability than the Poisson GLM.

a. The Poisson GLMM has $\hat{\alpha} = -3.69$ and $\hat{\beta} = 1.90$, with $\hat{\sigma} = 1.6$. Show that, for subjects at the mean of the random effects distribution, the estimated expected responses are 0.167 for blacks and 0.025 for whites.

b. Interpret $\hat{\beta}$.

10.23 A crossover study compares two drugs on a binary response variable. The study classifies subjects by age as under 60 or over 60. In a GLMM, these two age groups have the same conditional effect comparing the drugs, but the older group has a much larger variance component for its random effects. For the corresponding marginal model, explain why the drug effect for the older group will be smaller than that for the younger group.

10.24 True, or false? In a logistic regression model containing a random effect as a way of modeling within-subject correlation in repeated measures studies, the greater the estimate $\hat{\sigma}$ for the random effects distribution, the greater the heterogeneity of the subjects, and the larger in absolute value the estimated effects tend to be compared with the marginal model approach (with effects averaged over subjects, rather than conditional on subjects).

CHAPTER 11

A Historical Tour of Categorical Data Analysis*

We conclude by providing a historical overview of the evolution of methods for categorical data analysis (CDA). The beginnings of CDA were often shrouded in controversy. Key figures in the development of statistical science made groundbreaking contributions, but these statisticians were often in heated disagreement with one another.

11.1 THE PEARSON–YULE ASSOCIATION CONTROVERSY

Much of the early development of methods for CDA took place in the UK. It is fitting that we begin our historical tour in London in 1900, because in that year Karl Pearson introduced his chi-squared statistic (X^2). Pearson's motivation for developing the chi-squared test included testing whether outcomes on a roulette wheel in Monte Carlo varied randomly and testing statistical independence in two-way contingency tables.

Much of the literature on CDA in the early 1900s consisted of vocal debates about appropriate ways to summarize association. Pearson's approach assumed that continuous bivariate distributions underlie cross-classification tables. He argued that one should describe association by approximating a measure, such as the correlation, for the underlying continuum. In 1904, Pearson introduced the term *contingency* as a "measure of the total deviation of the classification from independent probability," and he introduced measures to describe its extent and to estimate the correlation.

George Udny Yule (1871–1951), an English contemporary of Pearson's, took an alternative approach in his study of association between 1900 and 1912. He believed that many categorical variables are inherently discrete. Yule defined measures, such as the odds ratio, directly using cell counts without assuming an underlying continuum. Discussing one of Pearson's measures that assumes underlying normality,

An Introduction to Categorical Data Analysis, Second Edition. By Alan Agresti

Yule stated "at best the normal coefficient can only be said to give us in cases like these a hypothetical correlation between supposititious variables. The introduction of needless and unverifiable hypotheses does not appear to me a desirable proceeding in scientific work." Yule also showed the potential discrepancy between marginal and conditional associations in contingency tables, later noted by E. H. Simpson in 1951 and now called *Simpson's paradox*.

Karl Pearson did not take kindly to criticism, and he reacted negatively to Yule's ideas. For example, Pearson claimed that the values of Yule's measures were unstable, since different collapsings of $I \times J$ tables to 2×2 tables could produce quite different values. In 1913, Pearson and D. Heron filled more than 150 pages of *Biometrika*, a journal he co-founded and edited, with a scathing reply to Yule's criticism. In a passage critical also of Yule's well-received book *An Introduction to the Theory of Statistics*, they stated

> If Mr. Yule's views are accepted, irreparable damage will be done to the growth of modern statistical theory. . . . [His measure] has never been and never will be used in any work done under his [Pearson's] supervision. . . . We regret having to draw attention to the manner in which Mr. Yule has gone astray at every stage in his treatment of association, but criticism of his methods has been thrust on us not only by Mr Yule's recent attack, but also by the unthinking praise which has been bestowed on a text-book which at many points can only lead statistical students hopelessly astray.

Pearson and Heron attacked Yule's "half-baked notions" and "specious reasoning" and concluded that Yule would have to withdraw his ideas "if he wishes to maintain any reputation as a statistician."

Half a century after the Pearson–Yule controversy, Leo Goodman and William Kruskal of the University of Chicago surveyed the development of measures of association for contingency tables and made many contributions of their own. Their 1979 book, *Measures of Association for Cross Classifications*, reprinted their four influential articles on this topic. Initial development of many measures occurred in the 1800s, such as the use of the relative risk by the Belgian social statistician Adolphe Quetelet in 1849. The following quote from an article by M. H. Doolittle in 1887 illustrates the lack of precision in early attempts to quantify *association* even in 2×2 tables.

> Having given the number of instances respectively in which things are both thus and so, in which they are thus but not so, in which they are so but not thus, and in which they are neither thus nor so, it is required to eliminate the general quantitative relativity inhering in the mere thingness of the things, and to determine the special quantitative relativity subsisting between the thusness and the soness of the things.

11.2 R. A. FISHER'S CONTRIBUTIONS

Karl Pearson's disagreements with Yule were minor compared with his later ones with Ronald A. Fisher (1890–1962). Using a geometric representation, in 1922 Fisher introduced *degrees of freedom* to characterize the family of chi-squared distributions. Fisher claimed that, for tests of independence in $I \times J$ tables, X^2 had $df = (I - 1)(J - 1)$. By contrast, in 1900 Pearson had argued that, for any application of his statistic, df equalled the number of cells minus 1, or $IJ - 1$ for two-way tables. Fisher pointed out, however, that estimating hypothesized cell probabilities using estimated row and column probabilities resulted in an additional $(I - 1) + (J - 1)$ constraints on the fitted values, thus affecting the distribution of X^2.

Not surprisingly, Pearson reacted critically to Fisher's suggestion that his formula for df was incorrect. He stated

> I hold that such a view [Fisher's] is entirely erroneous, and that the writer has done no service to the science of statistics by giving it broad-cast circulation in the pages of the *Journal of the Royal Statistical Society*. . . . I trust my critic will pardon me for comparing him with Don Quixote tilting at the windmill; he must either destroy himself, or the whole theory of probable errors, for they are invariably based on using sample values for those of the sampled population unknown to us.

Pearson claimed that using row and column sample proportions to estimate unknown probabilities had negligible effect on large-sample distributions. Fisher was unable to get his rebuttal published by the Royal Statistical Society, and he ultimately resigned his membership.

Statisticians soon realized that Fisher was correct. For example, in an article in 1926, Fisher provided empirical evidence to support his claim. Using 11,688 2×2 tables randomly generated by Karl Pearson's son, E. S. Pearson, he found a sample mean of X^2 for these tables of 1.00001; this is much closer to the 1.0 predicted by his formula for $E(X^2)$ of $df = (I - 1)(J - 1) = 1$ than Pearson's $IJ - 1 = 3$. Fisher maintained much bitterness over Pearson's reaction to his work. In a later volume of his collected works, writing about Pearson, he stated "If peevish intolerance of free opinion in others is a sign of senility, it is one which he had developed at an early age."

Fisher also made good use of CDA methods in his applied work. For example, he was also a famed geneticist. In one article, Fisher used Pearson's goodness-of-fit test to test Mendel's theories of natural inheritance. Calculating a summary P-value from the results of several of Mendel's experiments, he obtained an unusually large value $(P = 0.99996)$ for the right-tail probability of the reference chi-squared distribution. In other words X^2 was so small that the fit seemed *too* good, leading Fisher in 1936 to comment "the general level of agreement between Mendel's expectations and his reported results shows that it is closer than would be expected in the best of several thousand repetitions. . . . I have no doubt that Mendel was deceived by a gardening assistant, who knew only too well what his principal expected from each trial made." In a letter written at the time, he stated "Now, when data have been faked, I know very well how generally people underestimate the frequency of wide chance deviations, so that the tendency is always to make them agree too well with expectations."

In 1934 the fifth edition of Fisher's classic text *Statistical Methods for Research Workers* introduced "Fisher's exact test" for 2×2 contingency tables. In his 1935 book *The Design of Experiments*, Fisher described the tea-tasting experiment

(Section 2.6.2) based on his experience at an afternoon tea break while employed at Rothamsted Experiment Station. Other CDA-related work of his included showing how to (i) find ML estimates of parameters in the probit model (an iterative weighted least squares method today commonly called *Fisher scoring*), (ii) find ML estimates of cell probabilities satisfying the homogeneous association property of equality of odds ratios between two variables at each level of the third, and (iii) assign scores to rows and columns of a contingency table to maximize the correlation.

11.3 LOGISTIC REGRESSION

The mid-1930s finally saw some model building for categorical responses. For instance, Chester Bliss popularized the probit model for applications in toxicology dealing with a binary response. See Chapter 9 of Cramer (2003) for a survey of the early origins of binary regression models.

In a book of statistical tables published in 1938, R. A. Fisher and Frank Yates suggested $\log[\pi/(1-\pi)]$ as a possible transformation of a binomial parameter for analyzing binary data. In 1944, the physician and statistician Joseph Berkson introduced the term "logit" for this transformation. Berkson showed that the logistic regression model fitted similarly to the probit model, and his subsequent work did much to popularize logistic regression. In 1951, Jerome Cornfield, another statistician with a medical background, showed the use of the odds ratio for approximating relative risks in case–control studies with this model.

In the early 1970s, work by the Danish statistician and mathematician Georg Rasch sparked an enormous literature on item response models. The most important of these is the logit model with subject and item parameters, now called the *Rasch model* (Section 10.2.5). This work was highly influential in the psychometric community of northern Europe (especially in Denmark, the Netherlands, and Germany) and spurred many generalizations in the educational testing community in the United States.

The extension of logistic regression to multicategory responses received occasional attention before 1970, but substantial work after that date. For nominal responses, early work was mainly in the econometrics literature. In 2000, Daniel McFadden won the Nobel Prize in economics for his work in the 1970s and 1980s on the discrete-choice model (Section 6.1.5). Cumulative logit models received some attention starting in the 1960s and 1970s, but did not become popular until an article by Peter McCullagh in 1980 provided a Fisher scoring algorithm for ML fitting of a more general model for cumulative probabilities allowing a variety of link functions.

Other major advances with logistic regression dealt with its application to case–control studies in the 1970s and the conditional ML approach to model fitting for those studies and others with numerous nuisance parameters. Biostatisticians Norman Breslow and Ross Prentice at the University of Washington had a strong influence on this. The conditional approach was later exploited in small-sample exact inference in a series of papers by Cyrus Mehta, Nitin Patel, and colleagues at Harvard.

Perhaps the most far-reaching contribution was the introduction by British statisticians John Nelder and R. W. M. Wedderburn in 1972 of the concept of *generalized linear models*. This unifies the logistic and probit regression models for binomial responses with loglinear models for Poisson or negative binomial responses and with long-established regression and ANOVA models for normal responses.

More recently, attention has focused on fitting logistic regression models to correlated responses for clustered data. One strand of this is marginal modeling of longitudinal data, proposed by Kung-Yee Liang and Scott Zeger at Johns Hopkins in 1986 using generalized estimating equations (GEE). Another strand is generalized linear mixed models, including multi-level models.

11.4 MULTIWAY CONTINGENCY TABLES AND LOGLINEAR MODELS

In the early 1950s, William Cochran published work dealing with a variety of important topics in CDA. He introduced a generalization (Cochran's Q) of McNemar's test for comparing proportions in several matched samples. He showed how to partition chi-squared statistics and developed sample size guidelines for chi-squared approximations to work well for the X^2 statistic. He proposed a test of conditional independence for $2 \times 2 \times K$ tables, similar to the one proposed by Mantel and Haenszel in 1959. Although best known for the Cochran–Mantel–Haenszel test, Nathan Mantel himself made a variety of contributions to CDA, including a trend test for ordinal data and work on multinomial logit modeling and on logistic regression for case–control data.

The 1950s and early 1960s saw an abundance of work on association and interaction structure in multiway tables. These articles were the genesis of research work on loglinear models between about 1965 and 1975.

At the University of Chicago, Leo Goodman wrote a series of groundbreaking articles, dealing with such topics as partitionings of chi-squared, models for square tables (e.g., quasi-independence), latent class models, and specialized models for ordinal data. Goodman also wrote a stream of articles for social science journals that had a substantial impact on popularizing loglinear and logit methods for applications. Over the past 50 years, Goodman has been the most prolific contributor to the advancement of CDA methodology. In addition, some of Goodman's students, such as Shelby Haberman and Clifford Clogg, also made fundamental contributions.

Simultaneously, related research on ML methods for loglinear-logit models occurred at Harvard University by students of Frederick Mosteller (such as Stephen Fienberg) and William Cochran. Much of this research was inspired by problems arising in analyzing large, multivariate data sets in the National Halothane Study. That study investigated whether halothane was more likely than other anesthetics to cause death due to liver damage. A presidential address by Mosteller to the American Statistical Association (Mosteller, *J. Amer, Statist. Assoc.*, **63**: 1–28, 1968) describes early uses of loglinear models for smoothing multidimensional discrete data sets. A landmark book in 1975 by Yvonne Bishop, Stephen Fienberg, and Paul Holland, *Discrete Multivariate Analysis*, helped to introduce loglinear models to the general statistical community.

Karl Pearson

G. Udny Yule

Ronald A. Fisher

Leo Goodman

Figure 11.1. Four leading figures in the development of categorical data analysis.

Research at the University of North Carolina by Gary Koch and several colleagues was highly influential in the biomedical sciences. Their research developed weighted least squares (WLS) methods for categorical data models. An article in 1969 by Koch with J. Grizzle and F. Starmer popularized this approach. In later articles, Koch and colleagues applied WLS to problems for which ML methods are difficult to implement, such as the analysis of repeated categorical measurement data. For large samples with fully categorical data, WLS estimators have similar properties as ML.

Certain loglinear models with conditional independence structure provide *graphical models* for contingency tables. These relate to the conditional independence graphs that Section 7.4.1 used. An article by John Darroch, Steffen Lauritzen, and Terry Speed in 1980 was the genesis of much of this work.

11.5 FINAL COMMENTS

Methods for CDA continue to be developed. In the past decade, an active area of new research has been the modeling of clustered data, such as using GLMMs. In particular, multilevel (hierarchical) models have become increasingly popular.

The development of Bayesian approaches to CDA is an increasingly active area. Dennis Lindley and I. J. Good were early proponents of the Bayesian approach for categorical data, in the mid 1960s. Recently, the Bayesian approach has seen renewed interest because of the development of methods for numerically evaluating posterior distributions for increasingly complex models. See O'Hagan and Forster (2004).

Another active area of research, largely outside the realm of traditional modeling, is the development of algorithmic methods for huge data sets with large numbers of variables. Such methods, often referred to as *data mining*, deal with the handling of complex data structures, with a premium on predictive power at the sacrifice of simplicity and interpretability of structure. Important areas of application include genetics, such as the analysis of discrete DNA sequences in the form of very high-dimensional contingency tables, and business applications such as credit scoring and tree-structured methods for predicting behavior of customers.

The above discussion provides only a sketchy overview of the development of CDA. Further details and references for technical articles and books appear in Agresti (2002).

Appendix A: Software for Categorical Data Analysis

All major statistical software has procedures for categorical data analyses. This appendix has emphasis on SAS. For information about other packages (such as S-plus, R, SPSS, and Stata) as well as updated information about SAS, see the web site www.stat.ufl.edu/~aa/cda/software.html.

For certain analyses, specialized software is better than the major packages. A good example is StatXact (Cytel Software, Cambridge, MA, USA), which provides exact analysis for categorical data methods and some nonparametric methods. Among its procedures are small-sample confidence intervals for differences and ratios of proportions and for odds ratios, and Fisher's exact test and its generalizations for $I \times J$ and three-way tables. Its companion program LogXact performs exact conditional logistic regression.

SAS FOR CATEGORICAL DATA ANALYSIS

In SAS, the main procedures (PROCs) for categorical data analyses are FREQ, GENMOD, LOGISTIC, and NLMIXED. PROC FREQ computes chi-squared tests of independence, measures of association and their estimated standard errors. It also performs generalized CMH tests of conditional independence, and exact tests of independence in $I \times J$ tables. PROC GENMOD fits generalized linear models, cumulative logit models for ordinal responses, and it can perform GEE analyses for marginal models. PROC LOGISTIC provides ML fitting of binary response models, cumulative logit models for ordinal responses, and baseline-category logit models for nominal responses. It incorporates model selection procedures, regression diagnostic

An Introduction to Categorical Data Analysis, Second Edition. By Alan Agresti

options, and exact conditional inference. PROC NLMIXED fits generalized linear mixed models (models with random effects).

The examples below show SAS code (version 9), organized by chapter of presentation. For convenience, some of the examples enter data in the form of the contingency table displayed in the text. In practice, one would usually enter data at the subject level. Most of these tables and the full data sets are available at www.stat.ufl.edu/~aa/cda/software.html. For more detailed discussion of the use of SAS for categorical data analyses, see specialized SAS publications such as Allison (1999) and Stokes et al. (2000). For application of SAS to clustered data, see Molenberghs and Verbeke (2005).

CHAPTER 2: CONTINGENCY TABLES

Table A.1 uses SAS to analyze Table 2.5. The @@ symbol indicates that each line of data contains more than one observation. Input of a variable as characters rather than numbers requires an accompanying $ label in the INPUT statement.

Table A.1. SAS Code for Chi-Squared, Measures of Association, and Residuals with Party ID Data in Table 2.5

```
data table;
    input gender $ party $ count @@;
datalines;
female dem   762  female indep 327  female repub 468
male dem     484  male inidep  239  male repub   477
  ;
proc freq  order=data; weight count;
  tables gender*party / chisq expected measures cmh1;
proc genmod  order=data;  class gender party;
    model count = gender party / dist=poi link=log residuals;
```

PROC FREQ forms the table with the TABLES statement, ordering row and column categories alphanumerically. To use instead the order in which the categories appear in the data set (e.g., to treat the variable properly in an ordinal analysis), use the ORDER=DATA option in the PROC statement. The WEIGHT statement is needed when one enters the contingency table instead of subject-level data. PROC FREQ can conduct chi-squared tests of independence (CHISQ option), show its estimated expected frequencies (EXPECTED), provide a wide assortment of measures of association and their standard errors (MEASURES), and provide ordinal statistic (2.10) with a "nonzero correlation" test (CMH1). One can also perform chi-squared tests using PROC GENMOD (using loglinear models discussed in the Chapter 7 section of this Appendix), as shown. Its RESIDUALS option provides cell residuals. The output labeled "StReschi" is the standardized residual (2.9).

Table A.2 analyzes Table 2.8 on tasting tea. With PROC FREQ, for 2×2 tables the MEASURES option in the TABLES statement provides confidence intervals for the

Table A.2. SAS Code for Fisher's Exact Test and Confidence Intervals for Odds Ratio for Tea-Tasting Data in Table 2.8

```
data fisher;
input poured guess count @@;
datalines;
1 1 3    1 2 1    2 1 1    2 2 3
;
proc freq;  weight count;
   tables poured*guess / measures riskdiff;
   exact fisher or / alpha=.05;
```

odds ratio (labeled "case-control" on output) and the relative risk, and the RISKDIFF option provides intervals for the proportions and their difference. For tables having small cell counts, the EXACT statement can provide various exact analyses. These include Fisher's exact test and its generalization for $I \times J$ tables, treating variables as nominal, with keyword FISHER. The OR keyword gives the odds ratio and its large-sample and small-sample confidence intervals. Other EXACT statement keywords include binomial tests for 1×2 tables (keyword BINOMIAL), exact trend tests for $I \times 2$ tables (TREND), and exact chi-squared tests (CHISQ) and exact correlation tests for $I \times J$ tables (MHCHI).

CHAPTER 3: GENERALIZED LINEAR MODELS

PROC GENMOD fits GLMs. It specifies the response distribution in the DIST option ("poi" for Poisson, "bin" for binomial, "mult" for multinomial, "negbin" for negative binomial) and specifies the link in the LINK option. Table A.3 illustrates for binary regression models for the snoring and heart attack data of Table 3.1. For binomial grouped data, the response in the model statements takes the form of the number of "successes" divided by the number of cases. Table A.4 fits Poisson and negative binomial loglinear models for the horseshoe crab data of Table 3.2.

Table A.3. SAS Code for Binary GLMs for Snoring Data in Table 3.1

```
data glm;
input snoring disease total @@;
datalines;
0 24 1379    2 35 638    4 21 213    5 30 254
;
proc genmod; model disease/total = snoring / dist=bin link=identity;
proc genmod; model disease/total = snoring / dist=bin link=logit;
```

PROC GAM fits *generalized additive models*. These can smooth data, as illustrated by Figure 3.5.

Table A.4. SAS Code for Poisson Regression, Negative Binomial Regression, and Logistic Regression Models with Horseshoe Crab Data of Table 3.2

```
data crab;
input  color  spine  width  satell  weight;
if satell>0 then y=1; if satell=0 then y=0;
datalines;
2  3  28.3  8  3.05
...
2  2  24.5  0  2.00
;
proc genmod;
  model satell = width / dist=poi link=log;
proc genmod;
  model satell = width / dist=negbin link=log;
proc genmod descending; class color;
  model y = width color / dist=bin link=logit lrci type3 obstats;
  contrast 'a-d' color 1  0  0  -1;
proc logistic descending;
  model y = width;
  output out = predict  p = pi_hat lower = LCL  upper = UCL;
proc print  data = predict;
proc logistic descending; class color spine / param=ref;
  model y = width weight color spine / selection=backward lackfit outroc=classif1;
proc plot data=classif1; plot _sensit_ * _1mspec_  ;
```

CHAPTERS 4 AND 5: LOGISTIC REGRESSION

PROC GENMOD and PROC LOGISTIC can fit logistic regression. In GENMOD, the LRCI option provides confidence intervals based on the likelihood-ratio test. The ALPHA = option can specify an error probability other than the default of 0.05. The TYPE3 option provides likelihood-ratio tests for each parameter. In GENMOD or LOGISTIC, a CLASS statement for a predictor requests that it be treated as a qualitative factor by setting up indicator variables for it. By default, in GENMOD the parameter estimate for the last category of a factor equals 0. In LOGISTIC, estimates sum to zero. That is, indicator variables take the coding $(1, -1)$ of 1 when in the category and -1 when not, for which parameters sum to 0. The option PARAM=REF in the CLASS statement in LOGISTIC requests (1, 0) dummy variables with the last category as the reference level.

Table A.4 shows logistic regression analyses for Table 3.2. The models refer to a constructed binary variable Y that equals 1 when a horseshoe crab has satellites and 0 otherwise. With binary data entry, GENMOD and LOGISTIC order the levels alphanumerically, forming the logit with (1, 0) responses as $\log[P(Y = 0)/P(Y = 1)]$. Invoking the procedure with DESCENDING following the PROC name reverses the order. The CONTRAST statement provides tests involving contrasts of parameters, such as whether parameters for two levels of a factor are identical. The statement shown contrasts the first and fourth color levels. For PROC LOGISTIC, the INFLUENCE option provides residuals and diagnostic measures. Following the first LOGISTIC model statement, it requests predicted probabilities and lower and upper 95% confidence limits for the probabilities. LOGISTIC has options for stepwise selection of variables, as the final model statement shows. The LACKFIT option yields the Hosmer–Lemeshow statistic. The CTABLE option gives a classification table, with cutoff point specified by PPROB. Using the OUTROC option, LOGISTIC can output a data set for plotting a ROC curve.

Table A.5 uses GENMOD and LOGISTIC to fit a logit model with qualitative predictors to Table 4.4. In GENMOD, the OBSTATS option provides various "observation statistics," including predicted values and their confidence limits. The RESIDUALS option requests residuals such as the standardized residuals (labeled "StReschi"). In LOGISTIC, the CLPARM=BOTH and CLODDS=BOTH options provide Wald and likelihood-based confidence intervals for parameters and odds ratio effects of explanatory variables. With AGGREGATE SCALE=NONE in the model statement, LOGISTIC reports Pearson and deviance tests of fit; it forms groups by aggregating data into the possible combinations of explanatory variable values.

Table A.5. SAS Code for Logit Modeling of HIV Data in Table 4.4

```
data aids;
input race $ azt $ y n @@;
datalines;
 White Yes 14 107   White No 32 113   Black Yes 11 63   Black No 12 55
;
proc genmod; class race azt;
  model y/n = azt race / dist=bin type3 lrci residuals obstats;
proc logistic; class race azt / param=ref;
  model y/n = azt race / aggregate scale=none clparm=both clodds=both;
```

Exact conditional logistic regression is available in PROC LOGISTIC with the EXACT statement. It provides ordinary and mid P-values as well as confidence limits for each model parameter and the corresponding odds ratio with the ESTIMATE=BOTH option.

CHAPTER 6: MULTICATEGORY LOGIT MODELS

PROC LOGISTIC fits baseline-category logit models using the LINK=GLOGIT option. The final response category is the default baseline for the logits. Table A.6 fits a model to Table 6.1.

Table A.6. SAS Code for Baseline-category Logit Models with Alligator Data in Table 6.1

```
data gator;
input length choice $ @@;
datalines;
1.24 I  1.30 I  1.30 I  1.32 F  1.32 F  1.40 F  1.42 I  1.42 F
...
3.68 O  3.71 F  3.89 F
;
proc logistic;
    model choice = length / link=glogit aggregate scale=none;
run;
```

PROC GENMOD can fit the proportional odds version of cumulative logit models using the DIST=MULTINOMIAL and LINK=CLOGIT options. Table A.7 fits it to Table 6.9. When the number of response categories exceeds two, by default PROC LOGISTIC fits this model. It also gives a score test of the proportional odds assumption of identical effect parameters for each cutpoint.

Table A.7. SAS Code for Cumulative Logit Model with Mental Impairment Data of Table 6.9

```
data impair;
input mental ses life;
datalines;
1 1 1
....
4 0 9
;
proc genmod ;
   model mental = life ses / dist=multinomial link=clogit lrci type3;
proc logistic;
   model mental = life ses;
```

One can fit adjacent-categories logit models in SAS by fitting equivalent baseline-category logit models (e.g., see Table A.12 in the Appendix in Agresti, 2002). With the CMH option, PROC FREQ provides the generalized *CMH* tests of conditional independence. The statistic for the "general association" alternative treats X and Y as nominal, the statistic for the "row mean scores differ" alternative treats X as nominal and Y as ordinal, and the statistic for the "nonzero correlation" alternative treats X and Y as ordinal.

CHAPTER 7: LOGLINEAR MODELS FOR CONTINGENCY TABLES

Table A.8 uses GENMOD to fit model (AC, AM, CM) to Table 7.3. Table A.9 uses GENMOD to fit the linear-by-linear association model (7.11) to Table 7.15 (with column scores 1,2,4,5). The defined variable "assoc" represents the cross-product of row and column scores, which has β parameter as coefficient in model (7.11).

Table A.8. SAS Code for Fitting Loglinear Models to Drug Survey Data of Table 7.3

```
data drugs;
input a c m count @@;
datalines;
1 1 1 911    1 1 2 538    1 2 1 44    1 2 2 456
2 1 1   3    2 1 2  43    2 2 1  2    2 2 2 279
;
proc genmod; class a c m;
  model count = a c m a*m a*c c*m / dist=poi link=log lrci type3 obstats;
```

Table A.9. SAS Code for Fitting Association Models to GSS Data of Table 7.15

```
data sex;
input premar birth u v count @@;  assoc = u*v ;
datalines;
1 1 1 1  38   1 2 1 2  60   1 3 1 4 68   1 4 1 5 81
...
;
proc genmod; class premar birth;
  model  count = premar birth assoc / dist=poi link=log;
```

CHAPTER 8: MODELS FOR MATCHED PAIRS

Table A.10 analyzes Table 8.1. The AGREE option in PROC FREQ provides the McNemar chi-squared statistic for binary matched pairs, the X^2 test of fit of the symmetry model (also called *Bowker's test*), and Cohen's kappa and weighted kappa with SE values. The MCNEM keyword in the EXACT statement provides a small-sample binomial version of McNemar's test. PROC CATMOD can provide the confidence interval for the difference of proportions. The code forms a model for the marginal proportions in the first row and the first column, specifying a matrix in the model statement that has an intercept parameter (the first column) that applies to both proportions and a slope parameter that applies only to the second; hence the second parameter is the difference between the second and first marginal proportions. (It is also possible to get the interval with the GEE methods of Chapter 9, using PROC GENMOD with the REPEATED statement and identity link function.)

Table A.10. SAS Code for McNemar's Test and Comparing Proportions for Matched Samples in Table 8.1

```
data matched;
input taxes living count @@;
datalines;
 1 1 227   1 2 132   2 1 107   2 2 678
;
proc freq; weight count;
    tables taxes*living / agree; exact mcnem;
proc catmod; weight count;
    response marginals;
    model taxes*living =  (1 0 ,
                            1 1 ) ;
```

Table A.11 shows a way to test marginal homogeneity for Table 8.5 on coffee purchases. The GENMOD code expresses the I^2 expected frequencies in terms of parameters for the $(I-1)^2$ cells in the first $I-1$ rows and $I-1$ columns, the cell in the last row and last column, and $I-1$ marginal totals (which are the same for rows

Table A.11. SAS Code for Testing Marginal Homogeneity with Coffee Data of Table 8.5

```
data migrate;
input first $ second $ count m11 m12 m13 m14 m21 m22 m23 m24
    m31 m32 m33 m34 m41 m42 m43 m44 m55 m1 m2 m3 m4;
datalines;
high high 93  1  0  0  0  0  0  0  0  0  0  0  0  0  0  0  0  0 0 0 0 0
high tast 17  0  1  0  0  0  0  0  0  0  0  0  0  0  0  0  0  0 0 0 0 0
high sank 44  0  0  1  0  0  0  0  0  0  0  0  0  0  0  0  0  0 0 0 0 0
high nesc  7  0  0  0  1  0  0  0  0  0  0  0  0  0  0  0  0  0 0 0 0 0
high brim 10 -1 -1 -1 -1  0  0  0  0  0  0  0  0  0  0  0  0  0 1 0 0 0
...
nesc nesc 15  0  0  0  0  0  0  0  0  0  0  0  0  0  0  0  1  0 0 0 0 0
nesc brim  2  0  0  0  0  0  0  0  0  0  0  0  0 -1 -1 -1 -1  0 0 0 0 1
brim high 10 -1  0  0  0 -1  0  0  0 -1  0  0  0 -1  0  0  0  0 1 0 0 0
brim tast  4  0 -1  0  0  0 -1  0  0  0 -1  0  0  0 -1  0  0  0 0 1 0 0
brim sank 12  0  0 -1  0  0  0 -1  0  0  0 -1  0  0  0 -1  0  0 0 0 1 0
brim nesc  2  0  0  0 -1  0  0  0 -1  0  0  0 -1  0  0  0 -1  0 0 0 0 1
brim brim 27  0  0  0  0  0  0  0  0  0  0  0  0  0  0  0  0  1 0 0 0 0
      ;
proc genmod;
    model count = m11 m12 m13 m14 m21 m22 m23 m24 m31 m32 m33 m34 m41
        m42 m43 m44 m55 m1 m2 m3 m4
        / dist=poi  link=identity obstats residuals;
```

and columns). Here, m11 denotes expected frequency μ_{11}, m1 denotes $\mu_{1+} = \mu_{+1}$, and so forth. This parameterization uses formulas such as $\mu_{15} = \mu_{1+} - \mu_{11} - \mu_{12} - \mu_{13} - \mu_{14}$ for terms in the last column or last row. The likelihood-ratio test statistic for testing marginal homogeneity is the deviance statistic for this model.

Table A.12 shows analyses of Table 8.6. First the data are entered as a 4×4 table, and the loglinear model fitted is quasi independence. The "qi" factor invokes

Table A.12. SAS Code Showing Square-table Analyses of Tables 8.6

```
data square;
input recycle  drive  qi  count @@;
datalines;
1 1 1 12   1 2 5 43   1 3 5 163     1 4  5 233
2 1 5  4   2 2 2 21   2 3 5  99     2 4  5 185
3 1 5  4   3 2 5  8   3 3 3  77     3 4  5 230
4 1 5  0   4 2 5  1   4 3 5  18     4 4  4 132
;
proc genmod; class drive recycle qi;
  model count = drive recycle qi / dist=poi link=log; * quasi indep;
data square2;
input score below above @@; trials = below + above;
datalines;
1 4 43   1 8 99    1 18 230    2 4 163    2 1 185    3 0 233
;
proc genmod data=square2;
  model above/trials = / dist=bin link=logit noint;
proc genmod data=square2;
  model above/trials = score / dist=bin link=logit noint;
```

the δ_i parameters in equation (8.12). It takes a separate level for each cell on the main diagonal, and a common value for all other cells. The bottom of Table A.12 fits logit models for the data entered in the form of pairs of cell counts (n_{ij}, n_{ji}). These six sets of binomial count are labeled as "above" and "below" with reference to the main diagonal. The variable defined as "score" is the distance $(u_j - u_i) = j - i$. The first model is symmetry and the second is ordinal quasi symmetry. Neither model contains an intercept (NOINT). The quasi-symmetry model can be fitted using the approach shown next for the equivalent Bradley–Terry model.

Table A.13 uses GENMOD for logit fitting of the Bradley–Terry model to Table 8.9 by forming an artificial explanatory variable for each player. For a given observation, the variable for player i is 1 if he wins, -1 if he loses, and 0 if he is not one of the players for that match. Each observation lists the number of wins ("wins") for the player with variate-level equal to 1 out of the number of matches ("n") against the player with variate-level equal to -1. The model has these artificial variates, one of which is redundant, as explanatory variables with no intercept term. The COVB option provides the estimated covariance matrix of parameter estimators.

Table A.13. SAS Code for Fitting Bradley–Terry Model to Tennis Data in Table 8.9

```
data tennnis;
input win  n agassi federer henman hewitt roddick ;
datalines;
0  6  1 -1   0  0  0
0  0  1  0  -1  0  0
...
3  5  0  0   0  1 -1
;
proc genmod;
model win/n = agassi federer henman hewitt roddick / dist=bin link=logit noint covb;
```

CHAPTER 9: MODELING CORRELATED, CLUSTERED RESPONSES

Table A.14 uses GENMOD to analyze the depression data in Table 9.1 using GEE. The REPEATED statement specifies the variable name (here, "case") that identifies the subjects for each cluster. Possible working correlation structures are TYPE=EXCH for exchangeable, TYPE=AR for autoregressive, TYPE=INDEP for independence, and TYPE=UNSTR for unstructured. Output shows estimates and standard errors under the naive working correlation and incorporating the empirical dependence. Alternatively, the working association structure in the binary case can use the log odds ratio (e.g., using LOGOR=EXCH for exchangeability). The type3 option with the GEE approach provides score-type tests about effects. See Stokes et al. (2000, Section 15.11) for the use of GEE with missing data. See Table A.22 in Agresti (2002) for using GENMOD to implement GEE for a cumulative logit model for the insomnia data in Table 9.6.

Table A.14. SAS Code for Marginal and Random Effects Modeling of Depression Data in Table 9.1

```
data depress;
input case diagnose drug time outcome @@; * outcome=1 is normal;
datalines;
 1   0  0  0  1       1  0  0  1  1       1  0  0  2  1
...
340  1  1  0  0     340  1  1  1  0     340  1  1  2  0
;
proc genmod descending;  class case;
  model outcome = diagnose drug time drug*time / dist=bin link=logit type3;
  repeated subject=case / type=exch corrw;
proc nlmixed;
  eta = u + alpha + beta1*diagnose + beta2*drug  + beta3*time + beta4*drug*time;
  p = exp(eta)/(1 + exp(eta));
  model outcome ~ binary(p);
  random u ~ normal(0, sigma*sigma) subject = case;
```

CHAPTER 10: RANDOM EFFECTS: GENERALIZED LINEAR MIXED MODELS

PROC NLMIXED extends GLMs to GLMMs by including random effects. Table A.23 in Agresti (2002) shows how to fit the matched pairs model (10.3). Table A.15 analyzes the basketball data in Table 10.2. Table A.16 fits model (10.5) to Table 10.4 on abortion questions. This shows how to set the number of quadrature points for Gauss–Hermite quadrature (e.g., QPOINTS =) and specify initial parameter values (perhaps based on an initial run with the default number of quadrature points). Table A.14 uses NLMIXED for the depression study of Table 9.1. Table A.22 in Agresti (2002) uses NLMIXED for ordinal modeling of the insomnia data in Table 9.6.

Table A.15. SAS Code for GLMM Analyses of Basketball Data in Table 10.2

```
data basket;
input player $ y n @@;
datalines;
 yao 10 13      frye 9  10      camby 10  15      okur 9  14
....
;
proc nlmixed;
   eta = alpha + u;  p = exp(eta) / (1 + exp(eta));
   model y ~ binomial(n,p);
   random u ~ normal(0,sigma*sigma) subject=player;
   predict p out=new;
proc print data=new;
```

Table A.16. SAS Code for GLMM Modelling of Opinion in Table 10.4

```
data new;
input sex poor single any count;
datalines;
1 1 1 1 342
...
2 0 0 0 457
;
data new;  set new;
   sex = sex-1; case = _n_;
   q1=1; q2=0; resp = poor; output;
   q1=0; q2=1; resp = single; output;
   q1=0; q2=0; resp = any;  output;
drop poor single any;
proc nlmixed  qpoints = 50;
   parms alpha=0  beta1=.8  beta2=.3  gamma=0  sigma=8.6;
   eta = u + alpha + beta1*q1 + beta2*q2 + gamma*sex;
   p = exp(eta)/(1 + exp(eta));
   model resp ~ binary(p);
   ramdom u ~ normal(0,sigma*sigma) subject = case;
   replicate count;
```

Appendix B: Chi-Squared Distribution Values

	Right-Tail Probability						
df	0.250	0.100	0.050	0.025	0.010	0.005	0.001
1	1.32	2.71	3.84	5.02	6.63	7.88	10.83
2	2.77	4.61	5.99	7.38	9.21	10.60	13.82
3	4.11	6.25	7.81	9.35	11.34	12.84	16.27
4	5.39	7.78	9.49	11.14	13.28	14.86	18.47
5	6.63	9.24	11.07	12.83	15.09	16.75	20.52
6	7.84	10.64	12.59	14.45	16.81	18.55	22.46
7	9.04	12.02	14.07	16.01	18.48	20.28	24.32
8	10.22	13.36	15.51	17.53	20.09	21.96	26.12
9	11.39	14.68	16.92	19.02	21.67	23.59	27.88
10	12.55	15.99	18.31	20.48	23.21	25.19	29.59
11	13.70	17.28	19.68	21.92	24.72	26.76	31.26
12	14.85	18.55	21.03	23.34	26.22	28.30	32.91
13	15.98	19.81	22.36	24.74	27.69	29.82	34.53
14	17.12	21.06	23.68	26.12	29.14	31.32	36.12
15	18.25	22.31	25.00	27.49	30.58	32.80	37.70
16	19.37	23.54	26.30	28.85	32.00	34.27	39.25
17	20.49	24.77	27.59	30.19	33.41	35.72	40.79
18	21.60	25.99	28.87	31.53	34.81	37.16	42.31
19	22.72	27.20	30.14	32.85	36.19	38.58	43.82
20	23.83	28.41	31.41	34.17	37.57	40.00	45.32
25	29.34	34.38	37.65	40.65	44.31	46.93	52.62
30	34.80	40.26	43.77	46.98	50.89	53.67	59.70
40	45.62	51.80	55.76	59.34	63.69	66.77	73.40
50	56.33	63.17	67.50	71.42	76.15	79.49	86.66
60	66.98	74.40	79.08	83.30	88.38	91.95	99.61
70	77.58	85.53	90.53	95.02	100.4	104.2	112.3
80	88.13	96.58	101.8	106.6	112.3	116.3	124.8
90	98.65	107.6	113.1	118.1	124.1	128.3	137.2
100	109.1	118.5	124.3	129.6	135.8	140.2	149.5

Source: Calculated using *StaTable*, Cytel Software, Cambridge, MA, USA.

An Introduction to Categorical Data Analysis, Second Edition. By Alan Agresti

Bibliography

Agresti, A. (2002). *Categorical Data Analysis*, 2nd edn. New York: Wiley.

Allison, P. (1999). *Logistic Regression Using the SAS System*. Cary, NC: SAS Institute.

Bishop, Y. M. M., S. E. Fienberg, and P. W. Holland (1975). *Discrete Multivariate Analysis*. Cambridge, MA: MIT Press.

Breslow, N. and N. E. Day (1980). *Statistical Methods in Cancer Research, Vol. I: The Analysis of Case–Control Studies*. Lyon: IARC.

Christensen, R. (1997). *Log-Linear Models and Logistic Regression*. New York: Springer.

Collett, D. (2003). *Modelling Binary Data*, 2nd edn. London: Chapman & Hall.

Cramer, J. S. (2003). *Logit Models from Economics and Other Fields*. Cambridge: Cambridge University Press.

Diggle, P. J., P. Heagerty, K.-Y. Liang, and S. L. Zeger (2002). *Analysis of Longitudinal Data*, 2nd edn. Oxford: Oxford University Press.

Fahrmeir, L. and G. Tutz (2001). *Multivariate Statistical Modelling Based on Generalized Linear Models*, 2nd edn. Berlin: Springer.

Fitzmaurice, G., N. Laird, and J. Ware (2004). *Applied Longitudinal Analysis*. New York: Wiley.

Fleiss, J. L., B. Levin, and M. C. Paik (2003). *Statistical Methods for Rates and Proportions*, 3rd edn. New York: Wiley.

Goodman, L. A. and W. H. Kruskal (1979). *Measures of Association for Cross Classifications*. New York: Springer.

Grizzle, J. E., C. F. Starmer, and G. G. Koch (1969). Analysis of categorical data by linear models. *Biometrics*, **25**: 489–504.

Hastie, T. and R. Tibshirani (1990). *Generalized Additive Models*. London: Chapman and Hall.

Hensher, D. A., J. M. Rose, and W. H. Greene (2005). *Applied Choice Analysis: A Primer*. Cambridge: Cambridge University Press.

Hosmer, D. W. and S. Lemeshow (2000). *Applied Logistic Regression*, 2nd edn. New York: Wiley.

Lloyd, C. J. (1999). *Statistical Analysis of Categorical Data*. New York: Wiley.

Mehta, C. R. and N. R. Patel (1995). Exact logistic regression: Theory and examples. *Statist. Med.*, **14**: 2143–2160.

Molenberghs, G. and G. Verbeke (2005). *Models for Discrete Longitudinal Data*. New York: Springer.

O'Hagan, A. and J. Forster (2004). *Kendall's Advanced Theory of Statistics: Bayesian Inference*. London: Arnold.

Raudenbush, S. and A. Bryk (2002). *Hierarchical Linear Models*, 2nd edn. Thousand Oaks, CA: Sage.

Santner, T. J. and D. E. Duffy (1989). *The Statistical Analysis of Discrete Data*. Berlin: Springer.

Simonoff, J. S. (2003). *Analyzing Categorical Data*. New York: Springer.

Snijders, T. A. B. and R. J. Bosker (1999). *Multilevel Analysis*. London: Sage.

Stokes, M. E., C. S. Davis, and G. G. Koch (2000). *Categorical Data Analysis Using the SAS System*, 2nd edn. Cary, NC: SAS Institute.

Whittaker, J. (1990). *Graphical Models in Applied Multivariate Statistics*. New York: Wiley.

Index of Examples

An Introduction to Categorical Data Analysis, Second Edition. By Alan Agresti

Subject Index

An Introduction to Categorical Data Analysis, Second Edition. By Alan Agresti

Brief Solutions to Some Odd-Numbered Problems

CHAPTER 1

1. Response variables:

a. Attitude toward gun control,

b. Heart disease,

c. Vote for President,

d. Quality of life.

3. a. Binomial, $n = 100$, $\pi = 0.25$.

b. $\mu = n\pi = 25$ and $\sigma = \sqrt{[n\pi(1-\pi)]} = 4.33$. 50 correct responses is surprising, since 50 is $z = (50 - 25)/4.33 = 5.8$ standard deviations above mean.

7. a. $(5/6)^6$.

b. Note $Y = y$ when $y - 1$ successes and then a failure.

9. a. Let π = population proportion obtaining greater relief with new analgesic. For H_0: $\pi = 0.50$, $z = 2.00$, P-value $= 0.046$.

b. Wald CI is (0.504, 0.696), score CI is (0.502, 0.691).

11. $0.86 \pm 1.96(0.0102)$, or (0.84, 0.88).

13. a. $(1 - \pi_0)^{25}$ is binomial probability of $y = 0$ in $n = 25$ trials.

b. The maximum of $(1 - \pi)^{25}$ occurs at $\pi = 0.0$.

c. $-2\log(\ell_0/\ell_1) = -2\log[(0.50)^{25}/1.0] = 34.7$, P-value < 0.0001.

d. $-2\log(\ell_0/\ell_1) = -2\log[(0.926)^{25}/1.0] = 3.84$. With $df = 1$, chi-squared P-value $= 0.05$.

15. a. $\sigma(p)$ equals binomial standard deviation $\sqrt{n\pi(1-\pi)}$ divided by sample size n.

b. $\sigma(p)$ takes maximum at $\pi = 0.50$ and minimum at $\pi = 0$ and 1.

17. a. Smallest possible P-value is 0.08, so never reject H_0 and therefore never commit a type I error.

b. If $T = 9$, mid-P value $= 0.08/2 = 0.04$, so reject H_0. Probability of this happening is $P(T = 9) = 0.08 = P(\text{type I error})$.

c. (**a**) $P(\text{type I error}) = 0.04$, (**b**) $P(\text{type I error}) = 0.04$. Mid-$P$ test can have $P(\text{type I error})$ either below 0.05 (conservative) or above 0.05 (liberal).

CHAPTER 2

1. a. $P(-|C) = 1/4$, $P(\bar{C}|+) = 2/3$.

b. Sensitivity $= P(+|C) = 1 - P(-|C) = 3/4$.

c. $P(C,+) = 0.0075$, $P(C,-) = 0.0025$, $P(\bar{C},+) = 0.0150$, $P(\bar{C},-) = 0.9750$.

d. $P(+) = 0.0225$, $P(-) = 0.9775$.

e. $1/3$.

3. a. (i) 0.000061, (ii) $62.4/1.3 = 48$. (**b**) Relative risk.

5. a. Relative risk.

b. (i) $\pi_1 = 0.55\pi_2$, so $\pi_1/\pi_2 = 0.55$. (ii) $1/0.55 = 1.82$.

7. a. Quoted interpretation is that of relative risk.

b. Proportion $= 0.744$ for females, 0.203 for males.

c. $R = 0.744/0.203 = 3.7$.

9. (Odds for high smokers)/(Odds for low smokers) $= 26.1/11.7$.

11. a. Relative risk: Lung cancer, 14.00; Heart disease, 1.62.
Difference of proportions: Lung cancer, 0.00130; Heart disease, 0.00256.
Odds ratio: Lung cancer, 14.02; Heart disease, 1.62.

b. Difference of proportions describes excess deaths due to smoking. If $N =$ number of smokers in population, predict $0.00130N$ fewer deaths per year from lung cancer if they had never smoked, and $0.00256N$ fewer deaths per year from heart disease.

13. a. $\hat{\pi}_1 = 0.814$, $\hat{\pi}_2 = 0.793$. CI is $0.0216 \pm 1.645(0.024)$, or $(-0.018, 0.061)$.

b. CI for log odds ratio $0.137 \pm 1.645(0.1507)$, so CI for odds ratio is (0.89, 1.47).
c. $X^2 = 0.8$, $df = 1$, P-value $= 0.36$.

15. $\log(0.0171/0.0094) \pm 1.96\sqrt{(0.0052 + 0.0095)}$ is (0.360, 0.835), which translates to (1.43, 2.30) for relative risk.

17. **a.** $X^2 = 25.0$, $df = 1$, $P < 0.0001$.
b. $G^2 = 25.4$, $df = 1$.

19. **a.** $G^2 = 187.6$, $X^2 = 167.8$, $df = 2$ ($P < 0.0001$).
b. Standardized residuals of -11.85 for white Democrats and -11.77 for black Republicans show extremely strong evidence of fewer people in these cells than if party ID were independent of race. Standardized residuals of 11.85 for black Democrats and 11.77 for white Republicans show extremely strong evidence of more people in these cells than expected.
c. $G^2 = 24.1$ for comparing races on (Democrat, Independent) choice, and $G^2 = 163.5$ for comparing races on (Dem. + Indep., Republican) choice.

21. **a.** No, samples in different columns are dependent, because subjects can select as many columns as they wish.
b.

	A	
Gender	Yes	No
Men	60	40
Women	75	25

23. Extremely strong evidence of association. Strong evidence of tendency for those with less than high school education to be fundamentalist, and those with bachelor degree or higher to be liberal in religious beliefs.

25. **a.** Total of estimated expected frequencies in row i equals

$$\sum_j (n_{i+}n_{+j}/n) = (n_{i+}/n) \sum_j n_{+j} = n_{i+}$$

b. Odds ratio $= (n_{1+}n_{+1}/n)(n_{2+}n_{+2}/n)/(n_{1+}n_{+2}/n)(n_{2+}n_{+1}/n) = 1$.

27. **a.** $X^2 = 8.9$, $df = 6$, $P = 0.18$; nominal test with ordinal data.
b. Aspirations tend to be higher when family income is higher.
c. Ordinal test gives $M^2 = 4.75$, $df = 1$, $P = 0.03$.

29. Table has entries (7, 8) in row 1 and (0, 15) in row 2. $P = 0.003$.

31. **a.** P-value = 0.638.

b. 0.243.

33. **b.** 0.67 for white victims and 0.79 for black victims.

c. 1.18; yes.

35. Age distribution is relatively higher in Maine.

37. **a.** 0.18 for males and 0.32 for females.

b. 0.21.

39. (**a**) T, (**b**) T, (**c**) F, (**d**) T, (**e**) F.

CHAPTER 3

3. **a.** $\hat{\pi} = 0.00255 + 0.00109(\text{alcohol})$.

b. Estimated probability of malformation increases from 0.00255 at $x = 0$ to 0.01018 at $x = 7$. Relative risk = 0.01018/0.00255 = 4.0.

5. Fit of linear probability model is (i) $0.018 + 0.018(\text{snoring})$, (ii) $0.018 + 0.036(\text{snoring})$, (iii) $-0.019 + 0.036(\text{snoring})$. Slope depends on distance between scores; doubling distance halves slope estimate. Fitted values are identical for any linear transformation.

7. **a.** $\hat{\pi} = -0.145 + 0.323(\text{weight})$; at weight = 5.2, $\hat{\pi} = 1.53$, much higher than upper bound of 1.0 for a probability.

c. $\text{logit}(\hat{\pi}) = -3.695 + 1.815(\text{weight})$; at 5.2 kg, predicted logit = 5.74, and $\log(0.9968/0.0032) = 5.74$.

9. **a.** $\text{logit}(\hat{\pi}) = -3.556 + 0.0532x$.

11. **b.** $\log(\hat{\mu}) = 1.609 + 0.588x$. $\exp(\hat{\beta}) = \hat{\mu}_B/\hat{\mu}_A = 1.80$.

c. Wald $z = 3.33$, $z^2 = 11.1$ ($df = 1$), $P < 0.001$. LR statistic = 11.6 with $df = 1$, $P < 0.001$; higher defect rate for B.

d. Exponentiate 95% CI for β of $0.588 \pm 1.96(0.176)$ to get (1.27, 2.54).

13. **a.** $\log(\hat{\mu}) = -0.428 + 0.589(\text{weight})$.

b. 2.74.

c. $0.589 \pm 1.96(0.065) = (0.462, 0.717)$; CI for multiplicative effect on mean is (1.59, 2.05).

d. $z^2 = (0.589/0.065)^2 = 82.2$.

e. LR statistic = 71.9, $df = 1$.

15. **a.** $\exp(-2.38 + 1.733) = 0.522$ for blacks and $\exp(-2.38) = 0.092$ for whites.
b. Exponentiate endpoints of $1.733 \pm 1.96(0.147)$, which gives $(e^{1.44}, e^{2.02})$.
c. CI based on negative binomial model, because overdispersion for Poisson model.
d. Poisson is a special case of negative binomial with dispersion parameter $= 0$. Here, there is strong evidence that dispersion parameter > 0, because the estimated dispersion parameter is almost 5 standard errors above 0.

17. CI for log rate is $2.549 \pm 1.96(0.04495)$, so CI for rate is (11.7, 14.0).

19. **a.** Difference between deviances $= 11.6$, with $df = 1$, gives strong evidence Poisson model with constant rate inadequate.
b. $z = \hat{\beta}/SE = -0.0337/0.0130 = -2.6$ (or $z^2 = 6.7$ with $df = 1$).
c. $[\exp(-0.060), \exp(-0.008)]$, or (0.94, 0.99), quite narrow around point estimate of $e^{-0.0337} = 0.967$.

21. $\mu = \alpha t + \beta(tx)$, form of GLM with identity link, predictors t and tx, no intercept term.

CHAPTER 4

1. **a.** $\hat{\pi} = 0.068$.
b. $\hat{\pi} = 0.50$ at $-\hat{\alpha}/\hat{\beta} = 3.7771/0.1449 = 26$.
c. At LI $= 8$, $\hat{\pi} = 0.068$, rate of change $= 0.1449(0.068)(0.932) = 0.009$.
d. $e^{\hat{\beta}} = e^{0.1449} = 1.16$.

3. **a.** Proportion of complete games estimated to decrease by 0.07 per decade.
b. At $x = 12$, $\hat{\pi} = -0.075$, an impossible value.
c. At $x = 12$, $\text{logit}(\hat{\pi}) = -2.636$, and $\hat{\pi} = 0.067$.

5. **a.** $\text{logit}(\hat{\pi}) = 15.043 - 0.232x$.
b. At temperature $= 31$, $\hat{\pi} = 0.9996$.
c. $\hat{\pi} = 0.50$ at $x = 64.8$ and $\hat{\pi} > 0.50$ at $x < 64.8$. At $x = 64.8$, $\hat{\pi}$ decreases at rate 0.058.
d. Estimated odds of thermal distress multiply by $\exp(-0.232) = 0.79$ for each $1°$ increase in temperature.
e. Wald statistic $z^2 = 4.6$ ($P = 0.03$) and LR statistic $= 7.95$ ($df = 1$, $P = 0.005$).

7. **a.** $\text{logit}(\hat{\pi}) = -0.573 + 0.0043(\text{age})$. LR statistic $= 0.55$, Wald statistic $= 0.54$, $df = 1$; no evidence of age effect.

b. Age values more disperse when kyphosis absent.

c. $\text{logit}(\hat{\pi}) = -3.035 + 0.0558(\text{age}) - 0.0003(\text{age})^2$. LR statistic for $(\text{age})^2$ term equals 6.3 ($df = 1$), showing strong evidence of effect.

9. a. $\text{logit}(\hat{\pi}) = -0.76 + 1.86c_1 + 1.74c_2 + 1.13c_3$. The estimated odds a medium-light crab has a satellite are $e^{1.86} = 6.4$ times estimated odds a dark crab has a satellite.

b. LR statistic = 13.7, $df = 3$, P-value = 0.003.

c. For color scores 1,2,3,4, $\text{logit}(\hat{\pi}) = 2.36 - 0.71c$.

d. LR statistic = 12.5, $df = 1$, P-value = 0.0004.

e. Power advantage of focusing test on $df = 1$. But, may not be linear trend for color effect.

11. Odds ratio for spouse vs others = 2.02/1.71 = 1.18; odds ratio for \$10,000 – 24,999 vs \$25,000 + equal 0.72/0.41 = 1.76.

13. a. Chi-squared with $df = 1$, so P-value = 0.008.

b. Observed count = 0, expected count = 1.1.

15. a. *CMH* statistic = 7.815, P-value = 0.005.

b. Test $\beta = 0$ in model, $\text{logit}(\pi) = \alpha + \beta x + \beta_i^D$, where x = race. ML fit (when $x = 1$ for white and 0 for black) has $\hat{\beta} = 0.791$, with $SE = 0.285$. Wald statistic = 7.69, P-value = 0.006.

c. Model gives information about size of effect. Estimated odds ratio between promotion and race, controlling for district, is $\exp(0.791) = 2.2$.

17. a. $e^{-2.83}/(1 + e^{-2.83}) = 0.056$.

b. $e^{0.5805} = 1.79$.

c. $(e^{0.159}, e^{1.008}) = (1.17, 2.74)$.

d. $1/1.79 = 0.56$, CI is $(1/2.74, 1/1.17) = (0.36, 0.85)$.

e. H_0: $\beta_1 = 0$, H_a: $\beta_1 \neq 0$, LR statistic = 7.28, $df = 1$, P-value = 0.007.

19. a. $\exp(\hat{\beta}_1^G - \hat{\beta}_2^G) = 1.17$.

b. (i) 0.27, (ii) 0.88.

c. $\hat{\beta}_1^G = 0.16$, estimated odds ratio = $\exp(0.16) = 1.17$.

d. $\hat{\beta}_1^G = 0.08$, $\hat{\beta}_2^G = -0.08$.

21. a. Odds of obtaining condoms for educated group estimated to be 4.04 times odds for noneducated group.

b. $\text{logit}(\hat{\pi}) = \hat{\alpha} + 1.40x_1 + 0.32x_2 + 1.76x_3 + 1.17x_4$, where $x_1 = 1$ for educated and 0 for noneducated, $x_2 = 1$ for males and 0 for females, $x_3 = 1$ for high SES and 0 for low SES, and x_4 = lifetime number of partners. Log odds ratio = 1.40 has CI (0.16, 2.63). CI is $1.40 \pm 1.96(SE)$, so CI has width $3.92(SE)$, and $SE = 0.63$.

c. CI corresponds to one for log odds ratio of (0.207, 2.556); 1.38 is the midpoint of CI, suggesting it may be estimated log odds ratio, in which case $\exp(1.38) = 3.98$ = estimated odds ratio.

23. **a.** $R = 1$: $\text{logit}(\hat{\pi}) = -6.7 + 0.1A + 1.4S$. $R = 0$: $\text{logit}(\hat{\pi}) = -7.0 + 0.1A + 1.2S$. *YS* conditional odds ratio = $\exp(1.4) = 4.1$ for blacks and $\exp(1.2) = 3.3$ for whites. Coefficient of cross-product term, 0.22, is difference between log odds ratios 1.4 and 1.2.

b. The coefficient of S of 1.2 is log odds ratio between Y and S when $R = 0$ (whites), in which case *RS* interaction does not enter equation. P-value of $P < 0.01$ for smoking represents result of test that log odds ratio between Y and S for whites = 0.

25. **a.** Derive the four equations from overall equation

$$\text{logit}(\hat{\pi}) = -5.854 + 4.101c_1 - 4.186c_2 - 15.66c_3 + 0.200x - 0.094(c_1 \times x) + 0.218(c_2 \times x) + 0.658(c_3 \times x)$$

b. LR statistic = 4.4 ($df = 3$), $P = 0.22$.

27. **a.** −0.41 and 0.97 are coefficients for standardized versions of predictors for which standard deviation is 1.0.

b. For $c = 4$ (dark crabs), $\text{logit}(\hat{\pi}) = -12.11 + 0.458x$. Estimated probability changes from 0.33 to 0.64 when x changes from 24.9 to 27.7.

29. For main effects model, estimated conditional odds ratios = 3.7 for race and 1.9 for gender.

31. Model with main effects has estimated conditional odds ratios 17.3 between marijuana use and cigarette use and 19.8 between marijuana use and alcohol use.

35. **a.** Exponential term maximized when exponent equals 0, which is when $x = -\alpha/\beta$.

b. 24.8.

c. $0.40(0.302) = 0.12$.

37. (**a**) T, (**b**) F, (**c**) T, (**d**) F, (**e**) T.

CHAPTER 5

1. **a.** $\text{logit}(\hat{\pi}) = -9.35 + 0.834(\text{weight}) + 0.307(\text{width})$.

b. LR statistic = 32.9 ($df = 2$), $P < 0.0001$.

c. Wald statistics = 1.55 and 2.85 ($df = 1$), for P-values 0.21 and 0.09. Predictors are highly correlated (Pearson correlation = 0.887), so problem of multicollinearity.

3. a. Test statistic = 3.2 ($df = 3$). Yes, can remove it.

b. Change in deviance is smallest, 0.0 on $df = 2$, when remove S*W term.

c. Take out C*W term, as model W + C*S has larger P-value.

d. Yes, change in deviance = 9.0 ($df = 6$), which has P-value = 0.17.

e. Model C + S + W has smallest AIC.

5. Model with only four main effect terms has smallest AIC.

7. a. One intercept term, four main effect terms, six two-factor interaction terms, and four three-factor interaction terms, so numbers of parameters in models are $1, 1 + 4 = 5, 1 + 4 + 6 = 11, 1 + 4 + 6 + 4 = 15$.

b. AIC values are $1130.23 + 2(1) = 1132.23$, $1124.86 + 2(5) = 1134.86$, $1119.87 + 2(11) = 1141.87$, $1116.47 + 2(15) = 1146.47$. Best model has intercept only.

c. No; for example, expect c around 0.50 just by chance.

9. a. No, deviance can check fit only for categorical predictors.

b. LR statistic for testing that parameter for quadratic term is zero equals 3.9, with $df = 1$. P-value is about 0.05.

c. Derivative of linear predictor with respect to LI is $0.9625 - 2(0.016)\text{LI}$, which is >0 when $\text{LI} < 0.9625/0.032 = 30.1$. So, $\hat{\pi}$ increases as LI increases up to about 30.

d. Simpler model with linear effect on logit seems adequate.

11. Model seems adequate. A reference for this type of approach is the article by A. Tsiatis (*Biometrika*, **67**: 250–251, 1980).

15. Logit model with additive factor effects has $G^2 = 0.1$ and $X^2 = 0.1$, $df = 2$. Estimated odds of females still being missing are $\exp(0.38) = 1.46$ times those for males, given age. Estimated odds considerably higher for those aged at least 19 than for other age groups, given gender.

17. a. For death penalty response with main effect for predictors, $G^2 = 0.38$, $df = 1$, $P = 0.54$. Model fits adequately.

b. Each standardized residual is 0.44 in absolute value, showing no lack of fit.

c. Estimated conditional odds ratio = $\exp(-0.868) = 0.42$ for defendant's race and $\exp(2.404) = 11.1$ for victims' race.

19. **a.** $\text{logit}(\pi) = \alpha + \beta_1 d_1 + \cdots + \beta_6 d_6$, where $d_i = 1$ for department i and $d_i = 0$ otherwise.

b. Model fits poorly.

c. Only lack of fit in Department 1, where more females were admitted than expected if the model lacking gender effect truly holds.

d. -4.15, so fewer males admitted than expected if model lacking gender effect truly holds.

e. Males apply in relatively greater numbers to departments that have relatively higher proportions of acceptances.

27. $z_{\alpha/2} = 2.576$, $z_\beta = 1.645$, and $n_1 = n_2 = 214$.

29. $\text{logit}(\hat{\pi}) = -12.351 + 0.497x$. Probability at $x = 26.3$ is 0.674; probability at $x = 28.4$ (i.e., one standard deviation above mean) is 0.854. Odds ratio is 2.83, so $\lambda = 1.04$, $\delta = 5.1$. Then $n = 75$.

CHAPTER 6

1. **a.** $\log(\hat{\pi}_R/\hat{\pi}_D) = -2.3 + 0.5x$. Estimated odds of preferring Republicans over Democrats increase by 65% for every $10,000 increase.

b. $\hat{\pi}_R > \hat{\pi}_D$ when annual income >$46,000.

c. $\hat{\pi}_I = 1/[1 + \exp(3.3 - 0.2x) + \exp(1 + 0.3x)]$.

3. **a.** *SE* values in parentheses

Logit	Intercept	Size $\leq$ 2.3	Hancock	Oklawaha	Trafford
$\log(\pi_I/\pi_F)$	-1.55	1.46(0.40)	-1.66(0.61)	0.94(0.47)	1.12(0.49)
$\log(\pi_R/\pi_F)$	-3.31	-0.35(0.58)	1.24(1.19)	2.46(1.12)	2.94(1.12)
$\log(\pi_B/\pi_F)$	-2.09	-0.63(0.64)	0.70(0.78)	-0.65(1.20)	1.09(0.84)
$\log(\pi_O/\pi_F)$	-1.90	0.33(0.45)	0.83(0.56)	0.01(0.78)	1.52(0.62)

5. **a.** Job satisfaction tends to increase at higher x_1 and lower x_2 and x_3.

b. $x_1 = 4$ and $x_2 = x_3 = 1$.

7. **a.** Two cumulative probabilities to model and hence 2 intercept parameters. Proportional odds have same predictor effects for each cumulative probability, so only one effect reported for income.

b. Estimated odds of being at low end of scale (less happy) decrease as income increases.

c. LR statistic = 0.89 with *df* = 1, and *P*-value = 0.35. It is plausible that income has no effect on marital happiness.

d. Deviance = 3.25, *df* = 3, *P*-value = 0.36, so model fits adequately.

e. 1 − cumulative probability for category 2, which is 0.61.

9. a. There are four nondegenerate cumulative probabilities. When all predictor values equal 0, cumulative probabilities increase across categories, so logits increase, as do parameters that specify logits.

b. (i) Religion = none, (ii) Religion = Protestant.

c. For Protestant, 0.09. For None, 0.26.

d. (i) $e^{-1.27} = 0.28$; that is, estimated odds that Protestant falls in relatively more liberal categories (rather than more conservative categories) is 0.28 times estimated odds for someone with no religious preference. (ii) Estimated odds ratio comparing Protestants to Catholics is 0.95.

11. a. $\hat{\beta} = -0.0444$ ($SE = 0.0190$) suggests probability of having relatively less satisfaction decreases as income increases.

b. $\hat{\beta} = -0.0435$, very little change. If model holds for underlying logistic latent variable, model holds with same effect value for every way of defining outcome categories.

c. Gender estimate of −0.0256 has $SE = 0.4344$ and Wald statistic = 0.003 ($df = 1$), so can be dropped.

13. a. Income effect of 0.389 ($SE = 0.155$) indicates estimated odds of higher of any two adjacent job satisfaction categories increases as income increases.

b. Estimated income effects are −1.56 for outcome categories 1 and 4, −0.64 for outcome categories 2 and 4, and −0.40 for categories 3 and 4.

c. (**a**) Treats job satisfaction as ordinal whereas (**b**) treats job satisfaction as nominal. Ordinal model is more parsimonious and simpler to interpret, because it has one income effect rather than three.

17. Cumulative logit model with main effects of gender, location, and seat-belt has estimates 0.545, −0.773, and 0.824; for example, for those wearing a seat belt, estimated odds that the response is below any particular level of injury are $e^{0.824} = 2.3$ times the estimated odds for those not wearing seat belts.

21. For cumulative logit model of proportional odds form with Y = happiness and x = marital status (1 = married, 0 = divorced), $\hat{\beta} = -1.076$ ($SE = 0.116$). The model fits well (e.g., deviance = 0.29 with $df = 1$).

CHAPTER 7

1. a. $G^2 = 0.82$, $X^2 = 0.82$, $df = 1$.

b. $\hat{\lambda}_1^Y = 1.416$, $\hat{\lambda}_2^Y = 0$. Given gender, estimated odds of belief in afterlife equal $e^{1.416} = 4.1$.

3. a. $G^2 = 0.48$, $df = 1$, fit is adequate.

b. 2.06 for PB association, 4.72 for PH association, 1.60 for BH association.

c. H_0 model is (PH, BH). Test statistic = 4.64, $df = 1$, P-value = 0.03.

d. $\exp[0.721 \pm 1.96(0.354)] = (1.03, 4.1)$.

5. a. 0.42.

b. 1.45.

c. $G^2 = 0.38$, $df = 1$, $P = 0.54$, model fits adequately.

d. Logit model with main effects of defendant race and victim race, using indicator variable for each.

7. a. Difference in deviances = 2.21, $df = 2$; simpler model adequate.

b. $\exp(-1.507, -0.938) = (0.22, 0.39)$.

c. $e^{1.220} = 3.39$, $\exp(0.938, 1.507)$ is $(1/0.39, 1/0.22)$, which is $(2.55, 4.51)$.

9. a. Estimated odds ratios are 0.9 for conditional and 1.8 for marginal. Men apply in greater numbers to departments (1, 2) having relatively high admissions rates and women apply in greater numbers to departments (3, 4, 5, 6) having relatively low admissions rates.

b. Deviance $G^2 = 20.2$ ($df = 5$), poor fit. Standardized residuals show lack of fit only for Department 1.

c. $G^2 = 2.56$, $df = 4$, good fit.

d. Logit model with main effects for department and gender has estimated conditional odds ratio = 1.03 between gender and admissions. Model deleting gender term fits essentially as well, with $G^2 = 2.68$ ($df = 5$); plausible that admissions and gender are conditionally independent for these departments.

11. a. Injury has estimated conditional odds ratios 0.58 with gender, 2.13 with location, and 0.44 with seat-belt use. Since no interaction, overall most likely case for injury is females not wearing seat belts in rural locations.

13. a. $G^2 = 31.7$, $df = 48$.

b. $\log(\mu_{11cl}\mu_{33cl}/\mu_{13cl}\mu_{31cl})$
$= \log(\mu_{11cl}) + \log(\mu_{33cl}) - \log(\mu_{13cl}) - \log(\mu_{31cl})$
Substitute model formula, and simplify. Estimated odds ratio = $\exp(2.142)$ = 8.5. The 95% CI is $\exp[2.142 \pm 1.96(0.523)]$, or (3.1, 24.4).

c. 2.4 for C and L, 6.5 for H and L, 0.8 for C and H, 0.9 for E and L, 3.3 for C and E. Associations seem strongest between E and H and between H and L.

17. Logistic model more appropriate when one variable a response and others are explanatory. Loglinear model may be more appropriate when at least two variables are response variables.

19. b. The λ^{XY} term is not in the model, so X and Y are conditionally independent. All terms in the saturated model that are not in model (WXZ, WYZ) involve X and Y, and so permit XY conditional association.

21. a. $G^2 = 31.7$, $df = 48$. The model with three-factor terms has $G^2 = 8.5$, $df = 16$; the change is 23.1, $df = 32$, not a significant improvement.

b. (ii) For the result at the beginning of Section 7.4.4, identify set $B = \{E, L\}$ and sets A and C each to be one of other variables.

23. (**a**) No.

(**b**) Yes; in the result in Section 7.4.4, take $A = \{Y\}$, $B = \{X_1, X_2\}$, $C = \{X_3\}$.

25. a. Take $\beta = 0$.

b. LR statistic comparing this to model (XZ, YZ).

d. No, this is a heterogeneous linear-by-linear association model. The XY odds ratios vary according to the level of Z, and there is no longer homogeneous association. For scores $\{u_i = i\}$ and $\{v_j = j\}$, local odds ratio equals $\exp(\beta_k)$.

27. (**a**) T, (**b**) F, (**c**) T.

CHAPTER 8

1. $z = 2.88$, two-sided P-value $= 0.004$; there is strong evidence that MI cases are more likely than MI controls to have diabetes.

3. a. Population odds of belief in heaven estimated to be 2.02 times population odds of belief in hell.

b. For each subject, odds of belief in heaven estimated to equal 62.5 times odds of belief in hell.

5. a. This is probability, under H_0, of observed or more extreme result, with more extreme defined in direction specified by H_a.

b. Mid P-value includes only half observed probability, added to probability of more extreme results.

c. When binomial parameter $= 0.50$, binomial is symmetric, so two-sided P-value $= 2$(one-sided P-value) in (**a**) and (**b**).

7. 0.022 ± 0.038, or $(-0.016, 0.060)$, wider than for dependent samples.

9. $\hat{\beta} = \log(132/107) = 0.21$.

11. 95% CI for β is $\log(132/107) \pm 1.96\sqrt{1/132 + 1/107}$, which is (−0.045, 0.465). The corresponding CI for odds ratio is (0.96, 1.59).

13. **a.** More moves from (2) to (1), (1) to (4), (2) to (4) than if symmetry truly held.
b. Quasi-symmetry model fits well.
c. Difference between deviances = 148.3, with $df = 3$. P-value < 0.0001 for H_0: marginal homogeneity.

15. **a.** Subjects tend to respond more in always wrong direction for extramarital sex.
b. $z = -4.91/0.45 = -10.9$, extremely strong evidence against H_0.
c. Symmetry fits very poorly but quasi symmetry fits well. The difference of deviances = 400.8, $df = 3$, gives extremely strong evidence against H_0: marginal homogeneity (P-value < 0.0001).
d. Also fits well, not significantly worse than ordinary quasi symmetry. The difference of deviances = 400.1, $df = 1$, gives extremely strong evidence against marginal homogeneity.
e. From model formula in Section 8.4.5, for each pair of categories, a more favorable response is much more likely for premarital sex than extramarital sex.

19. $G^2 = 4167.6$ for independence model ($df = 9$), $G^2 = 9.7$ for quasi-independence ($df = 5$). QI model fits cells on main diagonal perfectly.

21. $G^2 = 13.8$, $df = 11$; fitted odds ratio = 1.0. Conditional on change in brand, new brand plausibly independent of old brand.

23. **a.** $G^2 = 4.3$, $df = 3$; prestige ranking: 1. *JRSS-B*, 2. *Biometrika*, 3. *JASA*, 4. *Commun. Statist.*

25. **a.** $e^{1.45-0.19}/(1 + e^{1.45-0.19}) = 0.78$.
b. Extremely strong evidence (P-value < 0.0001) of at least one difference among $\{\beta_i\}$. Players do not all have same probability of winning.

27. **a.** $\log(\pi_{ij}/\pi_{ji}) = \log(\mu_{ij}/\mu_{ji}) = (\lambda_i^X - \lambda_i^Y) - (\lambda_j^X - \lambda_j^Y)$. Take $\beta_i = (\lambda_i^X - \lambda_i^Y)$.
b. Under this constraint, $\mu_{ij} = \mu_{ji}$.
c. Under this constraint, model adds to independence model a term for each cell on main diagonal.

CHAPTER 9

1. a. Sample proportion yes = 0.86 for A, 0.66 for C, and 0.42 for M.

b. $\text{logit}[P(Y_t = 1)] = \beta_1 z_1 + \beta_2 z_2 + \beta_3 z_3$, where $t = 1, 2, 3$ refers to A, C, M, and $z_1 = 1$ if $t = 1$, $z_2 = 1$ if $t = 2$, $z_3 = 1$ if $t = 3$ (0 otherwise); for example, e^{β_1} is the odds of using alcohol. Marginal homogeneity is $\beta_1 = \beta_2 = \beta_3$.

3. a. Marijuana: For $S_1 = S_2 = 0$, the linear predictor takes greatest value when $R = 1$ and $G = 0$ (white males). For alcohol, $S_1 = 1$, $S_2 = 0$, the linear predictor takes greatest value when $R = 1$ and $G = 1$ (white females).

b. Estimated odds for white subjects $\exp(0.38) = 1.46$ times estimated odds for black subjects.

c. For alcohol, estimated odds ratio $= \exp(-0.20 + 0.37) = 1.19$; for cigarettes, $\exp(-0.20 + 0.22) = 1.02$; for marijuana, $\exp(-0.20) = 0.82$.

d. Estimated odds ratio $= \exp(1.93 + 0.37) = 9.97$.

e. Estimated odds ratio $= \exp(1.93) = 6.89$.

7. a. Subjects can select any number of sources, so a given subject could have anywhere from zero to five observations in the table. Multinomial distribution does not apply to these 40 cells.

b. Estimated correlation is weak, so results not much different from treating five responses by a subject as if from five independent subjects. For source A the estimated size effect is 1.08 and highly significant (Wald statistic = 6.46, $df = 1$, $P < 0.0001$). For sources C, D, and E size effect estimates are all roughly -0.2.

11. With constraint $\beta_4 = 0$, ML estimates of item parameters $\{\beta_j\}$ are $(-0.551, -0.603, -0.486, 0)$. The first three estimates have absolute values greater than five standard errors, providing strong evidence of greater support for increased government spending on education than other items.

13. $\text{logit}[\hat{P}(Y_t = 1)] = 1.37 + 1.148 y_{t-1} + 1.945 y_{t-2} + 0.174 s - 0.437 t$. So, y_{t-2} does have predictive power.

b. Given previous responses and child's age, estimated effect of maternal smoking weaker than when use only previous response as predictor, but still positive. LR statistic for testing maternal smoking effect is 0.72 ($df = 1$, $P = 0.40$).

17. Independent *conditional* on Y_{t-1}, but not independent marginally.

CHAPTER 10

1. a. Using PROC NLMIXED in SAS, $(\hat{\beta}, SE, \hat{\sigma}, SE) = (4.135, 0.713, 10.199, 1.792)$ for 1000 quadrature points.

b. For given subject, estimated odds of belief in heaven are $\exp(4.135) = 62.5$ times estimated odds of belief in hell.

c. $\hat{\beta} = \log(125/2) = 4.135$ with $SE = \sqrt{(1/125) + (1/2)} = 0.713$.

3. a. (i) 0.038, (ii) 0.020, (iii) 0.070.

b. Sample size may be small in each county, and GLMM analysis borrows from whole.

5. a. 0.4, 0.8, 0.2, 0.6, 0.6, 1.0, 0.8, 0.4, 0.6, 0.2.

b. $\text{logit}(\pi_i) = u_i + \alpha$. ML estimates $\hat{\alpha} = 0.259$ and $\hat{\sigma} = 0.557$. For average coin, estimated probability of head = 0.56.

c. Using PROC NLMIXED in SAS, predicted values are 0.52, 0.63, 0.46, 0.57, 0.57, 0.68, 0.63, 0.52, 0.57, 0.46.

7. a. Strong associations between responses inflates GLMM estimates relative to marginal model estimates.

b. Loglinear model focuses on strength of association between use of one substance and use of another, given whether or not one used remaining substance. The focus is not on the odds of having used one substance compared with the odds of using another.

c. If $\hat{\sigma} = 0$, GLMM has the same fit as loglinear model (A, C, M), since conditional independence of responses given random effect translates to conditional independence marginally also.

9. For $\hat{\beta}_A = 0$, $\hat{\beta}_B = 1.99$ $(SE = 0.35)$, $\hat{\beta}_C = 2.51$ $(SE = 0.37)$, with $\hat{\sigma} = 0$.

11. a. For given department, estimated odds of admission for female are $\exp(0.173) = 1.19$ times estimated odds of admission for male.

b. For given department, estimated odds of admission for female are $\exp(0.163) = 1.18$ times estimated odds of admission for male.

c. The estimated mean log odds ratio between gender and admissions, given department, is 0.176, corresponding to odds ratio = 1.19. Because of extra variance component, estimate of β is not as precise.

d. Marginal odds ratio of $\exp(-0.07) = 0.93$ in different direction, corresponding to odds of being admitted lower for females than males.

13. a. $e^{2.51} = 12.3$, so estimated odds of response in category $\leq j$ (i.e., toward "always wrong" end of scale) on extramarital sex for a randomly selected subject are 12.3 times estimated odds of response in those categories for premarital sex for another randomly selected subject.

b. Estimate of β much larger for GLMM, since a subject-specific estimate and variance component is large (recall Section 10.1.4).

17. a. At initial time, treatment effect $= 0.058$ (odds ratio $= 1.06$), so two groups have similar response distributions. At follow-up time, the treatment effect $= 0.058 + 1.081 = 1.139$ (odds ratio $= 3.1$).

b. LR statistic $= -2[-593.0 - (-621.0)] = 56$. Null distribution is equal mixture of degenerate at 0 and X_1^2, and P-value is half that of X_1^2 variate, and is 0 to many decimal places.

23. From Section 10.1.4, the effects in marginal models are smaller in absolute value than effects in GLMMs, with greater difference when $\hat{\sigma}$ is larger. Here, the effect for GLMM is the same for each age group, but diminishes more for the older age group in the marginal model because the older age group has much larger $\hat{\sigma}$ in GLMM.

WILEY SERIES IN PROBABILITY AND STATISTICS

ESTABLISHED BY WALTER A. SHEWHART AND SAMUEL S. WILKS

The ***Wiley Series in Probability and Statistics*** is well established and authoritative. It covers many topics of current research interest in both pure and applied statistics and probability theory. Written by leading statisticians and institutions, the titles span both state-of-the-art developments in the field and classical methods.

Reflecting the wide range of current research in statistics, the series encompasses applied, methodological and theoretical statistics, ranging from applications and new techniques made possible by advances in computerized practice to rigorous treatment of theoretical approaches.

This series provides essential and invaluable reading for all statisticians, whether in academia, industry, government, or research.

† ABRAHAM and LEDOLTER · Statistical Methods for Forecasting
AGRESTI · Analysis of Ordinal Categorical Data
AGRESTI · An Introduction to Categorical Data Analysis, *Second Edition*
AGRESTI · Categorical Data Analysis, *Second Edition*
ALTMAN, GILL, and McDONALD · Numerical Issues in Statistical Computing for the Social Scientist
AMARATUNGA and CABRERA · Exploration and Analysis of DNA Microarray and Protein Array Data
ANDĚL · Mathematics of Chance
ANDERSON · An Introduction to Multivariate Statistical Analysis, *Third Edition*
* ANDERSON · The Statistical Analysis of Time Series
ANDERSON, AUQUIER, HAUCK, OAKES, VANDAELE, and WEISBERG · Statistical Methods for Comparative Studies
ANDERSON and LOYNES · The Teaching of Practical Statistics
ARMITAGE and DAVID (editors) · Advances in Biometry
ARNOLD, BALAKRISHNAN, and NAGARAJA · Records
* ARTHANARI and DODGE · Mathematical Programming in Statistics
* BAILEY · The Elements of Stochastic Processes with Applications to the Natural Sciences
BALAKRISHNAN and KOUTRAS · Runs and Scans with Applications
BALAKRISHNAN and NG · Precedence-Type Tests and Applications
BARNETT · Comparative Statistical Inference, *Third Edition*
BARNETT · Environmental Statistics
BARNETT and LEWIS · Outliers in Statistical Data, *Third Edition*
BARTOSZYNSKI and NIEWIADOMSKA-BUGAJ · Probability and Statistical Inference
BASILEVSKY · Statistical Factor Analysis and Related Methods: Theory and Applications
BASU and RIGDON · Statistical Methods for the Reliability of Repairable Systems
BATES and WATTS · Nonlinear Regression Analysis and Its Applications

*Now available in a lower priced paperback edition in the Wiley Classics Library.
†Now available in a lower priced paperback edition in the Wiley–Interscience Paperback Series.

BECHHOFER, SANTNER, and GOLDSMAN · Design and Analysis of Experiments for Statistical Selection, Screening, and Multiple Comparisons
BELSLEY · Conditioning Diagnostics: Collinearity and Weak Data in Regression
† BELSLEY, KUH, and WELSCH · Regression Diagnostics: Identifying Influential Data and Sources of Collinearity
BENDAT and PIERSOL · Random Data: Analysis and Measurement Procedures, *Third Edition*
BERRY, CHALONER, and GEWEKE · Bayesian Analysis in Statistics and Econometrics: Essays in Honor of Arnold Zellner
BERNARDO and SMITH · Bayesian Theory
BHAT and MILLER · Elements of Applied Stochastic Processes, *Third Edition*
BHATTACHARYA and WAYMIRE · Stochastic Processes with Applications
BILLINGSLEY · Convergence of Probability Measures, *Second Edition*
BILLINGSLEY · Probability and Measure, *Third Edition*
BIRKES and DODGE · Alternative Methods of Regression
BLISCHKE AND MURTHY (editors) · Case Studies in Reliability and Maintenance
BLISCHKE AND MURTHY · Reliability: Modeling, Prediction, and Optimization
BLOOMFIELD · Fourier Analysis of Time Series: An Introduction, *Second Edition*
BOLLEN · Structural Equations with Latent Variables
BOLLEN and CURRAN · Latent Curve Models: A Structural Equation Perspective
BOROVKOV · Ergodicity and Stability of Stochastic Processes
BOULEAU · Numerical Methods for Stochastic Processes
BOX · Bayesian Inference in Statistical Analysis
BOX · R. A. Fisher, the Life of a Scientist
BOX and DRAPER · Response Surfaces, Mixtures, and Ridge Analyses, *Second Edition*
* BOX and DRAPER · Evolutionary Operation: A Statistical Method for Process Improvement
BOX and FRIENDS · Improving Almost Anything, *Revised Edition*
BOX, HUNTER, and HUNTER · Statistics for Experimenters: Design, Innovation, and Discovery, *Second Editon*
BOX and LUCEÑO · Statistical Control by Monitoring and Feedback Adjustment
BRANDIMARTE · Numerical Methods in Finance: A MATLAB-Based Introduction
BROWN and HOLLANDER · Statistics: A Biomedical Introduction
BRUNNER, DOMHOF, and LANGER · Nonparametric Analysis of Longitudinal Data in Factorial Experiments
BUCKLEW · Large Deviation Techniques in Decision, Simulation, and Estimation
CAIROLI and DALANG · Sequential Stochastic Optimization
CASTILLO, HADI, BALAKRISHNAN, and SARABIA · Extreme Value and Related Models with Applications in Engineering and Science
CHAN · Time Series: Applications to Finance
CHARALAMBIDES · Combinatorial Methods in Discrete Distributions
CHATTERJEE and HADI · Regression Analysis by Example, *Fourth Edition*
CHATTERJEE and HADI · Sensitivity Analysis in Linear Regression
CHERNICK · Bootstrap Methods: A Practitioner's Guide
CHERNICK and FRIIS · Introductory Biostatistics for the Health Sciences
CHILÈS and DELFINER · Geostatistics: Modeling Spatial Uncertainty
CHOW and LIU · Design and Analysis of Clinical Trials: Concepts and Methodologies, *Second Edition*
CLARKE and DISNEY · Probability and Random Processes: A First Course with Applications, *Second Edition*
* COCHRAN and COX · Experimental Designs, *Second Edition*
CONGDON · Applied Bayesian Modelling
CONGDON · Bayesian Models for Categorical Data
CONGDON · Bayesian Statistical Modelling

*Now available in a lower priced paperback edition in the Wiley Classics Library.
†Now available in a lower priced paperback edition in the Wiley–Interscience Paperback Series.

CONOVER · Practical Nonparametric Statistics, *Third Edition*
COOK · Regression Graphics
COOK and WEISBERG · Applied Regression Including Computing and Graphics
COOK and WEISBERG · An Introduction to Regression Graphics
CORNELL · Experiments with Mixtures, Designs, Models, and the Analysis of Mixture Data, *Third Edition*
COVER and THOMAS · Elements of Information Theory
COX · A Handbook of Introductory Statistical Methods
* COX · Planning of Experiments
CRESSIE · Statistics for Spatial Data, *Revised Edition*
CSÖRGŐ and HORVÁTH · Limit Theorems in Change Point Analysis
DANIEL · Applications of Statistics to Industrial Experimentation
DANIEL · Biostatistics: A Foundation for Analysis in the Health Sciences, *Eighth Edition*
* DANIEL · Fitting Equations to Data: Computer Analysis of Multifactor Data, *Second Edition*
DASU and JOHNSON · Exploratory Data Mining and Data Cleaning
DAVID and NAGARAJA · Order Statistics, *Third Edition*
* DEGROOT, FIENBERG, and KADANE · Statistics and the Law
DEL CASTILLO · Statistical Process Adjustment for Quality Control
DeMARIS · Regression with Social Data: Modeling Continuous and Limited Response Variables
DEMIDENKO · Mixed Models: Theory and Applications
DENISON, HOLMES, MALLICK and SMITH · Bayesian Methods for Nonlinear Classification and Regression
DETTE and STUDDEN · The Theory of Canonical Moments with Applications in Statistics, Probability, and Analysis
DEY and MUKERJEE · Fractional Factorial Plans
DILLON and GOLDSTEIN · Multivariate Analysis: Methods and Applications
DODGE · Alternative Methods of Regression
* DODGE and ROMIG · Sampling Inspection Tables, *Second Edition*
* DOOB · Stochastic Processes
DOWDY, WEARDEN, and CHILKO · Statistics for Research, *Third Edition*
DRAPER and SMITH · Applied Regression Analysis, *Third Edition*
DRYDEN and MARDIA · Statistical Shape Analysis
DUDEWICZ and MISHRA · Modern Mathematical Statistics
DUNN and CLARK · Basic Statistics: A Primer for the Biomedical Sciences, *Third Edition*
DUPUIS and ELLIS · A Weak Convergence Approach to the Theory of Large Deviations
EDLER and KITSOS · Recent Advances in Quantitative Methods in Cancer and Human Health Risk Assessment
* ELANDT-JOHNSON and JOHNSON · Survival Models and Data Analysis
ENDERS · Applied Econometric Time Series
† ETHIER and KURTZ · Markov Processes: Characterization and Convergence
EVANS, HASTINGS, and PEACOCK · Statistical Distributions, *Third Edition*
FELLER · An Introduction to Probability Theory and Its Applications, Volume I, *Third Edition,* Revised; Volume II, *Second Edition*
FISHER and VAN BELLE · Biostatistics: A Methodology for the Health Sciences
FITZMAURICE, LAIRD, and WARE · Applied Longitudinal Analysis
* FLEISS · The Design and Analysis of Clinical Experiments
FLEISS · Statistical Methods for Rates and Proportions, *Third Edition*
† FLEMING and HARRINGTON · Counting Processes and Survival Analysis
FULLER · Introduction to Statistical Time Series, *Second Edition*
† FULLER · Measurement Error Models

*Now available in a lower priced paperback edition in the Wiley Classics Library.
†Now available in a lower priced paperback edition in the Wiley–Interscience Paperback Series.

GALLANT · Nonlinear Statistical Models
GEISSER · Modes of Parametric Statistical Inference
GELMAN and MENG · Applied Bayesian Modeling and Causal Inference from Incomplete-Data Perspectives
GEWEKE · Contemporary Bayesian Econometrics and Statistics
GHOSH, MUKHOPADHYAY, and SEN · Sequential Estimation
GIESBRECHT and GUMPERTZ · Planning, Construction, and Statistical Analysis of Comparative Experiments
GIFI · Nonlinear Multivariate Analysis
GIVENS and HOETING · Computational Statistics
GLASSERMAN and YAO · Monotone Structure in Discrete-Event Systems
GNANADESIKAN · Methods for Statistical Data Analysis of Multivariate Observations, *Second Edition*
GOLDSTEIN and LEWIS · Assessment: Problems, Development, and Statistical Issues
GREENWOOD and NIKULIN · A Guide to Chi-Squared Testing
GROSS and HARRIS · Fundamentals of Queueing Theory, *Third Edition*
* HAHN and SHAPIRO · Statistical Models in Engineering
HAHN and MEEKER · Statistical Intervals: A Guide for Practitioners
HALD · A History of Probability and Statistics and their Applications Before 1750
HALD · A History of Mathematical Statistics from 1750 to 1930
† HAMPEL · Robust Statistics: The Approach Based on Influence Functions
HANNAN and DEISTLER · The Statistical Theory of Linear Systems
HEIBERGER · Computation for the Analysis of Designed Experiments
HEDAYAT and SINHA · Design and Inference in Finite Population Sampling
HEDEKER and GIBBONS · Longitudinal Data Analysis
HELLER · MACSYMA for Statisticians
HINKELMANN and KEMPTHORNE · Design and Analysis of Experiments, Volume 1: Introduction to Experimental Design
HINKELMANN and KEMPTHORNE · Design and Analysis of Experiments, Volume 2: Advanced Experimental Design
HOAGLIN, MOSTELLER, and TUKEY · Exploratory Approach to Analysis of Variance
* HOAGLIN, MOSTELLER, and TUKEY · Exploring Data Tables, Trends and Shapes
* HOAGLIN, MOSTELLER, and TUKEY · Understanding Robust and Exploratory Data Analysis
HOCHBERG and TAMHANE · Multiple Comparison Procedures
HOCKING · Methods and Applications of Linear Models: Regression and the Analysis of Variance, *Second Edition*
HOEL · Introduction to Mathematical Statistics, *Fifth Edition*
HOGG and KLUGMAN · Loss Distributions
HOLLANDER and WOLFE · Nonparametric Statistical Methods, *Second Edition*
HOSMER and LEMESHOW · Applied Logistic Regression, *Second Edition*
HOSMER and LEMESHOW · Applied Survival Analysis: Regression Modeling of Time to Event Data
† HUBER · Robust Statistics
HUBERTY · Applied Discriminant Analysis
HUBERTY and OLEJNIK · Applied MANOVA and Discriminant Analysis, *Second Edition*
HUNT and KENNEDY · Financial Derivatives in Theory and Practice, *Revised Edition*
HUSKOVA, BERAN, and DUPAC · Collected Works of Jaroslav Hajek—with Commentary
HUZURBAZAR · Flowgraph Models for Multistate Time-to-Event Data
IMAN and CONOVER · A Modern Approach to Statistics

*Now available in a lower priced paperback edition in the Wiley Classics Library.
†Now available in a lower priced paperback edition in the Wiley–Interscience Paperback Series.

† JACKSON · A User's Guide to Principle Components
JOHN · Statistical Methods in Engineering and Quality Assurance
JOHNSON · Multivariate Statistical Simulation
JOHNSON and BALAKRISHNAN · Advances in the Theory and Practice of Statistics: A Volume in Honor of Samuel Kotz
JOHNSON and BHATTACHARYYA · Statistics: Principles and Methods, *Fifth Edition*
JOHNSON and KOTZ · Distributions in Statistics
JOHNSON and KOTZ (editors) · Leading Personalities in Statistical Sciences: From the Seventeenth Century to the Present
JOHNSON, KOTZ, and BALAKRISHNAN · Continuous Univariate Distributions, Volume 1, *Second Edition*
JOHNSON, KOTZ, and BALAKRISHNAN · Continuous Univariate Distributions, Volume 2, *Second Edition*
JOHNSON, KOTZ, and BALAKRISHNAN · Discrete Multivariate Distributions
JOHNSON, KEMP, and KOTZ · Univariate Discrete Distributions, *Third Edition*
JUDGE, GRIFFITHS, HILL, LÜTKEPOHL, and LEE · The Theory and Practice of Econometrics, *Second Edition*
JUREČKOVÁ and SEN · Robust Statistical Procedures: Aymptotics and Interrelations
JUREK and MASON · Operator-Limit Distributions in Probability Theory
KADANE · Bayesian Methods and Ethics in a Clinical Trial Design
KADANE AND SCHUM · A Probabilistic Analysis of the Sacco and Vanzetti Evidence
KALBFLEISCH and PRENTICE · The Statistical Analysis of Failure Time Data, *Second Edition*
KARIYA and KURATA · Generalized Least Squares
KASS and VOS · Geometrical Foundations of Asymptotic Inference
† KAUFMAN and ROUSSEEUW · Finding Groups in Data: An Introduction to Cluster Analysis
KEDEM and FOKIANOS · Regression Models for Time Series Analysis
KENDALL, BARDEN, CARNE, and LE · Shape and Shape Theory
KHURI · Advanced Calculus with Applications in Statistics, *Second Edition*
KHURI, MATHEW, and SINHA · Statistical Tests for Mixed Linear Models
KLEIBER and KOTZ · Statistical Size Distributions in Economics and Actuarial Sciences
KLUGMAN, PANJER, and WILLMOT · Loss Models: From Data to Decisions, *Second Edition*
KLUGMAN, PANJER, and WILLMOT · Solutions Manual to Accompany Loss Models: From Data to Decisions, *Second Edition*
KOTZ, BALAKRISHNAN, and JOHNSON · Continuous Multivariate Distributions, Volume 1, *Second Edition*
KOVALENKO, KUZNETZOV, and PEGG · Mathematical Theory of Reliability of Time-Dependent Systems with Practical Applications
LACHIN · Biostatistical Methods: The Assessment of Relative Risks
LAD · Operational Subjective Statistical Methods: A Mathematical, Philosophical, and Historical Introduction
LAMPERTI · Probability: A Survey of the Mathematical Theory, *Second Edition*
LANGE, RYAN, BILLARD, BRILLINGER, CONQUEST, and GREENHOUSE · Case Studies in Biometry
LARSON · Introduction to Probability Theory and Statistical Inference, *Third Edition*
LAWLESS · Statistical Models and Methods for Lifetime Data, *Second Edition*
LAWSON · Statistical Methods in Spatial Epidemiology
LE · Applied Categorical Data Analysis
LE · Applied Survival Analysis
LEE and WANG · Statistical Methods for Survival Data Analysis, *Third Edition*
LEPAGE and BILLARD · Exploring the Limits of Bootstrap

*Now available in a lower priced paperback edition in the Wiley Classics Library.
†Now available in a lower priced paperback edition in the Wiley–Interscience Paperback Series.

LEYLAND and GOLDSTEIN (editors) · Multilevel Modelling of Health Statistics
LIAO · Statistical Group Comparison
LINDVALL · Lectures on the Coupling Method
LIN · Introductory Stochastic Analysis for Finance and Insurance
LINHART and ZUCCHINI · Model Selection
LITTLE and RUBIN · Statistical Analysis with Missing Data, *Second Edition*
LLOYD · The Statistical Analysis of Categorical Data
LOWEN and TEICH · Fractal-Based Point Processes
MAGNUS and NEUDECKER · Matrix Differential Calculus with Applications in Statistics and Econometrics, *Revised Edition*
MALLER and ZHOU · Survival Analysis with Long Term Survivors
MALLOWS · Design, Data, and Analysis by Some Friends of Cuthbert Daniel
MANN, SCHAFER, and SINGPURWALLA · Methods for Statistical Analysis of Reliability and Life Data
MANTON, WOODBURY, and TOLLEY · Statistical Applications Using Fuzzy Sets
MARCHETTE · Random Graphs for Statistical Pattern Recognition
MARDIA and JUPP · Directional Statistics
MASON, GUNST, and HESS · Statistical Design and Analysis of Experiments with Applications to Engineering and Science, *Second Edition*
McCULLOCH and SEARLE · Generalized, Linear, and Mixed Models
McFADDEN · Management of Data in Clinical Trials
* McLACHLAN · Discriminant Analysis and Statistical Pattern Recognition
McLACHLAN, DO, and AMBROISE · Analyzing Microarray Gene Expression Data
McLACHLAN and KRISHNAN · The EM Algorithm and Extensions
McLACHLAN and PEEL · Finite Mixture Models
McNEIL · Epidemiological Research Methods
MEEKER and ESCOBAR · Statistical Methods for Reliability Data
MEERSCHAERT and SCHEFFLER · Limit Distributions for Sums of Independent Random Vectors: Heavy Tails in Theory and Practice
MICKEY, DUNN, and CLARK · Applied Statistics: Analysis of Variance and Regression, *Third Edition*
* MILLER · Survival Analysis, *Second Edition*
MONTGOMERY, PECK, and VINING · Introduction to Linear Regression Analysis, *Fourth Edition*
MORGENTHALER and TUKEY · Configural Polysampling: A Route to Practical Robustness
MUIRHEAD · Aspects of Multivariate Statistical Theory
MULLER and STOYAN · Comparison Methods for Stochastic Models and Risks
MURRAY · X-STAT 2.0 Statistical Experimentation, Design Data Analysis, and Nonlinear Optimization
MURTHY, XIE, and JIANG · Weibull Models
MYERS and MONTGOMERY · Response Surface Methodology: Process and Product Optimization Using Designed Experiments, *Second Edition*
MYERS, MONTGOMERY, and VINING · Generalized Linear Models. With Applications in Engineering and the Sciences
† NELSON · Accelerated Testing, Statistical Models, Test Plans, and Data Analyses
† NELSON · Applied Life Data Analysis
NEWMAN · Biostatistical Methods in Epidemiology
OCHI · Applied Probability and Stochastic Processes in Engineering and Physical Sciences
OKABE, BOOTS, SUGIHARA, and CHIU · Spatial Tesselations: Concepts and Applications of Voronoi Diagrams, *Second Edition*
OLIVER and SMITH · Influence Diagrams, Belief Nets and Decision Analysis

*Now available in a lower priced paperback edition in the Wiley Classics Library.
†Now available in a lower priced paperback edition in the Wiley–Interscience Paperback Series.

PALTA · Quantitative Methods in Population Health: Extensions of Ordinary Regressions
PANJER · Operational Risk: Modeling and Analytics
PANKRATZ · Forecasting with Dynamic Regression Models
PANKRATZ · Forecasting with Univariate Box-Jenkins Models: Concepts and Cases
* PARZEN · Modern Probability Theory and Its Applications
PEÑA, TIAO, and TSAY · A Course in Time Series Analysis
PIANTADOSI · Clinical Trials: A Methodologic Perspective
PORT · Theoretical Probability for Applications
POURAHMADI · Foundations of Time Series Analysis and Prediction Theory
PRESS · Bayesian Statistics: Principles, Models, and Applications
PRESS · Subjective and Objective Bayesian Statistics, *Second Edition*
PRESS and TANUR · The Subjectivity of Scientists and the Bayesian Approach
PUKELSHEIM · Optimal Experimental Design
PURI, VILAPLANA, and WERTZ · New Perspectives in Theoretical and Applied Statistics
† PUTERMAN · Markov Decision Processes: Discrete Stochastic Dynamic Programming
QIU · Image Processing and Jump Regression Analysis
* RAO · Linear Statistical Inference and Its Applications, *Second Edition*
RAUSAND and HØYLAND · System Reliability Theory: Models, Statistical Methods, and Applications, *Second Edition*
RENCHER · Linear Models in Statistics
RENCHER · Methods of Multivariate Analysis, *Second Edition*
RENCHER · Multivariate Statistical Inference with Applications
* RIPLEY · Spatial Statistics
* RIPLEY · Stochastic Simulation
ROBINSON · Practical Strategies for Experimenting
ROHATGI and SALEH · An Introduction to Probability and Statistics, *Second Edition*
ROLSKI, SCHMIDLI, SCHMIDT, and TEUGELS · Stochastic Processes for Insurance and Finance
ROSENBERGER and LACHIN · Randomization in Clinical Trials: Theory and Practice
ROSS · Introduction to Probability and Statistics for Engineers and Scientists
ROSSI, ALLENBY, and McCULLOCH · Bayesian Statistics and Marketing
† ROUSSEEUW and LEROY · Robust Regression and Outlier Detection
* RUBIN · Multiple Imputation for Nonresponse in Surveys
RUBINSTEIN · Simulation and the Monte Carlo Method
RUBINSTEIN and MELAMED · Modern Simulation and Modeling
RYAN · Modern Experimental Design
RYAN · Modern Regression Methods
RYAN · Statistical Methods for Quality Improvement, *Second Edition*
SALEH · Theory of Preliminary Test and Stein-Type Estimation with Applications
* SCHEFFE · The Analysis of Variance
SCHIMEK · Smoothing and Regression: Approaches, Computation, and Application
SCHOTT · Matrix Analysis for Statistics, *Second Edition*
SCHOUTENS · Levy Processes in Finance: Pricing Financial Derivatives
SCHUSS · Theory and Applications of Stochastic Differential Equations
SCOTT · Multivariate Density Estimation: Theory, Practice, and Visualization
† SEARLE · Linear Models for Unbalanced Data
† SEARLE · Matrix Algebra Useful for Statistics
† SEARLE, CASELLA, and McCULLOCH · Variance Components
SEARLE and WILLETT · Matrix Algebra for Applied Economics
SEBER and LEE · Linear Regression Analysis, *Second Edition*
† SEBER · Multivariate Observations
† SEBER and WILD · Nonlinear Regression
SENNOTT · Stochastic Dynamic Programming and the Control of Queueing Systems

*Now available in a lower priced paperback edition in the Wiley Classics Library.
†Now available in a lower priced paperback edition in the Wiley–Interscience Paperback Series.

* SERFLING · Approximation Theorems of Mathematical Statistics
SHAFER and VOVK · Probability and Finance: It's Only a Game!
SILVAPULLE and SEN · Constrained Statistical Inference: Inequality, Order, and Shape Restrictions
SMALL and McLEISH · Hilbert Space Methods in Probability and Statistical Inference
SRIVASTAVA · Methods of Multivariate Statistics
STAPLETON · Linear Statistical Models
STAUDTE and SHEATHER · Robust Estimation and Testing
STOYAN, KENDALL, and MECKE · Stochastic Geometry and Its Applications, *Second Edition*
STOYAN and STOYAN · Fractals, Random Shapes and Point Fields: Methods of Geometrical Statistics
STYAN · The Collected Papers of T. W. Anderson: 1943–1985
SUTTON, ABRAMS, JONES, SHELDON, and SONG · Methods for Meta-Analysis in Medical Research
TAKEZAWA · Introduction to Nonparametric Regression
TANAKA · Time Series Analysis: Nonstationary and Noninvertible Distribution Theory
THOMPSON · Empirical Model Building
THOMPSON · Sampling, *Second Edition*
THOMPSON · Simulation: A Modeler's Approach
THOMPSON and SEBER · Adaptive Sampling
THOMPSON, WILLIAMS, and FINDLAY · Models for Investors in Real World Markets
TIAO, BISGAARD, HILL, PEÑA, and STIGLER (editors) · Box on Quality and Discovery: with Design, Control, and Robustness
TIERNEY · LISP-STAT: An Object-Oriented Environment for Statistical Computing and Dynamic Graphics
TSAY · Analysis of Financial Time Series, *Second Edition*
UPTON and FINGLETON · Spatial Data Analysis by Example, Volume II: Categorical and Directional Data
VAN BELLE · Statistical Rules of Thumb
VAN BELLE, FISHER, HEAGERTY, and LUMLEY · Biostatistics: A Methodology for the Health Sciences, *Second Edition*
VESTRUP · The Theory of Measures and Integration
VIDAKOVIC · Statistical Modeling by Wavelets
VINOD and REAGLE · Preparing for the Worst: Incorporating Downside Risk in Stock Market Investments
WALLER and GOTWAY · Applied Spatial Statistics for Public Health Data
WEERAHANDI · Generalized Inference in Repeated Measures: Exact Methods in MANOVA and Mixed Models
WEISBERG · Applied Linear Regression, *Third Edition*
WELSH · Aspects of Statistical Inference
WESTFALL and YOUNG · Resampling-Based Multiple Testing: Examples and Methods for *p*-Value Adjustment
WHITTAKER · Graphical Models in Applied Multivariate Statistics
WINKER · Optimization Heuristics in Economics: Applications of Threshold Accepting
WONNACOTT and WONNACOTT · Econometrics, *Second Edition*
WOODING · Planning Pharmaceutical Clinical Trials: Basic Statistical Principles
WOODWORTH · Biostatistics: A Bayesian Introduction
WOOLSON and CLARKE · Statistical Methods for the Analysis of Biomedical Data, *Second Edition*
WU and HAMADA · Experiments: Planning, Analysis, and Parameter Design Optimization
WU and ZHANG · Nonparametric Regression Methods for Longitudinal Data Analysis

*Now available in a lower priced paperback edition in the Wiley Classics Library.
†Now available in a lower priced paperback edition in the Wiley–Interscience Paperback Series.

YANG · The Construction Theory of Denumerable Markov Processes
YOUNG, VALERO-MORA, and FRIENDLY · Visual Statistics: Seeing Data with Dynamic Interactive Graphics
ZELTERMAN · Discrete Distributions—Applications in the Health Sciences
* ZELLNER · An Introduction to Bayesian Inference in Econometrics
ZHOU, OBUCHOWSKI, and McCLISH · Statistical Methods in Diagnostic Medicine